ROCK MECHANICS

SECOND EDITION

Other books published within the Series on Rock and Soil Mechanics

Reisner, W. & v. Eisenhart Rothe, M.:
Bins and Bunkers
for Handling Bulk Materials
1971

Dryer, W.:
The Science of Rock Mechanics
Part I: Strength Properties of Rocks
1972

Hanna, T.H.:
Foundation Instrumentation
1973

Gregory, C.E.:
Explosives
for North American Engineers
First Edition
1973

Reimbert, M. & A.:
Retaining Walls Volume I
Anchorages and Sheet Piling
1974

Vutukuri, V.S. & Lama, R.D.:
Handbook on Mechanical Properties
of Rocks
Volume I
1974

Reimbert, M. & A.:
Retaining Walls Volume II
Study of Passive Resistance
in Foundation Structures
1976

Hardy, H.R., Jr. & Leighton, F.W.:
First Conference on Acoustic
Emission/Microseismic Activity
in Geologic Structures and Materials
1977

Karafiath, L.L. & Nowatzki, E.A.:
Soil Mechanics
for Off-Road Vehicle Engineering
1977

**Baguelin, F., Jézéquel, J.F.
& Shields, D.H.:**
The Pressuremeter
and Foundation Engineering
1978

Lama, R.D. & Vutukuri, V.S.:
Handbook on Mechanical Properties
of Rocks
Volumes II, III & IV
1978

Gregory, C.E.:
Explosives
for North American Engineers
Second Edition
1979

Assonyi, Cs. & Richter, R.:
The Continuum Theory
of Rock Mechanics
1979

Hardy, H.R., Jr. & Leighton, F.W.:
Second Conference on Acoustic
Emission/Microseismic Activity
in Geologic Structures and Materials
1979

Jumikis, A.R.:
Rock Mechanics
1979

Hanna, T. H.:
Foundations in Tension
Ground Anchors
1982

Series on Rock and Soil Mechanics
Vol. 7 (1983)

ROCK MECHANICS
Second Edition

by
Alfreds R. Jumikis
Professor Emeritus of Rutgers University
The State University of New Jersey
USA

Second Edition
1983

TRANS TECH PUBLICATIONS

Rock Mechanics

Distributed in North America by
KARL DISTRIBUTORS
16 Bearskin Neck, Rockport, MA 01966, USA

and world wide by

TRANS TECH Publications
P.O. Box 266
D-3392 Clausthal-Zellerfeld
Federal Republic of Germany

and by

TRANS TECH S.A.
CH-4711 Aedermannsdorf, Switzerland

Copyright© of First Edition 1979
Copyright© of Second Edition 1983
by
Trans Tech Publications
D-3392 Clausthal-Zellerfeld
Federal Republic of Germany

All rights reserved

ISBN 0-87849-038-8
ISSN 0080-9004

Printed in the United States of America

No part of this publication may be reproduced, stored in a retrieval system, or transmitted, in any form or by any means, electronic, mechanical, photocopying, recording, or otherwise, without the written permission of the publisher.

PRINTED AND BOUND IN THE UNITED STATES OF AMERICA

FOREWORD TO THE SECOND EDITION

The favorable reception of the first edition of this book — Rock Mechanics — was so encouraging as to justify the preparation of the second edition of this volume. The philosophy of the second edition of this popular book is the same as that of the first edition, namely, the subject rock mechanics is presented as an introductory geotechnical engineering discipline to the education of civil engineering students and other interested readers.

The main chapters of the book describe and emphasize rock and its utilization as an engineering material by means of which, upon which, and within which engineers build structures.

Being an all-new, expanded and updated edition of the guide to Rock Mechanics, this edition is virtually a new book.

The major changes in this edition include:
  Rock classification
  Rock types and their description
  Laboratory rock testing equipment
  Rock properties
  Emphasis on effect of rock discontinuities and rock gouge on the stability of rock foundations
  Grouting of rocks
  Gunite and shotcrete
  Lugeon's water test of rocks.

Subject of current interest are topics for the understanding of natural and man-induced hazards in rock engineering:
  From stress concentrations in rock
    earthquakes
    faults
    water, and
    landslides and rockslides.

There are also some case histories described, such as the grouting of the Canelles Dam rock foundation; the failure of the Malpasset Dam, and the disaster of the Vaiont reservoir. These discussions should be of particular concern and interest to engineering geologists, civil engineers, students in geotechnical engineering, and other interested readers.

Included are also topics on rock bolting and prestressing; on pressure-grouted soil anchors, and rock slope stabilization.

Special features of this second edition are the selected bona fide references that are now added to the end of each chapter. These references provide a source to the reader for further detailed and specialized studies in rock mechanics.

To facilitate reading in this book, and because many readers do not use systematically a dictionary or encyclopedia of geologic and rock mechanics terms, a Glossary of Terms is appended to this book. An extended Glossary of Terms is of great value to readers in this book, particularly to engineering students in institutions where curricular objectives and constraints in the number of required courses exclude a course in engineering geology, or a course in physical geology, or even a course of an introductory geology. The Glossary of Terms is so much a part of the rock mechanics discipline that rock mechanics reading is unable to get far without the terms.

Increased emphasis is given to illustrative presentation. Photographs on rock structures; fractured rock specimens after their testing; the rupture surfaces in dry sand formed from various kinds of surface loading; large tunnels and oil storage caverns, and showing rock as a construction material — they all illustrate in a very graphic way as to how rock can be utilized for engineering construction purposes. The realism of the photographs and line drawings add greatly to the effectiveness of presentation of a subject such as rock mechanics where spatial, mental visualization is of paramount importance. The photographs in this book are refreshingly modern and unique, introducing the fundamental ideas in rock mechanics, and to provide the reader with an insight into the scope, meaning, and relationships used in rock mechanics.

May this book be interesting and beneficial to its readers.

A. R. JUMIKIS

Piscataway, New Jersey

Christmas 1982

FOREWORD TO THE FIRST EDITION

In this volume, the author presents an introductory segment to the relatively new civil engineering discipline known as engineering rock mechanics. This subject is presented here from the viewpoint of a civil engineer to civil engineers.

The content of this book deals with rock as an engineering construction material by means of which, upon which, and within which civil engineers build structures in rock. The discipline thus pertains to hydraulic structures engineering; to highway, railway, canal, foundation, and tunnel engineering; as well as to earthworks of, and substructures in, rock of all kinds in any way associated with engineering.

The main purpose of this book is to assist interested readers in understanding some of the basic rock mechanics principles as they apply to rock engineering. Hence, the book is developed basically as a guide in engineering rock mechanics.

Nowadays, rock mechanics develops very rapidly. The recent output of books, conference proceedings, research reports, and various articles on rock mechanics may be regarded as evidence of realization of the importance of the rock mechanics discipline in geotechnical engineering.

It should be said at this point that presently there are many published references and comprehensive discourses about various aspects of rock mechanics already available. As a matter of fact, rock mechanics literature indicates that such recently published material is literally soaked with new approaches in rock mechanics. However, in many instances some of the publications are not easy for a reader with limited time and/or knowledge to extract the basic information he needs; nor is it easy for him to evaluate and compare research papers and analyze the encountered engineering problems in rock. Hence, there is a need for abridging this condition. In other words, there is a need for a basic reading material such as the one presented in this book from which to learn about the nature of rock as an engineering construction material. These circumstances prompted the author to prepare this modest volume.

This book is written in simple language. It is easy to read. The basic concepts presented are easy to grasp and understand. Nevertheless, the text of the book is explicit. It concentrates on essentials, thus enabling the reader to obtain a certain amount of adequate knowledge for understanding and comprehending some basic principles, methods, and their use in rock mechanics, as well as the performance of rock under the influences of load, water, and temperature.

In essence, this unique volume emphasizes understanding. It gives a practical orientation to basic rock mechanics; provides a background as well as an outlook that motivates to further study; and will allow the reader to profit from his later studies of more comprehensive and complex publications on engineering rock mechanics than what is presented in this text.

May this book prove enjoyable and profitable to the reader.

ALFREDS R. JUMIKIS

College of Engineering
Piscataway, New Jersey

Christmas 1978

TABLE OF CONTENTS

FOREWORD TO THE SECOND EDITION 5
FOREWORD TO THE FIRST EDITION 7
PREFACE 19
ACKNOWLEDGMENTS TO THE SECOND EDITION 21
ACKNOWLEDGMENTS TO THE FIRST EDITION 25

Part 1
THE SUBJECT
Chapter 1
INTRODUCTION

1-1.	Some Historical Notes	31
1-2.	Manifestation for the Need of Rock Mechanics	34
	REFERENCES	35

Chapter 2
ROCKS

2-1.	Definition of Rock	37
	Geotechnics	40
	Bedrock	40
	Geologic Time Scale	41
2-2.	Classification of Rocks	44
	1. By Origin or Genesis	44
	2. Geological or Lithological Classification	62
	3. Engineering Classification of Intact Rock	62
	4. Rock Quality Designation (R.Q.D.)	64

2-3.	Rock as a Construction Material	64
	REFERENCES	71
	OTHER RELATED REFERENCES	72

Chapter 3
ROCK MECHANICS

3-1.	Definition of Rock Mechanics	73
3-2.	Some Features of Rock Mechanics	74
3-3.	Rock Mechanics Problems	75
3-4.	Objectives of Rock Mechanics	76
3-5.	Value of Rock Mechanics	78
3-6.	Theoretical Basis of Rock Mechanics	78
	General Notes	78
	Theory of Elasticity	79
	Theory of Plasticity	81
	Difference between the Engineering Discipline of Strength of Materials and Rock Mechanics	82
3-7.	Rock Engineering	84
3-8.	Ice Mechanics	84
	Permafrost	86
	Permacrete	87
3-9.	Work in Rock Engineering	88
	REFERENCES	89

Part 2
ROCK EXPLORATION

Chapter 4
METHODS OF ROCK EXPLORATION

4.1.	Need for Rock Exploration	93
4-2.	Exploration Methods	93
4-3.	Geological Exploration	94
	Subsurface Exploration	97
	Drilling Fluids	102
	Small Rock Samples versus Large Ones	113
4-4.	Hydrological Exploration	118
	Karst Regions	120
4-5.	Geophysical Exploration	120
4-6.	Thermal Exploration	121
4-7.	Evaluation of Rock as to its Diminution and Workability	122

TECHNOLOGICAL PROPERTIES OF ROCKS

4-8.	Properties	122
4-9.	Physical Properties	123
	Mineralogical Composition	124
	Rock Structure	124
	Rock Texture	124
	Specific Gravity	127
	Unit Weight	129
	Porosity of Rock	132
	Void Ratio	133
	Water in Rock	133
	Moisture Content	134
	Degree of Saturation	136
	Permeability	136
	Chemical Action	143
	Hardness of Water	143
	Gases	144
4-10.	Electrical Properties	146
	Radioactive Properties	147
4-11.	Thermal Properties of Rocks	147
	Heat	147
	Heat Capacity	150
	Specific Heat	150
	Latent Heat of Fusion	151
	Thermal Conductivity	152
	Thermal Diffusivity	152
4-12.	Coefficient of Thermal Expansion and Contraction	153

MECHANICAL PROPERTIES OF ROCKS

4-13.	General Properties	156
	Hardness	156
4-14.	Durability	158
4-15.	Elasticity	158
4-16.	Plasticity	159

CRITERIA OF ROCK FAILURE

4-17.	General Notes about Failure Criterion	161
4-18.	Various Criteria	162
	Maximum Tensile Stress Criterion for Rock	162
	Maximum Shear Stress Criterion	162
	Coulomb's Criterion of Failure	162

TABLE OF CONTENTS

Mohr's Criterion of Failure 166
Griffith's Criterion of Tensile Failure 168

DEFORMATION OF ROCKS

4-19. Definition 169
4-20. Mutual Dependence of Stress and Strain 169
4-21. Stress-Strain Diagrams 169
       Work and Strain Energy 171
       Factors Affecting Deformation 172
4-22. Stress-Strain Diagrams for Rocks 177
       Creep 180
4-23. Appearance of Deformed Rock Specimens 182
4-24. Modulus of Elasticity of Rocks 186
4-25. Poisson's Ratio 188
4-26. Strength Properties of Rocks 189
       Static Strength 191
4-27. Static Laboratory Compressive Strength 191
4-28. Tensile Strength of Rock 201
4-29. Shear Strength of Rock 206
       Methods of Shear Testing of Rocks in the Laboratory 207
       A. Direct Shear Strength Tests with Normal Stress on the Shear Plane Absent 210
       Single-Shear Test 210
       Double-Shear Test 210
       Punch Shear 210
       B. Direct Shear Strength Tests with Normal Stress on the Shear Plane Present 213
       Direct "Box" Shear Strength Test on Rock 213
       Direct Shear Strength Tests on Rock Cubes 216
       C. Torsion Tests 219
       Friction 221
4-30. Dynamic Properties of Rocks 222
4-31. Summary on Laboratory Testing of Rocks 224

IN-SITU STRENGTH OF ROCKS

4-32. Properties of In-Situ Rocks 225
4-33. Deformability 227
4-34. In-Situ Testing of Rocks 228
       Plate Loading Test 228
       Jacking Tests 231

	Cable Jacking Test	232
	Pressure Chamber Test	232
	Large-Scale Compression Strength Test	234
	Borehole Deformation Test	234
	In-Situ Tension Test	235
4-35.	In-Situ Shear Strength Test	235
	Torsion Shear Test	241
4-36.	Internal Stresses in a Rock Mass	242
	In-Situ Test of Stresses in a Rock Mass	243
	Direct Strain Measurement	244
	Dilatometer Test	244
	Flat-Jack Testing Method	245
	Overcoring Technique	247
	Hydraulic Fracturing Test of Rock	248
4-37.	Permeability	250
	REFERENCES	254
	OTHER RELATED REFERENCES	260

Chapter 5
ROCK MASS PROPERTIES

5-1.	Mechanical Defects of Rocks	265
5-2.	Fractures	267
5-3.	Joints	269
5-4.	Faults	272
5-5.	Folds	276
5-6.	Summary on Rock Mass Weaknesses	281
	Influence of Rock on Stability of Dams and Foundations	283
5-7.	Grouting	284
	Bituminous Grouting	288
	Clay Grouting	290
	Chemical Solidification of Soil	290
5-8.	Lugeon Water Pressure Test	292
5-9.	Gunite	293
5-10.	Shotcrete	293
	REFERENCES	296

Part 3
STRESSES IN ROCK ABOUT CIRCULAR UNDERGROUND OPENINGS

Chapter 6
STRESS FIELDS

6-1.	Force Field	301
6-2.	Primary Stresses in Sound Rock	302
6-3.	In-Situ State of Stress	306
6-4.	Stress Fields	307
	REFERENCES	310

Chapter 7
ELASTIC STRESS ANALYSIS IN ROCK ABOUT UNDERGROUND OPENINGS

7-1.	Underground Openings	311
7-2.	Secondary Stress Conditions in Rock	311
7-3.	Stresses in a Thick-Walled Cylinder	312
7-4.	Theoretical Basis for Stress Analysis in Rock	318
7-5.	Specialization of Stress Equations	326
7-6.	Discussion of Equations	328
7-7.	Elastic Deformations in Sound Rock upon Excavation of a Cavity	331
7-8.	Zone of Elastic Tangential Tensile Stresses	333
	REFERENCES	335

Chapter 8
PLASTIC ZONES IN ROCK AROUND UNDERGROUND OPENINGS

8-1.	Concept of Plastic Zone	337
8-2.	Derivation of Plasticity Condition in Rock	337
8-3.	Extent of Plastic Zone	341
8-4.	Discussion on Stresses in Elastic and Plastic Zones	344
8-5.	Zone of Disturbance	347
8-6.	General Notes about Slip Lines	348
8-7.	Form of Slip Lines	350
8-8.	Summary about Slip Lines	353
	REFERENCES	353

Chapter 9
STRESSES IN ELASTIC ROCK AROUND VERTICAL SHAFTS

9-1.	Primary Stress Conditions in the Elastic Zone	355
9-2.	Secondary Elastic Stress Conditions	356
9-3.	Discussion	358
9-4.	Secondary Stress Conditions in the Plastic Zone	361
9-5.	Other Forms of Openings and Modes of Stress Distribution in Rock	363
	REFERENCES	364

Chapter 10
SOME ENGINEERING PROBLEMS ASSOCIATED WITH WORK IN ROCK

10-1.	Kinds of Problems in Rock Engineering	365
	Rock Pressure	366
10-2.	Inferiorities of Rocks	367
10-3.	Subsurface Water	368
10-4.	Temperature	368
10-5.	Hazards in Various Kinds of Rocks	369
	Igneous Intrusive Rocks	369
	Igneous Extrusive Rocks	370
	Sedimentary Rocks	370
	Requirements in the Design in Rock	371
10-6.	Factors Contributing to Possible Hazards in Rock Engineering	372
10-7.	Stress Relaxation	373
10-8.	Hazards from Earthquakes and Faults	375
	Nature of Earthquakes	375
	Volcanic Earthquakes	376
	Tectonic Earthquakes	376
	Hazards of Earthquake Destruction	377
	Earthquake Intensity	378
	Earthquake Magnitude	382
	Hazards from Faults	390
	San Andreas Fault	391
	Codes	394
10-9.	Hazards from Water	396
	Water Movement Through Rock	396
	Hydraulic Fracturing	397
	Solution Cavities	397

Other Water Problems 398
Geological Hazards from Rock 398
Hydrostatic Uplift Pressure 400
Hoover Dam 403
Canelles Dam 404
10-10. Failure of the Malpasset Dam 408
10-11. Landslides 410
Nature of Landslides 410
Mud or Debris Flows 414
Creep 414
Solifluction 415
Quick-Clay Slides 415
Rock Falls 417
Niagara Waterfalls 417
Rock Slides 421
10-12. The Vaiont Reservoir Disaster 423
10-13. On the "Crushing Drama" 426
REFERENCES 427
OTHER RELATED REFERENCES 433

Part 4

STABILIZATION OF ROCK

Chapter 11

ROCK REINFORCEMENT

11-1. Anchoring 437
11-2. Rock Bolting 437
Requirements of Anchorage 440
11-3. Effect of Rock Bolting on Shear Stress 440
11-4. Rock Bolt Support 443
11-5. Determining Rock Bolt Support 445
11-6. Some Advantages and Disadvantages of Rock Bolting 448
11-7. The Williams Rock Bolt 448
11-8. Rock Caverns 451
Tunnel Beneath Oroville Dam 451
Washington, D.C.'s Metro 451
Wire Net 452
Underground Locomotive Maintenance Workshop of the
Norwegian State Railways 452
Underground Oil Storage Plant in Göteborg 455
World Trade Center Basement Foundation 455

11-9.	Prestressing	457
11-10.	Pressure-Grouted Soil Anchors	460
	REFERENCES	466
	OTHER RELATED REFERENCES	468

Chapter 12
ROCK SLOPES

12-1.	Definitions	471
12-2.	Factors Contributing to Slope Failure	472
12-3.	Stability of Rock Slopes	475
12-4.	Reinforcement of Rock Slopes	477
12-5.	Effect of Precipitation on Stability of Rock Slopes	482
12-6.	Rock-Slope Stability Based on Rock Strength	482
12-7.	Closing Remarks about Stability of Rock Slopes	485
12-8.	Stability Analyses of Soil Slopes	486
	Seepage Force	489
12-9.	Stability Analyses of Rock Slopes	493
12-10.	Sliding of a Rock Block on Geologically Predetermined Planar Discontinuity	493
12-11.	Anchoring of Rock Slopes	498
12-12.	Earthquake Effects	501
	General Notes	501
	Seismic Forces	501
	Earthquake Effect on Earth Retaining Wall	502
12-13.	Stabilization of Rock Slopes by Means of "Reticulated Root Piling"	504
12-14.	Methods of Remedy Against Rock Slides	507
	Rockfall Control by Means of Wire Mesh	509
	Rock Slope Treatment by Gabions and Wire Mesh	510
	REFERENCES	511
	OTHER RELATED REFERENCES	513

APPENDICES

Appendix 1	Greek Alphabet	517
Appendix 2	Key to Signs and Notations	519
Appendix 3	Glossary of Terms	529
Appendix 4	Rock Defects	561
Appendix 5	Conversion Factors for Units of Measurement	563
	A. Use of SI Units in Geotechnical Engineering	563
	B. Conversion Factors	564

TABLE OF CONTENTS

Appendix 6	Dynamic Viscosity Tables for Water	572
Appendix 7	Dynamic Viscosity Correction Factor Tables for Water	576

About the Author 581
INDEX 583
Author Index 585
Subject Index 591

PREFACE

The various aspects of engineering rock mechanics are concerned with engineering design and construction in rock.

In the study of engineering rock mechanics, rock is considered to be a construction material by means of which, upon which, and in which engineers build structures.

Generally, the geotechnical engineer must know the physical, mechanical, and strength properties of the materials with which he works.

From the civil engineering point of view, the increasing magnitude in size and scope of engineering projects designed and built in rock, and the subsequent resulting increase in responsibility to produce such projects, demand from the engineer qualitative as well as quantitative knowledge about rock as a construction material.

Rocks loaded by a structure may undergo displacements and deformations or, if overloaded, they may become damaged by cracking or crushing. Hence, the effects of loads on rock strata depend on the physical properties of such materials. In any subsurface excavation and construction, the strength properties of rocks must be known.

Although a commonly used construction material, rock is a very complex one. It is very difficult to describe, and most difficult to define. Hence, the study of rocks in engineering rock mechanics is an important branch of civil engineering in general, and in foundation engineering, hydraulic structures engineering, and subsurface construction engineering in particular.

In designing structural foundations in rock, engineers encounter problems which are frequently of a geological nature. It is so because all structures are laid on the ground.

Because geological and environmental factors expose difficulties when working in rock and, thus, influence the cost of construction, one's familiarization with the knowledge of physical geology as well as with the principles of engineering geology is helpful as well as profitable in engineering rock mechanics and rock engineering practice. However, where geology serves the purpose of construction, geology must be subordinated to the laws of engineering mechanics.

The basic idea in design of an engineering structure on or in rock is that the structure and the structure-supporting rock *en masse* as an integral unit should be considered as an inseparable, integral, functional and statical system, and that in this sense, every statical analysis of such rock structure systems should conform to and bear with this situation, and should be so dealt with.

This is so because the engineer has to accept the rock in which he builds as it occurs in its natural state with all of its own weaknesses, characterized by stratification, jointing, fisssures, faults and folds, variable rock properties, weathering, and the like. Of course, the strength properties of the rock may be better with increase in depth than at/or near the ground surface. Or else the strength of the rock may be increased by chemical injections and grouting, thus increasing the stability of the structure. These latter measures, however, are usually limited by some economical and/or technical restrictions.

If proper consideration is to be given to the most economical design and safe performance of a structure on or in rock, adequate knowledge about the engineering properties of the subsurface rocks must be available. Thus, because of todays large construction projects in rock, the engineer resorts to the theory of rock mechanics.

Although there is no theory of strength of rocks available as yet, in the context of rock mechanics, the latter may be in a way considered as the youngest branch of the subject of strength of materials dealing with rock as a construction material.

To conclude, it is apropos to say that the period of empiricism in handling rocks is now over. In its place, there enter scientific and new engineering concepts. The basic prerequisite for designing and constructing in rock, besides geological structures, stratigraphy and other pertinent information, is the knowledge of rock as an engineering material.

ACKNOWLEDGMENTS
TO THE SECOND EDITION

The author expresses once more his gratitude to the many colleagues in this country and also abroad for factual information, illustrative material, and for permissions to publish the material. The following colleagues, engineers, firms, government agencies, authorities and institutions generously provided photographs for the second edition of Rock Mechanics:

- Mr. JOHN ADAMS, Chief Engineer, Saint Lawrence Seaway Development Corporation, Massena, New York
- Mr. NEAL ANDERSEN, P.E., Chief, Office of Transportation Laboratory, Department of Transportation, Sacramento, California
- Mr. SVEN BYLUND, Svenska Entreprenad A.B. (SENTAB), Stockholm, Sweden
- Professor Dr. D. F. COATES, McGill University, Montreal, Canada, and Department of Energy, Mines, and Resources, Ottawa, Canada
- Mr. JULIO O. COTTA, Cimentaciones Especiales, S.A., Madrid, Spain
- Mr. DENNIS E. DEUSCHL, Director, Office of Communications and Consumer Affairs, Saint Lawrence Seaway Development Corporation, U.S. Department of Transportation, Washington, D.C.
- Mr. W. EBERT, Ingersoll-Rand Company, Rock Drill Division, Phillipsburg, New Jersey
- Mr. VERNON K. GARRET, Director, Office of Engineering, Washington Metropolitan Area Transit Authority (WMATA)
- Professor Dr. VICTOR GREENHUT, College of Engineering, Rutgers University
- Professor Dr. MARTHA M. HAMIL, Department of Geology, Rutgers University

Mr. H. Hartmark, NSB Hovedadministrasjonen. Norwegian State Railways, Oslo, Norway

Mr. Kurt Hellblom, NYA ASFALT A.B., Rock Division, Stockholm, Sweden

Engineer Andris A. Jumikis, P.E., Highland Park, New Jersey

Dr. J. L. Kulp, Vice President, Research and Development, Weyerhaeuser Company, Tacoma, Washington

Dr. Fernando Lizzi, Technical Director, FONDEDILE, Naples, Italy

Mr. Richard L. Marsh, Rock of Ages Corporation, Barre, Vermont

Mr. Paul J. Myatt, Washington Metropolitan Area Transit Authority

Mr. R. S. O'Neil, P.E., Consulting Engineer, Senior Vice President of De Leuw, Cather and Company, Washington, D.C.

Mr. Joseph L. Newtown, Security and Information Officer, Power Authority of the State of New York, St. Lawrence Power Project, Massena, New York.

Dr. W. C. Ormsby, Chief, Chemistry and Coatings Group, Federal Highway Administration, Washington, D.C.

Mr. W. W. Peak, Chief Engineering Geologist, California Division of Water Resources, Sacramento, California

Mr. Mario J. Pirastru, Regional Administrator, Niagara Frontier State Park and Recreation Commission, Niagara Falls, New York

Mr. Phil Portlock, Washington Metropolitan Area Transit Authority (WMATA)

Professor Dr. M. Rocha, President, Council of Engineering Research, Lisbon, Portugal

Professor G. Schnitter, Swiss Federal Institute of Technology, Zürich, Switzerland

Professor Dr. Edgar Schultze, Technical University Aachen, Aachen, Germany

Mr. Vasily Serpikov, P.E., Resident Engineer, Charles J. Kupper, Inc., Piscataway, New Jersey

Mr. Carl T. Sorrentino, Department of Highways, Denver, Colorado

Dr. Leif Viberg, Research Engineer, Swedish Geotechnical Institute, Linköping, Sweden

Mr. Thomas A. Wilkinson, District Geologist, Buffalo District, Corps of U.S. Army Engineers, Buffalo, New York

Mr. RON WILLIAMS, Jr., Vice President, Williams Form Engineering Corporation, Grand Rapids, Michigan

Mrs. MARY C. WOODS, Geologist and Editor-in-Chief of California Geology, California Division of Mines and Geology, Sacramento, California

Mr. DONALD L. YORK, Chief, Geotechnical Engineering. The Port Authority of New York and New Jersey

Aero Service Corporation, Philadelphia, Pennsylvania

California Department of Transportation

California Division of Mines and Geology

California Division of Water Resources

Canadian Consulate General, New York, N.Y.

Colorado Department of Highways

La Direzione di Esercizio del Traforo del Monte Bianco, Courmayeur, Italia

Division of Building Research, National Research Council of Canada, Ottawa

Kollbrunner-Rodio Foundation, Institute for Engineering Research, Zürich, Switzerland

Charles J. Kupper, Inc., Consulting Engineers and the Middlesex County Sewerage Authority, New Jersey

National Academy of Sciences, Washington, D.C.

National Park Service, U.S. Department of the Interior, Washington, D.C.

Ohio Department of Highways

Ontario Ministry of Industry and Tourism, Toronto, Canada

The Port Authority of New York and New Jersey, New York, N.Y.

Power Authority of the State of New York — St. Lawrence Power Project, Massena, New York

Rutgers University Press, New Brunswick, New Jersey

Rutgers University Library of Science and Medicine

Schweizerischer Wasserwirtschaftsverband, Baden, Switzerland

SCIENCE, a publication of the American Association for the Advancement of Science, Geology Section, Washington, D.C.

Swedish State Railways' Geotechnical Commission, Stockholm, Sweden

Swedish Geotechnical Institute, Linköping, Sweden

U.S. Corps of Engineers, Buffalo District, Buffalo, New York

U.S. Corps of Army Engineers, Philadelphia District, Philadelphia, Pa.

U.S. Department of Transportation, Saint Lawrence Seaway Development Corporation, Washington, D.C.

U.S. Geological Survey, Denver, Colorado.

Acknowledged is also the tangible assistance provided by Dr. E. G. NAWY, Chairman of the Department of Civil and Environmental Engineering of Rutgers University to the preparation of the Second Edition of this text. The typing and processing of the manuscript were done by Mrs. DORIS CLARK, and Mrs. TOSHIYE AOGAICHI.

A word of special appreciation goes also to the Editor and Publisher of TRANS TECH PUBLICATIONS, Dr. R. H. WÖHLBIER, for his many suggestions relative to the shaping of the book and to Dr. P. I. WELCH and R. WELCH, B.Ed., University of Cambridge, England, for carefully checking and proofreading the Second Edition.

Finally, the author is also grateful to his wife and son for their untiring support given to the author during the preparation of the Second Edition of Rock Mechanics.

A.R.J.

ACKNOWLEDGMENTS TO THE FIRST EDITION

The author expresses his cordial thanks to all those who contributed intellectually and materially, in words and in deeds, to the preparation of this manuscript for publication.

For tangible furtherance of this work, the author expresses his appreciation to Dr. E. H. DILL, Dean, College of Engineering, Rutgers, The State University of New Jersey, New Brunswick, New Jersey.

For providing secretarial services for typing this manuscript, the author expresses his sincere thanks to Dr. R. C. AHLERT, Executive Director of the Bureau of Engineering Research, and to Dr. J. WIESENFELD, Chairman, Department of Civil and Environmental Engineering, both at Rutgers University.

The author is especially grateful to Dr. EVERT HOEK for his furnishing to the author his great works on Rock Slope Engineering, and Underground Excavation Engineering, and for his constructive suggestions.

The author also acknowledges Dr. EDWARD J. CORDING's helpful written discussion about rock wedges formed in a circular sidewall of an underground opening.

The author is also most appreciative of the excellent library service of the Rutgers Library of Science and Medicine. Its efforts in procuring the many reference sources and books permitted personal examination and verification of facts in original, bona fide references.

To the authors credited in the documentation of this book, the writer owes a great deal, because some of their ideas and work helped in building up this work. Credit to these and other authors is given through references in the text as extensively as possible.

In addition to sources cited in the body of the text, the author acknowledges the many publications made available to him by various persons and organizations. Permissions to reproduce their published material were granted by the following individuals:

Dr. Arthur Casagrande, Harvard University
Dr E. Cording, Professor, University of Illinois
Dr. D. U. Deere, formerly Professor in the University of Illinois
Dr. E. Hoek, formerly Professor in the Imperial College of Science and Technology, London
Dr. E. R. Leeman, Assistant Director, National Mechanical Engineering Research Institute, South African Council for Scientific and Industrial Research, Pretoria, South Africa
Dr. M. Rocha, President, Council of Civil Engineering Research, Lisbon, Portugal.

The author is also glad to acknowledge the following publishers for the permissions granted for him to use their material:
Plenum Publishing Corporation, New York
McGraw-Hill Book Company, New York
E. and F. N. Spon Ltd., London
Springer Verlag, Berlin
John Wiley and Sons, Inc., New York.

This book has been prepared within the course of the author's academic, research, and practice activities relating to rocks and engineering rock mechanics and, in a way, is the outgrowth of a set of lecture notes prepared for a course entitled "Engineering Rock Mechanics," taught by the author in the College of Engineering at Rutgers, The State University of New Jersey.

A separate word of thanks is also expressed to the author's family. The author's son, Andris Alfreds Jumikis, P. E. in Civil Engineering, participated in research about the engineering properties of Triassic shale; helped in laboratory testing of granites, gneisses, and marbles; prepared a number of pencil sketches and all line drawings for this book; and assisted the author in his field work. The author's wife, Zelma Albertine Jumikis, deserves special credit for her patience in organizing and maintaining the voluminous files of notes pertaining to this work, and for her never-tiring moral support for this project.

Special thanks are conveyed herewith to Mrs. Erma M. Sutton for editing the author's handwritten draft of the manuscript for this book and for carefully typing and processing the manuscript. Her close attention to the work greatly expedited its preparation.

Likewise, appreciation is expressed to Mrs. DORIS CLARK for her patient typing of parts of the manuscript.

The publisher has lived up to its best tradition, and the author's cordial thanks go to all concerned in the production of this book.

A. R. J.

To
Zelma
and
Andris

PART 1
THE SUBJECT

CHAPTER 1

INTRODUCTION

1-1. Some Historical Notes

For the purpose of one's orientation, let us avail ourselves of a brief historical review about the development of rock mechanics, but more specifically of rock mechanics technology.

Since prehistoric times, the various strength properties of rocks have been recognized by primitive man. The caveman was aware of the stability of the roof of his cave when choosing his shelter. The Stone-Age man tried to choose optimum places to strike a piece of rock when carving tools and making flint modules for weapons. Also, by necessity, it was of interest throughout the history of human civilization to understand the performance characteristics of rock when subject to loads.

Likewise, the art of subsurface excavation has been known to mankind for many centuries. Subsurface mining, first in pits and later through shafts, began during the Stone Age some 15,000 years ago.

It is reported (7)* that tunneling started about 3,500 B.C. during the Bronze Age for copper ore mining, through adits on the mountain slopes in the Sinai Peninsula.

Limestone rock as an engineering material was used as a cutting edge for brick caissons in Egypt about 2,000 B.C., to drive a vertical shaft through conglomerate and sand layers until it reached the limestone bedrock. The cutting edge was made of a round limestone block with a vertical hole pierced through its middle (1).

It is interesting to observe that besides serving mining purposes, many tunnels in ancient times were driven also for water supply, public under-

---

* Numbers in parentheses refer to References at end of each part of the book.

ground passage, erecting of temples, building of tombs, military purposes, and other uses. For example, a road tunnel near Naples, Italy, driven about 36 B.C., was about 1,220 m long, 9.15 m high, and 7.60 m wide.

The first American nonmining tunnel, the Auburn Tunnel, was driven during the years 1818—1821, as part of a canal system in Pennsylvania.

The first American railway tunnel, the Allegheny Portage in Pennsylvania, dates back to 1831—1833.

The Mont Cenis Tunnel, about 11 km long, was the first to pierce the Alps (1857—1870).

The Mont Blanc Tunnel (~ 11 km long) was driven from 1959—1963.

All these and many other examples of tunneling merely point out the intuitive use of empirical knowledge for utilizing rock for engineering purposes long before science and engineering could supply the logics and theoretical background knowledge. Science moves slowly, creeping on from point to point. The same applies also to rock mechanics.

For the most part, European engineers and geologists in the past 25 years are credited with the development of the principles of rock mechanics, including the methods and techniques of rock testing in situ in connection with applications to a significant number of underground power plants and arch-type dams (Austria, France, Germany, Norway, Sweden, Switzerland, Portugal, Italy).

Use of rock mechanics in the United States was mainly limited to mining engineering until the early 1960s. At that time, geotechnical engineers found rock mechanics useful in hard-rock problems. This is somewhat comparable to the application of soil mechanics to problems of earthworks and foundation engineering.

Major organizations in the United States studying rock mechanics are: Society for Rock Mechanics, universities, U.S. Army Corps of Engineers laboratories, Department of Mines, U.S. Department of Transportation, Bureau of Reclamation, National Bureau of Standards, and some private consulting firms.

In 1964, the activities of the American Society for Testing and Materials were broadened to include testing of rocks.

Rock mechanics, first recognized as a distinct discipline only since about 1950, is now advancing in the accuracy, precision, and plausibility of its techniques, and in the rigorousness of its research methods. It can record a considerable progress in theory, experiment, and reliability in testing methods. Especially, progress has been made in determination of stresses in rock; shear strength testing; and reinforcement of rock, or rock bolting.

Reports on research about these topics were presented in three state-of--the-art papers at the 69th Annual Meeting of the ASTM, held June 26 to July 1, 1966, in Atlantic City, New Jersey (2).

The need for the new discipline Rock Mechanics was voiced at the 5th International Conference on Soil Mechanics and Foundation Engineering held in 1961 in Paris, France (9), and was emphasized at the 6th International Conference in 1965 held in Montreal, Canada (10).

No better proof is needed than the fact that the 1st International Congress on Rock Mechanics held in September 1966 at Lisbon, Portugal (11), resulted in three huge volumes of congress papers on rock mechanics dealing with exploration of rock masses; properties of rocks and rock masses; residual stresses in rock masses; comminution; natural and excavated slopes in rocks made in open cuts; underground excavations; deep borings; the behavior of rock masses as structural foundations; and support of massive dams and other engineering structures. Also, brought to the fore was the need for a uniform, standardized, static and dynamic strength test of rocks in situ as well as in the laboratory, and the correlation of such test results.

More than 800 educators, engineers, and engineering geologists from 42 countries attended this historical Congress to review failures of dams and rocks, their causes, and the need for systematically organized knowledge on theoretical, experimental, and applied engineering rock mechanics.

Thus the First International Congress on Rock Mechanics signified that for the first time in history there was a concerted awareness among engineers (11) and engineering geologists that complex geomechanical problems may respond to quantitative analyses, and that the benefit of such detailed analyses are those of better understanding and consequent better control of natural phenomena encountered in rock engineering.

The U.S. National Research Council's study in 1967 on the value of and need for rock mechanics (6) recommended an acceleration of the total effort in rock mechanics research.

Since 1966, many symposia, conferences, congresses, short courses, and courses on rock mechanics within the course for continued education have been held here and abroad.

International Congresses on Rock Mechanics have been held in Lisbon, Portugal (1966, ref. 11); Belgrad, Yugoslavia (1970, ref. 12); Denver, Colorado, USA 1974, (ref. 13); and Montreux, Switzerland (1979, ref. 14).

The first international congress on engineering geology was held in 1970 in Paris, France; the second one in 1974 in São Paulo, Brazil; the third one in 1978 in Madrid, Spain, and the fourth one in 1982 in New Delhi, India.

1-2. Manifestation for the Need of Rock Mechanics

For today's megalopolis, the designing and constructing in rock of underground openings for vehicular and/or aqueous tunnels, underground garages, underpinnings of structures, shelters, storage spaces, underground hydroelectric power plants in connection with requirements for more electrical energy and utilization of "white coal" for that purpose, earth and massive dams, foundations, and other civil engineering structures require the turning over of an enormous amount of earth and rockworks under various and difficult geologic and topographic conditions. Tunnels driven under rivers frequently prove to be more economical than bridges.

As a consequence of the development of our age of specialization, urbanization, industrialization, and conservation of water resources, subsurface excavations in rock are often many times greater now than in the past, and rock slopes associated with highway, railway, causeway, and open-pit engineering are now often several hundred feet high.

Also, it has been said that some phases in progress in engineering, in adding to the store of man's knowledge, have been the results of intensive research on failures of some engineering structures, such as failures of some massive dams founded on rock.

Past failures of rock as a dam-supporting foundation (3, 4, 5) induced the profession to make a collective effort to try to interpret, evaluate, and to control rock performance under load, water, and temperature (8, 15). Thus, the need for knowledge and experience in engineering rock mechanics is coming more and more to the fore. Large-scale work in rock has been handicapped by lack of understanding of the complex mechanical behavior of the wide variety of geological materials, among rocks. The tremendous scope of engineering projects today dealing with rock as a construction material on a scientific-technical basis requires solutions of engineering problems hardly known before. In the service of safety of life, such engineering activities require civil engineers to design and construct safe and reasonably economical structures. To cope with the unprecedented increase in size and scope of such structures in rock, and to handle strength problems in rocks, today imposes upon the engineer a demand for an extensive, appropriate knowledge not merely of soil mechanics but of rock mechanics as well.

REFERENCES

1. Anon., *Engineering News Record*, December 7, 1933, p. 675.
2. ASTM, *Proceedings of the 69th Annual Meeting of the ASTM*, held June 26 to July 1, 1966 in Atlantic City, New Jersey. Philadelphia, Pa., 1966.
3. JAEGER, C., 1963, "The Malpasset Report," *Water Power*, vol. 15, No. 2, February, 1963, pp. 55—61.
4. JAEGER, C., 1972, *Rock Mechanics and Engineering*. Cambridge at the University Press, 1972, pp. 323—339.
5. KIERSCH, G. E., 1964, "Vaiont Reservoir Disaster," Civil Engineering, March, 1964, vol. 34, No. 3, pp. 31—39.
6. National Academy of Sciences — National Academy of Engineering, National Research Council: *News* Report *XVII*, No. 2, February, 1967, p. 4.
7. NASIATKA, Th. M., 1968. *Tunneling Technology*, Washington, D.C.: U.S. Department of the Interior, Bureau of Mines, U.S. Government Printing Office. Information Circular 8375.
8. PÉQUIGNOT, C. A., (Editor), 1963, *Tunnels and Tunnelling*. London: Hutchinson and Co., Scientific and Technical Publishers LTD.
9. *Proceedings of the 5th International Conference on Soil Mechanics and Foundation Engineering*, held July 17 to 22, 1961, in Paris, France. Published by Dunod, Paris, 1961.
10. *Proceedings of the 6th International Conference on Soil Mechanics and Foundation Engineering*, held September 8 to 15, 1965 at Montreal, Canada. Published by the University of Toronto Press.
11. *Proceedings of the (First) Congress of the International Society of Rock Mechanics, 1966*, held September 25 to October 1, 1966 at Lisbon, Portugal (3 volumes: vols. 1 and 2 — 1966; vol. 3 — 1967).
12. Proceedings of the Second Congress of the International Society of Rock Mechanics, 1970, held September 21—26, 1970, at Belgrade, Yugoslavia. Published by the Institute for Development of Water Resources. Jugoslav CERNI in 1971 at Belgrade.
13. *Proceedings of the Third Congress of the International Society for Rock Mechanics, 1974*, held September 1 to 7, 1974, at Denver, Colorado. Published by the National Academy of Sciences, Washington, D.C.
14. *Proceedings of the Fourth Congress of the International Society for Rock Mechanics*, 1979, held September 2 to 8, 1979 at Montreux, Switzerland.
15. SMITH, N., 1971, *A History of Dams*. London: Peter Davies.

CHAPTER 2

ROCKS

2-1. Definition of Rock

Generally, to the geologist the term "rock" applies to all constituents of the earth's crust. He speaks about "consolidated" rock and "unconsolidated" rock (soil). To the civil engineer, especially the geotechnical engineer, the term "rock" is understood to apply to the hard and solid formations of the earth's crust. The derivatives of rocks, the weathering products, are soils (the unconsolidated sediments). One should be aware that the genesis, viz., formation of geological structures, was not designed by the handbook of some standards. The very heterogeneity of rocks does not lend access to their mathematical analysis and rigorous quantitative study. However, with the future development in rock mechanics which lies ahead of us, hopefully a more exact and quantitative treatment will be possible.

There are several definitions of rock. Some of them are given below.

A rock is a mixture of one or more different minerals. It has no definite chemical composition. Also, it is said that rock is an aggregate of fused or compressed discrete mineral particles. Thus, various minerals are combined in various proportions to form rock.

Broadly speaking, the term "rock" is a general geological term which comprehends any naturally occurring aggregates of minerals or mass of mineral matter, whether or not coherent, constituting an essential part of the earth's crust.

To the geologist, rocks are essential units of the earth's crust. To the civil engineer, the physical and strength properties of rocks are of utmost concern.

Rock is also considered to be mineral and organic (soft coal and anthracite, for example) matter that comprehends the solid part of the earth's crust, excluding soil. TERZAGHI (14) defines soil as "sediments and other unconsolidated accumulations of solid particles produced by the mechanical or chemical disintegration of rocks." Thus, the distinction between rock and soil is in the degree of consolidation and in the limit of the size of the particles.

A simple concept of rock has been presented by EMERY (4), according to whom "Rock is a granular material composed of 'grains and glue.' There is nothing else involved." The "glue" may be ferroginous, calcareous, argillaceous, or siliceous material which cements the grains. Thus, "rock may be described, then, as a granular aelotropic, heterogeneous technical substance which occurs naturally and which is composed of grains of varied polycrystalline or noncrystalline (amorphous) materials which are cemented together either by a *glue* or by a *mechanical bond,* but ultimately by atomic, ionic and molecular bonds within the grains and the glue and at every interface of bonding."

To the engineer who builds on, in, and by means of rock, the term "rock" signifies firm and coherent or consolidated substances that cannot normally be excavated by manual methods alone. To him, rock is a material having manifold properties, like any other material.

In designing in rock, engineers frequently assume rock as a homogeneous and isotropic medium. However, most rocks are not sound; hence they are neither homogeneous nor isotropic.

A *homogeneous* rock material is one in which the physical properties of all of its parts or elements are the same, as opposed to a heterogeneous material.

Heterogeneity (nonhomogeneity or inhomogeneity) is a characteristic of a material, or a force field signifying that the material has properties which vary with position within it (unlike qualities).

An *isotropic* material is one that has the same physical, viz. elastic, properties in every direction at any point. For example, in an elastic medium the velocities of elastic wave propagation are independent of direction.

An *anisotropic* (aeolotropic) material is one with certain of its properties varying with direction at any point. For example, many rocks have preferred particle and crystal orientation. Or, because of the state of the stress of the geological formations, sound waves are transmitted through them with different velocities in the horizontal and vertical directions. Hence, such rocks are said to be anisotropic. Therefore they react differently to forces in different directions, depending upon the degree of anisotropy. Thus, the term "anisotropic" is used to mean having different orientations of mechanical properties and fabric.

All these definitions of rock convey that mechanically, rock is a multiple-body system (see Fig. 2-1), and an extremely complex material difficult to work with.

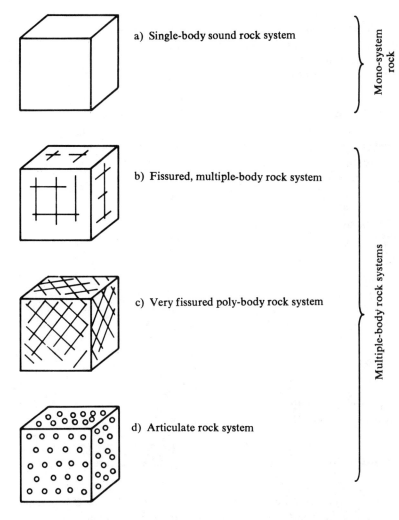

Fig. 2-1
Single and "multiple-body" rock systems

Sometimes rock is defined as a hard, compact material that under stress responds uniformly in deformation and in proportion to the force applied, within certain limits. Such a definition would imply an assumption that rock is an elastic material within these limits. Such an assumption depends to a large degree of course upon the composition and structure of the rock.

A *solid* (compact, fast, firm, rigid, stable, strong) is one which exerts a considerable resistance to separation of its individual particles and/or component parts.

A rock, unlike a steel which can be refined to a consistent internal state before its use, is a naturally occurring material and must be used in its natural state. Certain simplifying assumptions are justified to assist design processes, others are not, and to a large extent the basis for all assumptions lies in the composition and structure of the rock.

As geology has grown more quantitative, more of its parts have become simplified by the reduction of verbal descriptions to numerical measurements, and by the use of mathematics to express relationships among the numbers. However, large areas of geology have stubbornly resisted numerical treatment. The complexity of some problems and the incompleteness of the geologic record limit this approach, even in the age of computers.

Geotechnics

Geotechnics is the civil engineering discipline containing elements of engineering geology fortified with pertinent knowledge from other earth sciences and engineering such as foundation engineering, soil mechanics, rock mechanics, earthworks, hydraulics, hydrology, highway engineering, hydraulic structures engineering (power plants, waterfront structures, tunnels, underground mining openings, for example).

Bedrock

A formation at some depth beneath a mantle of soil is termed bedrock. It is formed from the crust of the earth. It is dry, solid rock exposed to the surface of the earth, or overlain by unconsolidated material.

A *competent rock* is one which is sufficiently strong to transmit a compressive force under given conditions. Rocks which are sufficiently plastic to deform without fracturing are *incompetent*.

Sandstone, quartzite, and igneous rocks are relatively competent under all conditions. Shale and slate are commonly incompetent. Limestone is likely to be competent at low temperature and under moderate pressure. At depth where high pressure and temperature favor recrystallization, rock is likely to be relatively incompetent.

Competent ground is one that does not require support when a tunnel is excavated through it.

An *intact rock* is a rock material which can be sampled and tested in the laboratory, and which is free of the larger-scale structural features such as joints, bedding planes, partings, and shear zones.

Regolith (mantle rock) is loose fragments of rock and soil that act as a cover for bedrock. The soil above the bedrock is called *overburden*.

Rock mass is the in situ rock made up of the rock substance plus the structural discontinuities. It is a much used term with an obvious meaning. It tends to imply something larger, less individualized, less distinctly marked off from its surroundings.

Rock substance is the solid part of the rock mass typically obtained as a drill rock core.

In rock mechanics, it is necessary to distinguish between the rock mass and rock material. *Rock material* may be defined as the aggregate of mineral particles together with rock pores, cavities, fissures and cracks termed voids. The voids may be isolated or interconnected. They may be filled with gas or air, or they may be fully or partially filled with water. Rock mass, in its turn, is the aggregate or regular or irregular blocks of rock material in situ. These blocks are separated by structural features such as bedding planes, joints, fissures, cavities, and other discontinuities.

Rock masses are heterogeneous and usually discontinuous assemblages of rock materials. This means that the scale of an experiment or test with a rock determines to some extent the result of the experiment or test.

Geologic Time Scale

To appreciate rock as a natural material with which the engineer must come to terms, it is desirable to familiarize with the geologic column and the geologic time scale of the earth's history (Table 2-1).

The geological column, or stratigraphical column, summarizes the earth's history, and defines the order in which sedimentary beds occur, and the periods during which igneous activity has taken place.

As is customary, as in Table 2-1, the oldest time intervals are known at the bottom, in the order up in which the rocks of those ages normally occur in the earth.

The term *geologic time scale* generally refers to the absolute calendar in which the progressive development of animals and plants may be related. The interval of earth's history is from the Cambrian period to the present.

Isotopic age determinations (radioactive decay measurements) on rocks of known stratigraphic age define an absolute time scale for earth's history.

The new geologic time scale after Kulp (10) is shown in Table 2-1.

The terms era, period, epoch and age mean divisions of time.

Era is one of the major divisions or units of geological time comprising one or more periods. This term suggests a time interval marked by a new and distinct order of things. Paleozoic, Mesozoic, and Cainozoic are well recognized eras. Eras are subdivided into periods. Eras are separated from each other by major breaks in the geologic record.

The term *period* means a fundamental unit of any length or of unspecified duration of the standard geologic time scale of the earth's history corresponding to a stratigraphical system. That is, a period is the time interval during which a group of rocks was deposited. Cambrian, Carboniferous, Cretaceous, Tertiary are examples of periods. The time unit period is smaller than era and larger than an epoch. The period names refer to regions where the rocks at that age were first studied.

In geological usage, the term *epoch* refers to a subdivision of a period. For example, the Eocene epoch of the Tertiary period; the Pennsylvanian epoch of the Carboniferous period.

Age refers to a fairly definite period of time, strongly dominated by a prominent feature. It may be said that age refers to the time in which a particular event occurred, for example, the Ice Age.

The term *Archeozoic era* is the era during which, or during the later part of which, the oldest system of rocks was formed.

The term Archean (Archaen) means ancient, and has been generally applied to the oldest rocks of the Precambrian period. All rocks formed before Cambrian time are now called Precambrian.

Precambrian rocks form the basement or foundation upon which youngest rocks lie. Outcrops are so patchy as to make a complete account of these strata difficult.

The approximate age of the oldest rocks discovered is 3000 million (3 billion) years. The approximate age of meteorites is 4.5 billion years.

Attempts to use radioactivity to determine the age of the earth ultimately lead to the presently accepted age of 4700 million = 4.7 billion years before the present when the formation of the earth's crust took place.

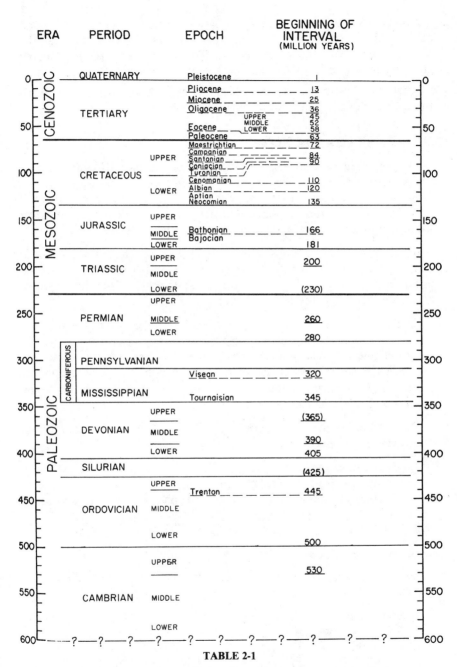

TABLE 2-1

Geologic Time Scale of Earth's History (Ref. 10).
Copyright April 14, 1961 by the American Association for the Advancement of Science.

However, it has been said that no physical evidence has ever been found of an original "crust" of the earth. How much older the earth is than the oldest rocks is a matter of conjecture. The age of the earth is probably greater than the estimated maximum age of the rocks.

Small but very instructive exposures of Precambrian rocks occur in the Inner Gorge of the Grand Canyon, in Arizona.

2-2. Classification of Rocks

The upper layers of the earth's crust which are accessible to underground and surface rock engineering are in their appearance, stratification, and form of extraordinary variety of petrographical nature and of structural diversity. The structure of rocks, especially, has a marked effect on their strength properties. Hence, there is a need for classification of rocks.

Rocks can be studied most effectively if they are classed according to certain principles into a definite system. A *system* is generally understood to be an ordered grouping of certain elements and/or facts in a field of knowledge according to certain principles.

By its very nature, classification is usually the very first and most important step of activities in the organization of any branch of a scientific and/or engineering discipline. Naturally, therefore, this applies also to rock mechanics in general, as well as to rock classification into certain systems in particular. Rock classification, hence, should be one of the necessary prerequisites in rock engineering.

Rocks may be classified according to several principles. Some of the classification systems are:

1. By origin or genesis,
2. Geological or lithological classification,
3. Engineering classification of intact rock on the basis of rock strength, or
4. A combination of several of these.

1. By Origin or Genesis

Based on genesis or mode of origin, rocks are grouped into three broad groups:

Igneous rocks,
Sedimentary rocks, and
Metamorphic rocks.

Igneous rocks form when hot molten silicate material from within the earth's crust solidifies.

Sedimentary rocks form from deposition and accumulation of sediments of other rocks, plant remains, and animal remains by wind, or water at the earth's surface, and their later solidification into rock.

Metamorphic rocks form when already existing rocks undergo changes by recrystallization in the solid state at high pressure, temperature, and chemical action at some time in their geological history.

The rocks record the history of the earth.

Igneous Rocks

Igneous rocks form when hot molten silicate material from within the earth solidifies.

When magma cools relatively quickly at shallow depth in the earth's crust, or where magma is extruded on the surface as lava, the solidified product is igneous rock having fine-grained crystals. However, upon slow cooling, the crystals of the rock are coarse-grained.

As to their occurrence, igneous rocks are classed into two major groups, namely:

a) intrusive or plutonic rocks, and
b) extrusive or volcanic rocks.

See Table 2-2.

Intrusive rocks are those that have cooled and crystallized within the earth's crust below the earth's surface. Hence, intrusive rocks have usually coarse-grained texture because of slow cooling.

TABLE 2-2

Igneous Rocks

Intrusive rocks	Extrusive rocks
1	2
Diorite	Andesite
Gabbro	Basalt
Granite	Diabase (trap rock)
Pegmatite (a coarse form of granite)	Obsidian
	Rhyolite
Syenite	Trachite

Extrusive rocks were formed by cooling of the molten material near or at the earth's surface. Hence, these rocks were cooled quickly resulting in a fine-grained to glassy textured material. The crystals of extrusive rocks are usually too fine to be distinguished by the naked eye.

All fresh, unweathered intrusive rocks are characterized by their high compressive and shear strengths. However, attention should be paid to jointing, faulting, and other defects and weaknesses in these rocks.

Most volcanic flows, particularly those of the acid type, show a great abundance of joints, which are closely spaced and cause the rock to break into many small angular fragments. None of the volcanic rock is used much as dimension stone.

The term *acid*, or *acidic*, or *silicic*, applies to rock with a relatively high proportion of the acid-forming radical, silica (SiO_2), whereas basic means relatively rich in iron and magnesium.

It must be emphasized that many rock types are transitional in character, and that they can be identified with certainty only after petrological examination of thin rock platelets under the petrological microscope.

The common igneous rock-making minerals such as quartz, feldspar, hornblende, biotite, augite, and olivine are called phenocrysts.

Short Description of Igneous Rocks

The description of rocks that follows, is given in alphabetical order.

Andesite

Andesite is a fine-grained igneous rock equivalent of diorite. It has no quartz or orthoclase, and is composed mostly of about 75% plagioclase feldspars and the balance of ferromagnesian silicates.

Basalt

Basalt is a dark-colored, fine-grained, heavy extrusive (volcanic) igneous rock. It is the most abundant lava-formed rock, and is composed mainly of pyroxene, plagioclase feldspar, and augite, with or without olivine. Basalts and andesites represent about 98% of all extrusive rocks. Basalt is approximately the fine-grained extrusive equivalent of gabbro. Basalt may be quite dense, or filled with many gas bubbles, depending upon the condition prevailing at the time of extrusion.

The physical properties of basalt vary considerably. Therefore basalts require a thorough, extended examination before their engineering characteristics can be evaluated.

Cellular basalts are used in the manufacture of lightweight building blocks. Because of their cellular nature, these lightweight blocks have excellent insulating properties against heat, cold, and sound.

Diabase

Diabase is also known by the name "trap rock". Diabase is a dark-colored, fine-grained, basic, extrusive igneous rock usually occurring in dikes or intrusive sheets. The essential minerals of diabase are the same as in basalt, and are mostly plagioclase feldspar, quartz, and augite with the plagioclase in long, narrow, lath-shaped crystals oriented in all directions, and the augite filling the interstices.

Diabase has been used occasionally for monumental and paving block purposes. It has a good strength and takes a high polish. It is, however, difficult to quarry in large blocks and difficult to work with. Diabase and basalt rock when not vesicular make excellent road metal, and are widely used in the construction field as crushed stone.

Diorite

Diorite is a black-and-white speckled coarse-grained, intrusive igneous granitoid rock whose essential minerals are hornblende and feldspar, with common accessory minerals of mica (biotite) and pyroxenes. Quartz may be present in considerable amounts, in which case the rock is called quartz diorite. The dark crystals are hornblende with a little biotite. The white crystals are mostly lime-soda feldspar or plagioclase.

Diorite is less common in occurrence than granite.

Dolerite

Dolerite is an older term for designating coarsely-crystalline basaltic igneous extrusive rocks. The term is interchangeable with diabase.

Dolerite is not used as a building stone, but has been used successfully as crushed rock because of its toughness and good resistance to abrasion.

Gabbro

Gabbro is a dark-colored, coarse-grained equiangular intrusive igneous rock. Sometimes gabbro is called the black granite. Gabbro is the extrusive equivalent of basalt. Gabbro makes up vast bodies of rock. It consists of plagioclase feldspar and quartz, and one or more dark minerals. The common dark minerals are hornblende, pyroxene, and olivine. Magnetite, ilmenite and apatite are accessory minerals.

Gabbros have been used more widely as ornamental stone than for construction purposes.

Granite

Granite is the most common intrusive igneous rock that has cooled at depths below the surface forming batholites, bosses, and dikes. Granite is a light-colored, visibly coarse-grained crystalline rock. Its interlocking texture of crystals has been formed by slow crystallization of the melt.

Granites are composed essentially of about 50% orthoclase, 25% quartz, plagioclase feldspars, mica, and hornblende. The rock colors are usually white, gray, pink, reddish, or greenish largely determined by the color of the feldspar. Commonly, mica is present either as muscovite or biotite.

Subsequently to the solidification of a mass of granite the rock becomes traversed by planes of fracture or joints. The direction of these joints follow definite patterns in some masses (see "Rock of Ages" granite quarry at Barre, Vermont, Fig. 2-2), while in others no preferred directions can be made out, the joints being oriented in a haphazard manner (1, 2, 16).

As a building material, granite has been used as a building stone, facing slab, and for the support of structures for a long time. The Granite City of Aberdeen, Scotland, is a well-known example where granite has been used for building practically an entire town.

Because of its durability, under normal conditions granites are competent to support high loads in ordinary construction practice. The technological properties of a granite are here of as much importance as its mineralogical composition. The two, of course, are interrelated.

Because of its mineral composition and interlocking of crystals, granite is usually very dense, compact, and as compared with some other rocks, it has a low porosity where water would be absorbed and retained, or seep. Hence, used for structural purposes, granite is impervious to change in temperature and weather, and has excellent frost resistance properties. However, a coarse-grained granite is more pervious to water than a fine-grained granite.

The resistance of granite to weathering varies with the climatic conditions, of course. Spalling of granite usually does not occur unless in the atmosphere around the granite building stone, smoke and sulfur prevails. In time, these gases and chemicals form hydrous lime sulfate mineral gypsum in the pores of the granite.

When subjected to excessively high temperatures, granite would spall and fracture.

Granite breaks down mechanically into arcosic sand.

Because of its mineral composition, granite is hard and resistant to impact and abrasion. The shear strength of granite is superior to that of most other rocks and building stones.

Granite, the giant among the rocks — rock of ages — is usually imagined as a symbol of strength. However, granite is also a less exposed example of its weakness. Recall that one of the essential minerals of granite is feldspar which is contained in abundance in all varieties of granite. The main weakness of granite lies in this single mineral feldspar, namely: under the action of rainwater and/or groundwater, feldspar alters to clay.

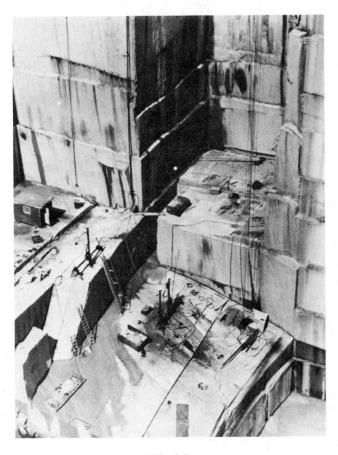

Fig. 2-2
"Rock of Ages" granite quarry at Barre, Vermont.
(Photo courtesy of R. L. MARCH, Rock of Ages Corporation, Barre, Vermont)

Also, the various inherent geological defects in granite such as jointing, fractures and faults, for example, constitute a hazard in hydraulic structures- and in foundation engineering.

For the permeability aspects of granite in dam engineering refer to Chapters 4, 5 and 10.

Obsidian

Obsidian is a dark-colored extrusive igneous rock of glassy texture. It is known as volcanic glass. Obsidian is the glassy equivalent of granite. The mineral content of obsidian is much like that of granite.

Pegmatite

Pegmatite has been called "giant granite" because it generally consists of very coarse crystalline aggregates of the same minerals as granite. Pegmatite is a porphyritic intrusive igneous rock. It is an irregularly textured coarse form of granite, composed of giant crystals or crystalline masses of quartz and potash feldspar, and frequently also of plagioclase.

Pegmatites occur in fractures near the borders of granite intrusions.

Some pegmatites are worked for mica.

Rhyolite

Rhyolite is a fine-grained extrusive igneous rock with the composition of granite. Consists of quartz, though potash feldspar and biotite mica are common.

Syenite

Syenite is a grained, wholly crystalline intrusive igneous rock. It is similar in texture like granite, but with less silica. Its chief constituent is orthoclase feldspar, but containing little or no quartz. Biotite, hornblende, and sometimes augite are commonly occurring accessory minerals.

Syenites are not as widely distributed as the granites.

The general properties of syenite are similar to those of granite. Because of their rare occurrence, syenites are little used as a construction material.

Trachite

Trachite — an extrusive igneous rock — is the equivalent of the intrusive rock syenite.

Sedimentary rocks

Sedimentary rocks are those formed by deposition and accumulation of sediments of other rocks, plant remains, and animal remains by water, or wind at the earth's surface, and their later solidification into rock. Thus, one distinguishes between sediments
   a) in water (aqueous deposits),
   b) from air (eolian deposits brought about by the wind),
   c) from glacial ice.

There are
1. *Mechanical sedimentary rocks* formed from the accumulation of pebbles, sand and clay, and fragments of other rocks. These are the clastic rocks formed from other fragmental rocks such as conglomerate, sandstone, and shale.
2. *Organic sedimentary rocks* formed from plant and animal remains, such as certain limestones and coal, and
3. *Chemical sedimentary rocks* formed from minerals that were once dissolved in water and then precipitated out. Examples: limestone, salt, gypsum.

Stratification is the single most characteristic feature of sedimentary rocks.

Some sedimentary rock deposits are composed of volcanic-ejected fragments and deposited on land or in water, such as breccia (coarse), and tuff (fine).

Breccia

Breccia is a rock consisting of consolidated coarse, angular rock fragments with sharp edges and unworn corners larger than sand grains, cemented together with a fine-grained siliceous, calcareous or other matrix. Not all breccias are of sedimentary origin. Many breccias have been formed by volcanic eruptions.

Conglomerate

Conglomerate is the consolidated equivalent of gravel. Conglomerate (popularly known as puddingstone) is formed from the cementing together of gravel, pebbles, and *rounded* fragments of other rocks set in an abundant fine sand and silica, calcium carbonate, or iron oxides precipitating out from circulating water through gravel deposits. The rock fragments are rounded and smoothed from transportation by water or from wave action.

Dolomite

Dolomite rocks or magnesium limestone rocks are made up wholly or in large part of the mineral dolomite [the double carbonate of calcium and magnesium — $CaMg(CO_3)_2$]. The dolomite rock is formed when magnesium replaces part of the calcium in limestone.

The mineral dolomite is somewhat harder, heavier and less soluble in hydrochloric acid than calcite. Dolomite effervesces in cold, diluted acid only when scratched or powdered. Dolomite will also react without scratching (to produce powder) with concentrated or with warm dilute hydrochloric acid.

As a rock, dolomite occurs in great masses especially in older sedimentary rocks.

Limestone

Limestone is a bedded carbonate rock, and consists predominantly of calcium carbonate ($CaCO_3$) which has been formed by either organic (biologic) or inorganic processes. If of biologic origin, limestone is formed from accumulation of lime shells from shell fish. Limestone responds to the hydrochloric acid test, i.e., limestone effervesces in diluted HCl-acid.

The color of limestones vary from white through varying shades of gray and black.

Most limestones have a clastic texture, but crystalline textures are common. The carbonate rocks, dolomite and limestone, constitute about 22% of the sedimentary rocks exposed above the sea level.

Limestones vary greatly in porosity, some being very impervious, some very porous, hence, pervious. Because carbonate rocks are relatively soluble, solution cavities in limestones may be abundant. One refers here to Karst topography. In limestone areas solution cavities and hence ready permeability should always be suspected until contrary evidence is obtained.

Limestone has a wide use in construction and in industry. It is also one of interior and exterior dimension stone, and it is also the basic ingredient in the manufacture of cement and lime.

Sandstone

Sandstone is a consolidated, porous and pervious rock composed mainly of sand particles — quartz grains — cemented together by clayey, siliceous, calcareous or limonitic material which fills the spaces between the grains.

A sandstone which is rich in quartz is stable over a wide range of temperature and pressure. A siliceous cement usually produces the strongest sandstone.

Calcareous cement may dissolve, and can be detected by means of the hydrochloric acid test upon which test the calcareous material will effervesce.

The flaking off of the sandstone at free surfaces, known as "spalling" occurs in course of time.

Sound sandstones are durable, and are used as a building stone and construction material in a good many different ways. Generally, dry sandstone is a competent load bearing rock. However, sandstone which allows water to ooze, is of doubtful stability.

Graywacke (greywacke)

Graywacke is a variety of sandstone. Graywacke is a gray, or green, or dark-colored, hard sandstone composed largely of angular grains of quartz, feldspar and a variety of rock and mineral fragments derived by rapid disintegration of fine-grained basic igneous rocks, slates and dark-colored rocks embedded in a compact matrix having the composition of clay-slate. It is also said that rocks with high amplibole and pyroxene contents are referred to as graywackes.

Some graywackes are massive and show no bedding. Others show marked graded bedding and are associated with slate.

Graywacke is similar in physical properties to coarse, hard sandstones, but is generally less resistant to weathering.

Generally, graywacke is a competent load-bearing rock.

Shale

Shale is a laminated, most abundantly occurring sedimentary rock on all continents. The constituent particles of shale are predominantly of the silt and clay grade. Silt and clay are changed into shale by the process of adhesion, compaction or densification, and cementation. The chief mineral in shale is kaolin. Sandstone is next in abundance, and limestone is least. Shale, sandstone and limestone make up about 99% of all sedimentary rocks.

The color of shale varies from light gray to black. Organic matter makes a black, carbonaceous shale. Red and green shales are also common. The red shales are due to ferric oxides.

The Triassic shale in New Jersey — the red Brunswick shale formation

— varies in its composition. This shale formation is formed by consolidation of silt and clay. The clay-size particles forming the bulk of the shale are reported to be mostly the clay mineral dioctahedral illite (5, 9, 13, 15), powdery hematite (13), and small amounts of kaolinite.

The Triassic shale deposit often lies next to and is interbedded with sandstone, which is abundant in northeastern New Jersey and in adjacent areas of the State of New York.

The dull-red color of the Triassic shale varies according to the Munsell color classification chart coordinates as follows (11):

in dry condition: from YR 5/4 to 10R 4/2;

in wet condition: from YR 3/4 to 5R 4/2.

The shale system may be classed into two major parts:

a) the sound, unweathered shale, and

b) the weathered part of the shale overlying the sound shale.

The sound shale is generally hard, and in its unweathered condition is usually a competent load-bearing formation for foundation support. However, it has been observed in practice that the strength, viz., bearing capacity, of the shale may vary because of various degrees of weathering. Also, a number of discontinuities generally exist in the bedrock shale.

Whereas the sound, unweathered shale is hard and excavation in it can be made most effectively by the use of explosives, or in the case of tunneling sound shale must be drilled by means of tunnel-boring equipment, the weathered shale may be excavated by first loosening the shale by rippers and rooters, for example. Then the loosened material is removed by means of a power shovel, bulldozer, or scraper.

Discontinuities in shale may take many forms, such as bedding planes, joints, faults, laminations. The Brunswick (Triassic) shale is also crisscrossed with fissures that have been filled with various minerals, such as calcite (Fig. 2-3), and small calcite pockets or white "peppery" spots (Figs. 2-4). Also, the joints may be open. Figure 2-5 shows some shale core specimens after testing for unconfined compressive strength (6, 7, 8).

It is theorized that as a result of great pressure and temperature changes which occurred during the geological time, the shale rock has been cut by numerous open joints and fault zones, and broken up into fragments.

Shales are fissible and break along planes parallel to the original bedding. The weathered shale can be easily scratched, and have a smooth feel. The lamination, or fissibility, is usually best displayed after weathering.

Shales are frequently very compressible, especially the weathered shale.

Fig. 2-3
Triassic shale
a) and b): calcite veins in shale
c): ~ 1 cm thick calcite vein

Fig. 2-4
Shale sample "peppered" with calcite spots

Fig. 2-5
Some shale core specimens after testing for unconfined compressive strength

Many clastic sediments loose their coherence when exposed to the action of alternate wetting and drying, freezing and thawing, and other weathering agents, reverting to the original clayey mass from which shales had been formed.

Compaction shales are unreliable for most engineering purposes, and should be carefully investigated prior to design and construction.

Affected by the elements, shale rock spalls in loose, thin bedding scales and laminae (Fig. 2-6a). The shale laminae are fissile and easy to break with the fingers into approximately parallel-sided platelets (notice the thin shadows in Fig. 2-6b).

The drying of shales brings about cracks on bedding and shear planes, and eventually reduces the material to chips, granules, platelets, or finer particles.

Flysch. Flysch consists of a rhythmic sequence of fine-grained sandstone or siltstone and shale (see Fig. 2-7). Least resistant layers are shale.

Unconformity. Gaps in the geologic record are known as *unconformities.* An unconformity is a surface of erosion or nondeposition that separates younger strata from older rocks. The plane in contact between the two beds or series of beds is the plane of *unconformity.*

The most conspicuous kind of unconformity is the *angular unconformity.* The angular unconformity is characterized by two groups of sedimentary rocks with different angles of dip in depositional contact, indicating that the lower sequence of strata was tilted or folded before being eroded and covered by the upper sequence.

An example of angular unconformity in the Moroccan Meseta about 50 km southeast of Rabat is shown in Fig. 2-8: Pliocene limestone overlying carboniferous flysch.

An unconformity that develops between massive igneous rocks that were exposed to erosion and then covered by sedimentary rocks in called *nonconformity.*

Conformity is the mutual relationship between sedimentary beds laid down in orderly sequence with little or no evidence of time lapses, and specifically without any evidence that the lower beds were folded, tilted, or eroded before the higher beds were laid down.

If the older rocks remained essentially horizontal during erosion, the surface separating them from the younger rocks is called a *disconformity.*

Obviously, these various arrangements of stratifications affect the distribution of structural loads, and affect the stability of rock and structure where the rock permeability is concerned.

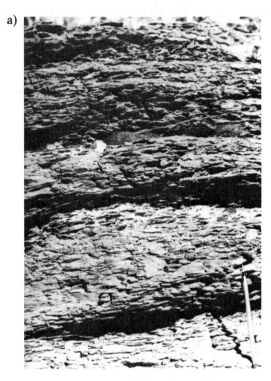

Fig. 2-6
Shale platelets.
a) Platelets in an exposed weathered bank of shale
b) Thin shale platelets placed upright (notice the thin shadows)
(Photos by ANDRIS A. JUMIKIS)

Fig. 2-7
Rhythmic sequence of flysch (Mississippian), Kiamichi Mountain, Oklahoma.
Least resistant layers are shale.
(Photo courtesy of Dr. MARTHA M. HAMIL, Professor of Geology, Rutgers University)

Fig. 2-8
Angular unconformity in the Moroccan Meseta about 50 km southeast of Rabat.
Pliocene limestone overlying Carboniferous flysch.
(Photo courtesy of Dr. MARTHA M. HAMIL, Professor of Geology, Rutgers University)

Metamorphic Rocks

Metamorphism or transformation is a process whereby consolidated rocks undergo physical or chemical changes, or both, in composition, texture or internal structure to achieve equilibrium with conditions other than those under which they were originally formed.

Metamorphism is the result of a changed geological environment in which the stability of the rock can be maintained only by a corresponding change in their make-up.

Metamorphism is characterized by the development of new textures, new minerals, or both, and these are often so unlike the former ones that it is frequently difficult to determine the nature of the original rock.

The factors in metamorphic processes of rocks are heat, pressure and chemically active fluids (hot liquids, vapors and gases). Heat from within the earth and from hot intrusive magmas as well as from pressure and friction of movement within the rock that is being deformed, speeds up chemical activity. The dominant process in metamorphism is recrystallization.

Because metamorphic rocks are the result of transformation, this means that metamorphic rocks were formed from older, pre-existing rock formations. Both igneous and sedimentary rocks have in the geological past been metamorphosed.

Some of the commonly known metamorphic rocks are gneiss, marble, quartzite, schist, and slate.

Through metamorphism,

basalt	changes to	schist
granite	changes to	gneiss
limestone	changes to	marble
sandstone	changes to	quartzite, and
shale	changes to	schist or slate

Gneiss

Gneiss, transformed from granite or conglomerate, is a coarse-grained, foliated (banded) crystalline metamorphic rock with alternating layers of light and dark minerals. The bands are rich in feldspar and quartz. Gneiss is most commonly of the same composition as granite, and might then be called granitic gneiss. Usually most gneisses are hard and have great compressive and shear strengths. Some gneisses are used as building and dimension stone. However, gneiss is not widely used. Seepage through gneiss underneath massive dams because of great hydrostatic pressure from impounded water affects the stability of hydraulic structures on gneiss, and has been and still is of great concern of the engineer.

Marble

Marble is formed either from limestone or dolomite. It is a nonfoliated, crystalline metamorphic rock of coarse, interlocking granular texture. Marble is composed mainly of calcite or dolomite or both. Therefore, marble responds to the hydrochloric acid test.

Pure marble is white. But depending upon impurities, the color of marble may be pink, yellow, brown, gray and black.

Marble is more dense than limestone, because the porosity of marble during the process of its metamorphism has been decreased by pressure and recrystallization. However, sometimes solution cavities in marble are encountered.

Marble is used as a cut stone for buildings and for monuments.

Quartzite

Quartzite is formed from recrystallization of quartz sandstone under heat and pressure. Quartzite is a nonporous crystalline metamorphic rock whose dominant mineral is quartz. The quartz-sandstone is cemented by interstitial quartz into a hard mass. Quartzite breaks through sand grains as contrasted to sandstone, which breaks around the sand grains.

Quartzite is one of the most hard, durable, strong, and wear-resistant rocks. The rock is mostly used in industry. In foundation engineering, quartzite is a competent load-bearing rock.

Schist

Schist is a crystalline, highly foliated rock metamorphosed from fine-grained sedimentary rocks such as shale, mudstone or volcanic ashes. Some schists are formed from fine-grained igneous rocks such as felsites and basalts. Schist is a rock largely or completely recrystallized, structurally characterized by fine-scale foliation resulting from the parallel disposition of fibrous or platy minerals, most commonly the micas. The closely spaced lamination tends to split readily into thin flakes or smooth platelets. This kind of separation is known as schistosity.

Because of their well-expressed cleavage, schists are very isotropic rocks. Many schists, talc and chlorite schists, for example, are soft ($H = 1.0$ to $H = 2.5$), and are therefore incompetent to carry high pressure. Shales may also be rather permeable in a direction parallel to schistosity.

Slate

Slate is another foliated rock metamorphosed from shale. Slate is a dense rock with a strongly developed foliation, and the texture is fine-grained. Slates are composed mainly of quartz, and secondary — mica, calcite, pyrite, and biotite. The color of slates range from black through gray, green and red.

Slate is used in industry, and in construction trades. In excavations, large slate blocks may become detached when undermined.

2. Geological or Lithological Classification

Lithology is the study of the physical character of rocks. The lithology of a rock pertains to its mineralogical composition and texture together with a descriptive term from some accepted rock classification system, i.e., granite, limestone or, say mica schist. Thus, the terms and classification are geological.

Although in several ways the lithological rock classification has some advantages, for engineering purposes the geologic term alone is insufficient for engineering classification of rocks.

3. Engineering Classification of Intact Rock

This classification system is here presented according to Deere's (3) lectures given at the ASCE Rock Mechanics Seminar April 9—10 and May 7—8, 1968.

The classification system pertains to intact rock and to rock in situ. Intact-rock classification presupposes rock-testing. The classification is based on two important engineering properties of rock, namely: the uniaxial compressive strength, and the tangent modulus of elasticity (taken at a stress level equal to one-half the ultimate compressive strength of the rock). Table 2-3 shows the engineering classification of intact rock on the basis of uniaxial compressive strength after Deere and Miller (3).

In this table, the rock-strength classes follow a geometric progression.

Table 2-4 shows the engineering classification of intact rock on the basis of modulus ratio $M_R = E_{t50}/\sigma_{ult}$ after Deere and Miller (3), where E_{t50} = tangent modulus at 50 % ultimate compressive strength of the rock, and σ_{ult} = uniaxial ultimate compressive strength.

Based on this classification, rocks are classified both by strength and modulus ratio as AM, BL, BH, and CM, for example. The classification is also sensitive relative to mineralogy, texture, and direction of anisotropy.

TABLE 2-3

Engineering Classification of Intact Rock
(after DEERE and MILLER, Ref. 3)

1. On the basis of strength

Class	Description	Uniaxial Compressive Strength psi	MN/m²	Rock Material
A	Very high strength	> 32,000	~ 220	Quartzite; diabase; dense basalts
B	High strength	16,000–32,000	~ 110 to ~ 220	Majority of igneous rocks; strong metamorphic rocks; weakly cemented sandstones; hard shales; majority of limestones; dolomites
C	Medium strength	8,000–16,000	~ 55 to ~ 110	Many shales; porous sandstones and limestones; schistose varieties of metamorphic rocks
D	Low strength	4,000–8,000	~ 28 to ~ 55	Porous low-density rocks; friable sandstone; tuff; clay shales; weathered and chemically altered rocks of any lithology
E	Very low strength	< 4,000	< 28	

TABLE 2-4

Engineering Classification of Intact Rock
(after DEERE and MILLER, Ref. 3)

2. On the basis of modulus ratio

Class	Description	Modulus ratio, E_{t50}/σ_{ult}
H	High	> 500
M	Average (medium)	200–500
L	Low	< 200

4. Rock Quality Designation (R.Q.D.)

According to Deere (3), the rock quality designation (R.Q.D.) is an index which indicates the recovery of rock cores obtained by adding up the lengths of rock core pieces the length of which are 10 cm and longer. This length is then divided by the total length of the recovered rock core to obtain the index of fracture frequency. The reliability of this index of fracture is, however, affected by 1) the diameter of the rock core, and 2) the workmanship experience and skill of the rock drilling crew. For this R.Q.D. test, a standardized diameter of 50 mm is desirable.

2-3. Rock as a Construction Material

Rock as a material has been ingrained in human culture since the beginning of history and has now become the "working material" in geotechnical and rock engineering.

To a civil engineer who lays foundations for structures on rocks and builds in rock, rock is an engineering material. As naturally occurring material, rocks have many uses and applications. For example, rock is used:

1. For laying structural foundations to support structures;
2. For constructing vehicular and equeous tunnels, underground power plants, and other underground openings and cavities;
3. For building moles or breakwaters;
4. For protective blanketing of earth dams and other earthworks against erosion by water in the form of riprap;
5. For ballast on railroads in the form of crushed rock;
6. As a base and subbase courses for roads and airfield runways in the form of crushed rock;
7. As an aggregate for concrete (crushed rock);
8. As a dimension stone;
9. As facing stone for buildings, bridges, and hydraulic structures to protect these structures from weathering;
10. For making artifical sand from rock in regions poor in sand but rich in suitable rocks.

These are only a few uses for rock as an engineering material. There may be many more.

Without natural stones, the building of highways, railroads, airfields, bridges, harbors, industrial structures, and residential and cultural edifices, the regulation of rivers and the entire field of concrete engineering would be impossible.

As engineering materials, rocks must be indentified, described, tested, and classified.

To utilize rock as a construction material in design and for building of safe and economical structures, its physical and mechanical properties must be known and stress conditions in rock *en masse* must be understood.

Rock as a construction material, or a rock mass, frequently contains large quantities of water and various tectonic discontinuities, and may be subjected to high *confining pressures*—all of which may radically alter its properties, which are normally based on small-scale laboratory test data.

Among the important questions which the geotechnical engineer has to consider in his work are:

1. The structure of rocks in relation to tunneling operations;
2. Dam, reservoir and building foundations;
3. Landslides (stability of slopes);
4. The nature or character of the common rocks in their use for building stone and road material; and
5. The geological conditions affecting and controlling underground water supplies.

Figure 2-9 shows an example as to how natural rock in situ is used as a material for carving of sculptures. Carved into a smooth-grained granite wall of the 1829 m ($\approx$ 6000 feet) Mount Rushmore of the Black Hills of South Dakota are the faces of the four American Presidents — Washington, Jefferson, Lincoln and Theodore Roosevelt facing the sun most of the day. Work on the impressive Mount Rushmore National Memorial began in 1927, and was finished, with frequent interruptions, in 1941. On the average, the heads measure 18.29 m (60 feet) from chin to top, with each nose 6.10 m (20 feet) long, each mouth 5.50 m (18 feet) wide, and the eyes 3.35 m (11 feet) across.

This memorial is among the most beautiful spots in the USA.

The finished gigantic sculpture must be inspected each year for cracks and damage. Repair work is done, if needed, at that time.

Figure 2-10 shows a steep-sided granite quarry at Barre, Vermont. Here vertical granite slabs are made by wire-saw cutting.

The use of traprock as sea-defense along part of the Atlantic coast of New Jersey is shown in Fig. 2-11.

Fig. 2-9
Mount Rushmore National Memorial, South Dakota.
(Photo courtesy of National Park Service, U.S. Dept. of the Interior)

Fig. 2-10
"Rock of Ages": A steep-sided granite quarry at Barre, Vermont.
Vertical slabs of granite are made by wire-saw cutting.
(Courtesy of R. L. MARCH, Rock of Ages Corporation, Barre, Vermont)

Fig. 2-11
Traprock used as sea-defense along part of the Atlantic coast of New Jersey

Underground Rock Caverns

Underground rock caverns have several practical and economical advantages. Rock caverns may be utilized as public underground air raid shelters for the purpose of civil defense in any eventual war, and in peace time crises; as automobile parking space; as state emergency storage; distribution- and bulk storage for oil refineries and oil companies, and consumption storage for power plants and industry; for storing of liquefied natural gas, compressed air, and even excess heat from thermal power plants (either nuclear or fossil heated) in the form of hot water.

Storage caverns and underground power plants in rock are economically feasible where the solid bedrock is near the ground surface. In Finland and Sweden, large caverns have been built in different kind of rocks such as crystalline limestone, diabase, diorite, gabbro, gneiss, granite, and schist.

In many places around the Swedish coast complete naval bases, military establishments and storage of strategic supplies of oil, have been placed underground in enormous rock caverns (18).

The advantages of storing practically all kinds of oil in underground rock caverns as compared with storage in the above-the-ground surface steel tank storage facilities are:

underground storage is fire proof;

high safety against sabotage and during any eventual hostility;

relatively safe from external explosions;

the waste-rock aggregate obtained from underground rock excavation is a valuable by-product which may be used economically for other kinds of construction;

low cost of initial investment, maintenance, and operation;

evaporation losses of light products are small;

the relatively low thermal conductivity of rock reduces the heating cost of heavy oils;

the storage structures in rock are not damaged by corrosion; hence the age of an underground surface plant is exceptionally long;

Fig. 2-12
The Rosslyn, Virginia, underground Metro passenger station of the Washington Metropolitan Area Transit System.
(Photo courtesy of PHIL PORTLOCK, Washington Metropolitan Area Transit Authority)

Fig. 2-13
Construction of oil storage caverns in granite in Sweden.
(Photo by Gösta Nordin. Photo courtesy of SVEN BYLUND, Svenska Entreprenad A.B., Stockholm)

70   ROCK MECHANICS

expensive surface sites in urban areas are preserved: the landscape remains unchanged because underground storage facilities cause less environmental disturbance and air pollution than surface storage.

Fig. 2-12 shows the Rosslyn, Virginia, Metro subway passenger station of the Washington, D.C. Metropolitan Area Transit Authority.

The tunnel of the station is of an arched vault design. It is 213.36 m (700 ft.) long, 13.72 m (45 feet) high, and 19.81 m (65 feet) wide, and is constructed in rock to a maximum depth of 36.58 m (120 feet) below street level. As of 1979, there are approximately 23 km of rock tunnels on the 96.56 km (60 miles) Metro system (12, 17).

Fig. 2-13 shows an underground cavity in rock developed for storage of oil in Sweden.

Fig. 2-14 is an underground opening in rock built for a locomotive maintenance workshop of the Norwegian State Railways in Oslo, Norway.

Fig. 2-14
Underground locomotive maintenance workshop of the
Norwegian State Railways in Oslo, Norway.
(Photo by Arne Svendsen. Photo courtesy of H. HARTMARK, Norwegian State Railways)

REFERENCES

1. BALDWIN, B., 1961, "The Barre Granite Quarries," Trip B-2, Section IV, *New England Intercollegiate Geological Conference Guidebook*, 53rd Annual Meeting.
2. The Barre Guild, *The Story of Granite*. Barre, Vermont: The Barre Granite Association (no date given).
3. DEERE, D. U., and R. P. MILLER, 1966, "Engineering Classification and Index Properties for Intact Rock," *Technical Report No. AFWL-TR-65-116*, Air Force Weapons Laboratory, Kirtland Air Force Base, New Mexico.
4. EMERY, C. L. 1966, "The Strain in Rocks in Relation to Highway Design", Washington, D. C.: National Research Council. HRB Record No. 135 on *Rock Mechanics,* pp. 1-9.
5. FREY, J. Q. and G. I., STOKES, 1901, *Shale Investigation*, graduate thesis. New Brunswick, New Jersey: Rutgers University.
6. JUMIKIS, A. R., 1966, "Some Engineering Aspects of Brunswick Shale," *Proceedings*, (1st) International Congress on Rock Mechanics, held Sept. 25 to Oct. 1, 1966, at Lisbon. Published by the Laboratório Nacional de Engenharia Civil, Lisbon, Portugal; Vol. 1, 1966, Paper No. 2.1, pp. 99—102.
7. JUMIKIS, A. R., and A. A. JUMIKIS, 1975, *Red Brunswick Shale and its Engineering Aspects*. New Brunswick, New Jersey: College of Engineering. Rutgers, The State University of New Jersey. Engineering Research Bulletin No. 55, 1975, pp. 3-19.
8. JUMIKIS, A. R., 1978, "Geotechnical Properties of Triassic Shale." *Proceedings of the 3rd International Congress on Engineering Geology,* held September 4-8, 1978 in Madrid, Spain. Paper 21. Section II, Vol. 1, pp. 211—217.
9. KENWORT, C. E., 1922, "A Scale of Grade and Class Terms for Plastic Sediments." *Journal of Geology*, vol. 30, 1922, pp. 377—392.
10. KULP, J. L., 1961, "Geologic Time Scale." *Science*, April 14, 1961, Vol. 133, No. 3459, Fig. 1, pp. 1105—1114.
11. Munsell Color Co., Inc., 1954, *Munsell Color Charts*. Baltimore, Maryland: 1954.
12. SELTZ-PETRASH, A., 1979, "Washington Metro: A People's Eye View." *Civil Engineering-ASCE,* June, 1979, Vol. 49, No. 6, pp. 59—63.
13. STURM, E., 1956, *Mineralogy and Petrography of the Newark Group Sediments of New Jersey*. Ph. D. Thesis New Brunswick, New Jersey: Rutgers University, 1956, p. 7.
14. TERZAGHI, K., 1943, *Theoretical Soil Mechanics,* New York, N.Y.: John Wiley and Sons, p. 1.
15. UGOLINI, F. C., 1960, *Soil Development on the Red Beds of New Jersey*. Ph. D. Thesis. New Brunswick, New Jersey: Rutgers University, 1960, p. 8.
16. WHITE, W. S., 1946, *Rock-Bursts in the Granite Quarries at Barre, Vermont. U. S. Geological Survey Circular 13, 1946.*
17. WILHOYT, E. E., Jr., 1974, "Contracting Washington D. C.'s Metro Subway," *Civil Engineering-ASCE,* November, 1974, pp. 74—77.
18. WINDOLF, G., 1976, "Sweden's Underground Millions." *Tunnels and Tunneling*. Sept.-Oct. 1976, Vol. 8, No. 6, pp. 24—26.

OTHER RELATED REFERENCES

ATTEWALL, P. B., and I. W. FARMER, 1976, *Principles of Engineering Geology.* London: Chapman and Hall and John Wiley and Sons, Inc.

BLYTH, F. G. H., and M. H. DE FREITAS, 1974, *A Geology for Engineers.* London: Edward Arnold, Ltd.

Channel Tunnel Study Group, 1960, "Report on the Channel Tunnel Study Group," Channel Tunnel Company, London.

COATES, D. F., 1970, *Rock Mechanics Principles.* Ottawa: Information Canada, Mines Branch Monograph 874.

DREYER, W., 1972, *The Science of Rock Mechanics.* Clausthal, Germany: Trans Tech Publications.

DUNKAN, N., 1969, *Engineering Geology and Rock Mechanics* (2 volumes). London: Leonard Hill.

FARMER, J. W., 1968, *Engineering Properties of Rocks.* London: E.&F.N. Spon Ltd.

The Geological Society of America (1950/55 and other years), *Application of Geology to Engineering Practice.* New York, N.Y.: The Geologic Society of America.

JAEGER, C., 1961, "Rock Mechanics and Hydro-Power Engineering," *Water Power,* Vol. 13, Nos. 9 and 10, September-October, 1961.

JAEGER, C., 1972, *Rock Mechanics and Engineering.* Cambridge University Press.

JAEGER, J. C., and N. G. W. COOK, 1971, *Fundamentals of Rock Mechanics.* London: Chapman and Hall Ltd.

JOHNSON, A. M., 1970, *Physical Processes in Geology.* San Francisco, California: Freeman, Cooper and Company.

KRYNINE, D. P., and W. R. JUDD, 1957, *Principles of Engineering Geology and Geotechnics.* New York, N.Y.: McGraw-Hill Book Company, Inc.

LEGGET, R. F., 1962, *Geology and Engineering* (2nd ed.). New York, N.Y.: McGraw-Hill Book Company, Inc.

MÜLLER, L., 1963, *Der Felsbau.* Stuttgart: Ferdinand Enke Verlag, vol. 1.

OBERT, L., and W. I. DUVALL, 1967, *Rock Mechanics and the Design of Structures in Rock.* New York, N.Y.: John Wiley and Sons, Inc.

SCHULTZ, J. R., and A. B. CLEAVES, 1955, *Geology in Engineering.* New York, N.Y.: John Wiley and Sons, Inc.

SLATER, H., and C. BARNETT, 1958, *The Channel Tunnel.* London: Allan Wingate.

STAGG, K. G., and O. C. ZIENKIEWICZ (editors), 1968, *Rock Mechanics in Engineering Practice.* New York, N.Y.: John Wiley and Sons, Inc.

STINI, J., 1950, *Tunnelbaugeologie.* Wien: Springer-Verlag.

SZÉCHY, K., 1966, *The Art of Tunnelling.* Budapest: Académiai Kaidó.

TALOBRE, J. A., 1957, *La Mécanique des Roches.* Paris: Dunod.

TREFETHEN, J. M., 1959, *Geology for Engineers* (2nd ed.). New York, N.Y.: D. Van Nostrand Company, Inc.

WAHLSTROM, E. E., 1974, *Dams, Dam Foundations and Reservoir Sites.* Amsterdam: Elsevier Scientific Publishing Company.

WALTERS, R. C. S., 1962, *Dam Geology.* London: Butterworths.

CHAPTER 3

ROCK MECHANICS

3-1. Definition of Rock Mechanics

Let us now fix the idea of rock mechanics and its position.

Rock mechanics is principally characterized by the fact that rock is not a continuum but a regulated discontinuum, also called articulate, grain-structure mechanics.

In a large, solid, sound (healthy) mass, rock may be considered to be a continuum. Rock mechanics is the mechanics of the discontinuum.

Whereas soil mechanics works in plane, rock mechanics works both in plane and in space.

According to the definition put forward by the Committee on Rock Mechanics of the Geological Society of America in 1964, viz., of the National Academy of Sciences Committee on Rock Mechanics in 1966 (18): "Rock Mechanics is the theoretical and applied behavior of rock; it is that branch of mechanics concerned with the response of rock to the force fields of its environment."

According to JUDD (10), every word in this definition was carefully evaluated to ensure a minimum of misunderstanding. No attempt was made to qualify the term "rock". However, by the logical contents of the matter, the subject rock mechanics pertains not merely to mechanics, but to the geologic materials as well that exist in nature or have been obtained from a geological environment. Although intimately concerned with the same natural, geological material, the new engineering discipline rock mechanics does not lie in the province of seeking explanation about the origin of geological structures as the scientific discipline geology does. Also, rock mechanics is an engineering field of its own, distinct form engineering geology. Engineering rock mechanics deals with rock as an engineering material. It also deals primarily with those changes in mechanical behavior in rock such as stress, strain, and motion of rocks brought about by engineering activities. Among other things, the subject of rock mechanics concerns itself with the design and stability of underground structures in rocks.

As an engineering discipline, rock mechanics is not an abstract study in mathematics and/or mechanics; it is a high-level applied engineering science. It can be successful only when connected with the real geological situation when designing and constructing foundations on rock and substructures in rock.

Thus, to the engineer, the subject of rock mechanics involves the analysis of static and/or dynamic loads or forces as applied to the rock, the analysis of the internal effects in terms of either stress, strain, or stored energy; and the analysis of the consequences of these internal effects, i.e. fracture, flow, or simply deformation of the rock. One, therefore, may say that rock mechanics is the youngest branch of the discipline strength of materials.

As in the field of soil mechanics and other mechanics, rock mechanics makes an extensive use of the theory of elasticity and the theory of plasticity, and studies rocks and structure-rock systems experimentally.

One sees that rock mechanics is a maturing engineering discipline, emphasizing its breath in both theoretical and applied areas. Rock mechanics is important within the scope of progress in the geotechnical sciences.

3-2. Some Features of Rock Mechanics

1. In a large, solid, sound rock mass, rock may be considered to be a continuum.

2. However, in its natural, geologic environment, rock is principally characterized by the fact that rock is not a continuum but a regulated discontinuum because of joints, fissures, shistosity, cracks, cavities, and other possible discontinuities. In this respect, it may be said that under certain conditions, rock mechanics is the mechanics of the discontinuum or, in other words, the mechanics of the structure of the rock.

3. Mechanically, rock is a "multiple-body" system.

4. Soil mechanics analyses work mostly in plane, whereas rock mechanics works both in plane and in space.

5. Rock mechanics has developed independently of soil mechanics. Of course, there is a certain overlap between these two engineering sciences.

A civil engineer in the fields of geotechnics and construction must know the material with which, upon which, and in which he works.

Rock, as one of the most commonly used construction materials, is very complex and difficult to describe and define. It is a material about which we know very little. To avoid structural failures associated with rock, it is an absolute necessity to understand it and to control it, especially in its natural environment under the influence of load, water, and temperature.

3-3. Rock Mechanics Problems

Some of the problems in rock mechanics associated with man's activity on and in rock, and the relationships that exist between a foundation on or an underground opening in rock, may be listed as follows:

How will rock react when put to man's use?
What is the bearing capacity of a rock on its surface and at various dephts to carry various loads?
What is the shear strength of rock?
How will rock perform under dynamic loads?
What is the effect of earthquakes on a rock-foundation system?
What is the modulus of elasticity of rocks?
What is the Poisson's ratio of rocks?
What are the effects of rock defects (jointing, bedding planes, schistosity, fissures, cavities, and other discontinuities) on its strength properties?
Which laboratory test method of rocks most nearly provides the actual in-situ properties of a foundation-supporting rock *en masse?*
How to account for joints and faults in design in rock?
How to cope with time-dependent deformations (creep) in rock?
What are the laws of plastic flow of rocks?
What is the effect of anisotropy of rock on stress distribution in rock?
How to correlate the results of rock strength tests made in situ with those made in the laboratory on prepared rock specimens?
What test method would most readily provide the actual in-situ conditions and properties of a rock?
What are the mechanisms of failure of rocks?
Can the state of stress in rock be accurately estimated, or can it even be measured?
What are the rock-slope design factors?
Will roof bolting in an underground opening in rock provide a satisfactorily safe permanent installation?

Also, there is the problem as to the factor of safety to use in design in rock.

The engineer is also interested in the water regimen in rock, in the mutual interaction between rock and foundation; suitability of rock as a construction material; effect of weathering on soil; and grouting of fissured rock to improve its performance relative to load and uplift pressures.

From these questions, one can see that "rock mechanics versus foundation" problems are attached to the rock, and that they should be studied as problems of deformation, strength, and stability.

The scope of the contents of rock mechanics, of course, overlaps several other disciplines of engineering and science (soil mechanics, engineering geology, geophysics, for example) which, in the past were not always considered to be too closely related.

The rock mechanics problems could be solved reasonably accurately:

1. If the stresses in a loaded rock were possible to indicate even approximately (for example, analytically), and

2. If for the induced stresses and stress distribution in rock to be analyzed, it were possible to indicate, even approximately, the corresponding factor of safety against rupture or dangerous deformations.

Although the technology strives for quantifying knowledge of the *en masse* properties of rock material and a plausilbe comprehension of stress distribution within it, unfortunately, even today, no exhaustive solution to these problems can be given as yet, mainly because of the inhomogeneity and anisotropy of the rock material and, in the words of STINI (29), because "Nature is different everywhere, and she does not follow textbooks." In this respect, rock mechanics has not reached a state of completeness as yet. These problems cannot be solved plausibly by any formulas as yet, but require some idealization of the rock material and the consideration of the effects of many environmental factors, without the possibility of an exact proof of stresses as a soft cushion of our conscience.

Besides, the problem of stress distribution in rock is a hint that the difficult problems of construction in rock cannot be simply burdened upon the geologist. Instead, the rock engineer should have the information on rock qualities as observed by the geologist, to evaluate numerically in his conclusions.

3-4. Objectives of Rock Mechanics

Some of the objectives of engineering rock mechanics are:

1. To perform engineering rock surveys;

2. To develop rational rock sampling, identification and classification methods;

3. To develop suitable rock testing devices and standard testing methods for compressive strength and shear strength of rock materials;

4. To collect and classify information on rocks and their physical properties in the light of fundamental knowledge of rock mechanics, foundation engineering and hydraulic structures engineering;

5. On the basis of test results, to study the physical, mechanical (static and dynamic), elastic, inelastic, plastic and rheological properties of rocks, and their mode of failure under static and dynamic loading;

6. To study rock performance under thermal conditions and water regimen;

7. To deal with statics and dynamics of structures in rock;

8. To develop methods of in-situ measurement of static and dynamic deformation properties of rock and residual stresses in rock under various environmental conditions of weathering, leaching, seismics, and tectonics;

9. To perform research on the mechanisms of failure of rocks;

10. To organize research on rock reinforcement and for in-situ stress measurements;

11. To replace by scientific methods the empirical ones of design used in rock engineering in the past, thus contributing to the advancement of the rock mechanics discipline;

12. To stimulate and disseminate knowledge on rocks and rock mechanics;

13. To apply the knowledge of rock mechanics for the solution of practical engineering problems;

14. To study performance of natural intact rock *(en masse)* under load and environmental conditions; and

15. To deal with statics of structures in rock; stability of rocks is of utmost importance from the viewpoint of both safety and economy.

Thus, by and large, the objective of engineering rock mechanics is its application to the solution of geotechnical problems.

International and national committees are busy planning *specifications* for *classification* of rocks, and for various laboratory and in-situ techniques which can be used to develop a better awareness of the action of rock under various environments.

3-5. Value of Rock Mechanics

Just as the value of soil mechanics is indisputable in soil engineering, so is the value of rock mechanics in rock engineering. Rock mechanics is one of the most useful aids to foundation engineers in their everyday professional activities.

3-6. Theoretical Basis of Rock Mechanics

Just as every other engineering discipline has a theoretical basis, so has rock mechanics. As is so often the case in a relatively new discipline, correct solutions to problems may require the engineer and researcher to borrow some general knowledge and theories from other disciplines and to adjust them to suit their particular field in forming a theoretical basis. The theoretical basis is needed for the ultimate understanding of the strength of the rock and its performance in situ. Inasmuch as every physical law has been derived from experiment, experiments with rocks play an important part in rock mechanics. Also, experiments with rocks serve for checking certain aspects of adopted theories from other disciplines. However, according to Sir ROGER BACON, experimental science does not receive knowledge from any higher sciences: experimental science is the mistress; all other sciences are her servants.

General Notes

Many rock mechanics and rock engineering problems of practical importance involve both elasticity and plasticity of rocks, especially when concerned with strength and safety of an underground opening in rock, or with the stability of rock as a material supporting structural foundations.

Rock mechanics, like any other engineering discipline, has a theoretical basis. In rock mechanics, frequently the theory of elasticity and the theory of plasticity are used as a part of that basis. Therefore, the acquaintance with some basic concepts of these theories is desirable.

Let us first recall that elasticity and plasticity are important universal properties of crystalline and amorphous solids, among rocks.

Frequently, a rock mechanics problem on hand calls for quantitative studies: first with all parts considered to be within the range of elasticity, and thereafter with some parts admitted to be in the range of plasticity (refer to Chapters 7, 8, and 9 in this book).

Both the theory of elasticity and the theory of plasticity deal with stresses and deformations in solid materials.

In their present state, the theory of elasticity and the theory of plasticity are concerned mostly with problems of loads and equilibrium within the realm of statics.

Theory of Elasticity

The essential store of all knowledge of structural theory, design, and construction practice is the mathematical theory of elasticity.

A review of the development of the theory of elasticity shows that during a period of over 250 years, the theory of elasticity had gradually developed into an exact discipline as a part of mechanics. Today, mechanics is the solid foundation for the design of engineering structures.

In the mathematical theory of elasticity, an ideal, homogeneous, elastic, isotropic medium (viz., material) is postulated. The theory of elasticity is based on HOOKE's law of linear proportionality E between stress σ and strain ε, i.e.,

$$\sigma = E \cdot \varepsilon \tag{3-1}$$

Ut tensio sic vis

Freely translated, HOOKE's law reads: As the stress, so the strain.

The coefficient of proportionality E is called the modulus of elasticity of the material.

The concepts of stress and strain are fundamental in the theory as well in the design of engineering structures.

It is also assumed in this theory that the deformation, viz., strains of a loaded material, are small quantities. Hence, they can be neglected in setting up equations of equilibrium.

It is further assumed in this theory that in the externally loaded elastic material, all strains are instantaneously and totally recoverable upon removal of all external load. A material is perfectly elastic if it recovers completely upon removal of all external stresses.

In rock mechanics, the object of a problem relating to elasticity is usually to ascertain the stress distribution in an elastic material. In some other cases, strains must be found at any point brought about by given body forces and given conditions at the boundary of that body. In such elasticity problems, one is dealing with infinitesimal strains and displacements.

According to NÁDAI (17), these assumptions are supported by actual measurements of the observed strains or displacements within the elastic range in metals, rocks, and other materials, to a sufficient degree for most practical applications.

But rock is not exactly an ideal, elastic material. An elastic medium is an idealization of actual material properties, for obviously, deformation of crystalline materials under the action of external forces is the result of distortion of the crystal lattice in which the atoms, ions, and/or molecules are rearranged, involving structural adjustments inside the rock material, which adjustments will be a finite process. Hence, there will be less than a total recovery of deformations after the removal of the externally applied load from the material.

Also, the concept of isotropy is applied in rock mechanics as a simplifying assumption used in continuum mechanics. Most crystals and rocks are not isotropic.

The main objections against the use of the theory of elasticity in rock mechanics are

1. Because of the presence of discontinuities and various rock defects in the rock in situ, thus affecting the mechanical properties of the rock (rock has a grained structure consisting of minerals and crystals of several kinds with many different orientations, and with irregular boundaries between the grains);

2. Because of the ever present internal (residual) stresses in the rock mass; and

3. Because of the dependence on time of the deformations (rheological aspects).

Unfortunately, for rocks, frequently the established functions to use are so complex and involved, that in attempting to seek a reasonable and acceptable solution one must resort to idealizations and simplifications; or to the method of trial and adjustment; or to graphoanalytical methods; or in-situ and laboratory testing of rocks; or experimental model testing of prototype structure.

Nevertheless, to a certain extent, the theory of elasticity and the theory of plasticity are used in the design of structures in rock. Here, simplified rock properties are assumed in order to assist analytical calculations based on the theory of elasticity and/or plasticity. According to FARMER (4), sometimes such assumptions are successful and sometimes not. It is therefore essential that the limits of applicability of the theory of elasticity to rock be clearly defined.

If the laws of HOOKE's linear elasticity, and those of ideal plasticity, are compared with the typical experimental features, one can determine the applicability and limitations of the mathematical formulations for rock. By observing near the origin of the stress-strain diagrams for rocks, one notices that the linear stress-strain relationship is generally good in nearly all cases.

Another point to make here in favor of applicability of the theory of elasticity to rock mechanics is as follows: the stress-strain relationships as derived on the basis of the theory of elasticity alone, are suitable for the theoretical verification of the development of the stress-free body around underground openings, and the stresses should be carried entirely by the rock surrounding the opening.

For rock engineering purposes, geological and rock mechanics data, results obtained from in-situ testing of rocks, and analyses of strength properties of rocks buttressed by model tests, coupled with experience and judicial judgement, naturally form the basis for the design and construction in rock.

The reader should not infer that the above made assumptions are a rigid framework limiting the scope and applicability of rock mechanics. Unless simplifications represent the real conditions in rock with sufficient accuracy, the physical laws governing the nature of the performance of the rock-structure-load system should not be arbitrarily twisted in order to obtain a simplified working system that could be treated easily mathematically, or be "in good agreement", or would render in advance a preconceived result.

Theory of Plasticity

The object of the theory of plasticity is to study, mathematically, stresses and displacements in plastically deforming materials. The basic feature of the theory of plasticity as compared with the theory of elasticity is the non-linear relationship between stress and strain.

In contradistinction to the theory of elasticity, in the theory of plasticity one considers bodies in which, under the action of large enough stresses, there set in permanent deformations which remain after the removal of the externally applied load. Actually, all materials possess a certain degree of plasticity, i.e., property of preserving a certain part of a deformation after relief of stress.

The theory of plasticity as applied to an ideal rock assumes that on a certain plane in rock there exists a state of failure or rupture at any point located on such a plane. Such a theory has been put forward by HENCKY (9), NÁDAI (17), PRANDTL (21), and others.

PRANDTL contributed to the theory of plastic equilibrium of a horizontally bounded hemispatial medium subjected to a vertical punching pressure (21).

Other authorities on plasticity and rupture are SAINT VENANT (24), VON MISES (16), KÁRMAN (14), TRESCA (32), and others.

Generally, the theory of plasticity is concerned with all deformations that are independent of time. The explanation of plastic flow in solid materials is the permanent displacement in the relative positions of the elements or atoms in the crystal lattice.

The theory of plasticity is confined to equilibrium within the stressed body. Thus, the theory can determine only the state of stress, both in the plastic and elastic zones, after all deformations have ceased to change with time.

Plastic deformation is physically anisotropic. Any initial isotropy which may have been present is usually destroyed by plastic deformation.

Within the range of plasticity, the relationships between stress and strain are more complex than within the elastic range. The strains are functions not only of the corresponding acting stress and temperature, but also of the past history of stresses, such as those which have been brought about by tectonic activities in the earth's crust.

The study of the strength properties of rocks is based on the conclusions of the theory of plasticity, because fracture, as a rule, precedes plastic deformations.

Materials with polycrystalline grain structure, such as minerals and brittle rocks, under certain conditions may be brought into the plastic state, in which permanent deformations may occur without fracture of the material. Solid materials in this condition are said to yield, or to deform plastically.

In describing the plastic states of stress in rock, extensive use is made of the slip surfaces (Chapter 8).

The theory of plasticity has yielded powerful general concepts that have become tools of enormous importance in the rational design of a wide range of structures, among them structures in rock. In civil engineering, the theory of plasticity is applied to soil mechanics and foundation engineering, and to rock mechanics as well. Plasticity is also of great interest to the mining industry, as well as to geology and to geophysics.

Difference between the Engineering Discipline of Strength of Materials and Rock Mechanics

Functionally, the ground (soil and/or rock) belongs to structure, and vice versa. The ground and the structure are a mutually interacting integral system called the rock-structure-load system. This system includes all the effects of discontinuities, viz., rock weaknesses or defects. In rock mechanics, therefore, consideration of relevant important engineering concepts should not be omitted from the presentation that follows. However, corresponding to the sense of content and intent of this book on rock mechanics, the engineering geology concepts are subordinated to some extent under the aegis of engineering mechanics within the scope of rock mechanics.

Particularly in the phase of mechanics within the discipline of rock mechanics, elements from mechanics of materials, theory of elasticity, theory of plasticity, theory of mechanics of fracture, rheology, and other branches of scientific-technical knowledge are involved.

In the wake of the above, naturally some questions arise, such as:

What is rock mechanics?
What is the difference between strength of materials and rock mechanics?

The definition of rock mechanics was given earlier in Chapter 3 of this book. However, popularly speaking, rock mechanics may be regarded as the strength of materials of rocks. The strength of the rock is its ability to resist externally applied loads.

In the discipline of strength of materials, called also the mechanics of materials, a load is applied to a *stress-free* material, and the so imparted stresses and strains in it are analyzed. Here, the external loading and the geometric dimensions of the loaded structural element are known.

In rock mechanics (refer to Section 7), the rock in situ is initially subjected to an *initial* or *primary stress field*. Upon introducing an underground opening in the rock mass, a new stress condition in rock is brought about, namely: stress is released locally (stress reduction or stress relaxation), resulting in a corresponding unloading deformation.

One notices that the basic difference between strength of materials and rock mechanics is the changed stress conditions and deformations in the body of the material brought about by an excavated underground opening. Thus, rock mechanics has for its purpose to analyze newly induced stresses and strains in the rock massif brought about by an underground opening (see Fig. 7-3).

The geometric dimensions of the region or zone of disturbance in the rock around the cavity that takes up the unloading deformation must first be determined (Fig. 7-3), which is difficult to do. It can thus be reasoned that the load and the load-bearing structural element cannot be precisely defined. Also, one bears in mind that rock as a construction material is not a homogeneous material manufactured artificially according to certain standards. On the contrary, rock is a natural, nonhomogeneous material. Besides, of special significance is the fact that the rock deformations, brought about because of the changed stress conditions, are seldom of elastic nature. In most cases they are of plastic nature, or at least partly elastic and partly plastic. Thus the discipline of practical engineering rock mechanics in situ pertains actually to the plastic zones in the rock. Because of their complexity, the plastic zones are difficult to comprehend analytically. Therefore, approximation or simplification of the more exact theory is made, so that the practicing engineer can utilize the predictions and results of the mathematical theory of plasticity.

Because the mechanical phenomenon of pressure on tunnels may have several different causes, there exist many methods of calculations of stress conditions around various forms of tunnels (5, 23, 31). Besides, the striving after a universal, omnipotential formula for computing the magnitude of the pressure on underground openings and lined and walled tunnels cannot conform exactly with the reality.

3-7. Rock Engineering

The discipline of engineering rock mechanics is the scientific basis for rock engineering. Rock engineering is the practical, technical use and engineering application of rock mechanics in the design of engineering structures such as various underground openings as set forth: power plants, storage space, protective shelters, fortifications, shafts, adits, mines, vehicular and aqueous tunnels; in the utilization of rock as a support of structural foundations; or in the design of roads, railroads, canals, concrete dams on rock, quarries, open excavations, design of rock slopes in cuts, anchorage of rocks.

Just as soil engineering was influenced decisively by new concepts in the field of soil mechanics, so it is with rock engineering: through research, experience, and accumulation of man's knowledge of the physical and strength properties of rocks, a new engineering discipline of rock mechanics and rock engineering is coming to the fore.

Although a relatively new discipline still in need of development, today rock mechanics and rock engineering are recognized as distinct geotechnical engineering disciplines in their own rights, just as soil mechanics and foundation engineering.

3-8. Ice Mechanics

The place of ice mechanics is between soil mechanics and rock mechanics. Geologically, ice is a rock (the solid phase of water). As a rock, ice has been recognized by the First Congress of the International Society of Rock Mechanics held September 25 to October 1, 1966, at Lisbon, Portugal.

Ice, H_2O (hydrogen oxide) is not normally thought of as a rock or a mineral, but it meets the definition of rock. Glacial ice is in every sense a metamorphic rock.

Because ice is solid at temperatures below 0°C at atmospheric pressure, it is abundant at the surface of the earth.

Ice is unique in many ways, and it is one of the substances that expands on freezing by 9% in volume. A tremendous ice force is produced upon freezing, particularly when the ice is confined. Ice pressure on hydraulic structures is a major design problem in regions affected by frost.

Ice is also an important, effective mechanical weathering, erosion and transportation agent of tremendous capacity. Next to water, ice is a spectacular physical agent in the weathering of rocks and soils.

Ice wedging by frost action in rock contributes to the widening of joints, cracks, and fissures. Water, filled in a crack of rock, freezes from the ground surface down because the water in the crack is cooled most at the surface. This confines the water in the crack. A continued freezing induces ice pressure because of expansion of the ice. This confined ice wedging against the rock brings about further cracking of the rock. The resulting products of mechanical weathering of rock include everything from the huge boulders to particles of the size of silt and clay.

A knowledge of the mechanical properties of ice is useful in organizing and adapting suitable drilling, blasting, excavation, and design methods.

Ice content and unfrozen water within cryogenic rocks and soils determine their physical and mechanical properties.

The average ultimate unconfined compressive strength of 12 cylindrical pure ice specimens 7.2 cm high and 3.3 cm in diameter, as tested by the author at a temperature of $-8.5\,°C$ was $\sigma_{ult\,\|} = 1.569\,MN/m^2$ (11,13).

According to BARNES (1), the allowable compressive strength of pure ice may be taken as $588\,kN/m^2$, and the allowable tensile stress as $294\,kN/m^2$.

According to SIPRE'S report (26), the tensile strength of ice is from 620 to $1034\,kN/m^2$, and the ultimate bending stresses are 1.344 to $1.682\,MN/m^2$, although for warm melting of ice these values vary from 262 to $779\,kN/m^2$.

The reported shear strength of ice is about $\tau_\perp = 779\,kN/m^2$, and $\tau_\| = 677\,kN/m^2$.

More about mechanical properties of single ice crystals have been written by GLEN (6), GOLD (7), and ÔURA (20).

In foundation, tunnel, shaft, dam and hydraulic structures engineering ice is utilized in temporary thermal soil stabilization, and for checking temporarily seepage in soil and or rock. This is accomplished by artificial freezing of soil, or rock. Artificial freezing in excavation operations is an expedient, efficient, time-tested and successful means of stabilizing the walls and bottoms of foundation pits, and around tunnels and shafts in permeable, waterlogged soils, in waterbearing layers of soil, and in sandy soils subject to "quick" condition (= a hydraulic condition).

In the method of thermal soil stabilization, the water in rock joints and fissures is temporarily converted into solid ice. This is done by artificial freezing of the ground. The frozen condition of the ground is maintained only until the foundation, or tunnel, or shaft work is completed. After the foundation work in a "frozen-out" ground is completed, the frozen ground is allowed to thaw.

Soil freezing is also applicable for underpinning of structures (3).

Artificial freezing of soil has been and is still used in the mining industry.

In 1906, during the construction of a section of the Metropolitan, the subway system of Paris, a tunnel under the Seine was driven partly through artificially frozen soil. The section is 1092 m long, running from Place du Châtelet to Place St. Michel, under the Ile de la Cité and both arms of the Seine (15).

During the construction of the 167.64 m (550 ft) high and 1271.93 m (4173 ft) long Grand Coulee multiple purpose gravity dam across the deep gorge of the Columbia River, Washington, soil and mud was frozen to form a temporary earth- and water retaining structure to prevent the mud and water from flowing into the foundation pit (8). The dam was built on granite overlain by extensive basalt lava.

Recently, here and abroad, many artificial soil freezing jobs have been completed successfully (22), among them the shaft for the Richmond water supply tunnel in Brooklyn, New York (11, 27, 28).

Fig. 3-1 shows an artificial soil freezing operation for the Richmond water supply tunnel shaft in Brooklyn, New York.

Permafrost

A natural phenomenon of interest in thermal geotechnics is permafrost in arctic, subarctic and antarctic regions. Permafrost, or permanently frozen ground, is defined exclusively on the basis of temperature below 0°C, irrespective of its lithological nature, texture, and water content.

Permafrost presents to civil engineers many complex design and construction problems associated with climatic extremes; minute changes in the temperature of frozen soils, viz., thermal disturbances in permafrost, and exposure to sun considerably influence the physical and mechanical properties of permafrost. The basic principle to adhere to in engineering operations in permafrost is that of preserving its natural frozen regimen (12, 19, 22, 25).

Fig. 3-1
Artificial soil freezing around a vertical circular shaft for the Richmond water supply tunnel in Brooklyn, New York.
Left: tarpaulin-covered soil freezer pipes
Right: lined-out vertical circular shaft

In the recently completed 1280 km (800 miles) long Alaska pipeline system — the largest engineering project in the north so far known — all thermal factors and thermal conditions had to be considered in its design.

Permacrete

The term "permacrete" is used to describe artificial concrete-like mixtures of soil materials, which, when cemented with ice, form a consolidated aggregate useful as a construction material in cold regions.

The use of the permacrete material in permafrost tunneling and mining has some advantages over concrete and timbering. Permacrete can be used for roof stabilization and direct subsurface construction (also as columns and arches). According to SWINZOW (12, 30), permacrete is inexpensive and safe, and strong full-storage containers can be constructed in almost any permafrost environment (not only in tunnels).

It should be said that once the frozen ground and permafrost problems are understood and correctly evaluated, their successful solution is, for the most part, a matter of common sense whereby the frost forces are utilized to play into the hand of the engineer and not against it.

3-9. Work in Rock Engineering

The following questions

What is the deformability of fissured and sound, massive rock in situ?

What is the correlation between the mechanical properties of rocks and their geological and petrographical data?

How to cope with porosity and permeability in fissured rock in situ?

How to stabilize weak rock masses?

How to perform measurements of stresses and strains in rock before and after excavation and construction?

point out that in rock engineering work, one must interpret geological data in physical and mechanical terms. Then rock mechanics can be of great help in designing safe foundations and underground structures.

Rock mechanics problems may be also broadly grouped as problems of

1. preliminary design and
2. final design

involving detailed studies of rocks involved.

Another kind of grouping of rock mechanics and rock engineering may be based on methods of rock exploration, namely:

1. methods of design in rock, and
2. methods of treatment of rock in situ to improve certain

properties of rock.

As to the province of activity of the geologist, he performs and helps to guide rock exploration work, and interprets the geological data collected. During the design and construction phases of the project, the geologist cooperates with the engineer, outlining the regional and site geology, the genesis and past history of the rocks, and studies and reports what is hidden and not yet known geologically about the construction site. The geologist points out possible pitfalls such as Karst topography, erosion, and past and possible future rock- and landslides, for example. It is that tectonics, stratigraphy, hydrology, rock types and rock defects at the site may indicate problems to be anticipated during the time of construction and service of the structure.

It is of importance to detect the presence of active faults; faults which can be activated by an earthquake; fault zones, and to learn about the gouge material contained in rock joints, fissures, cracks, crevices and in the shear planes.

The discipline of rock mechanics pursues also its research in tunneling, large openings in rocks such as underground power plants, storage space, shafts, and other kinds of subsurface structures.

REFERENCES

1. BARNES, H. T., 1928, *Ice Engineering*. Montreal, Canada: Renauf.
2. Cold Regions Specialty Conference on *Applied Techniques for Cold Environment*, held May 17—19, 1978 at Anchorage, Alaska.
   Vol. 1 published by the ASCE in 1978 (629 pages).
   Vol. 2 published by the ASCE in 1979 (pp. 631—1167).
3. DUMONT-VILLARES, A. D., 1956, "The Underpinning of the 26-storey Companhia Paulista de Seguros Building, São Paulo, Brazil." Géotechnique (London), Vol. 6, No. 1.
4. FARMER, I. W., 1968, *Engineering Properties of Rocks*. London: E. and F. N. Spon Ltd., p. 40.
5. FENNER, R., 1938, "Untersuchungen zur Erkenntnis des Gebirgsdrucks." *Glückauf*, No. 32, pp. 681—695. No. 33, pp. 705—715.
6. GLEN, J., 1974, *The Physics of Ice*. Cold Regions Science and Engineering Monograph 11-C 2 a. Hanover, New Hampshire: Cold Regions Research and Engineering Laboratory (CRREL), April, 1974.
7. GOLD, L. W., 1973, "Ice-Challenge to the Engineer," *Proceedings, Fourth Canadian Congress of Applied Mechanics,* held May 28 to June 1, 1973, at Montreal. Ottawa, Canada: Division of Building Research. Paper No. 395, 1973, pp. G-19 to G-36.
8. GORDON, G., 1937, "Arch Dam of Ice Stops Slide," *Engineering News-Record,* Vol. 118, 1937, p. 211.
9. HENCKY, H., 1923, "Über einige statisch bestimmte Fälle des Gleichgewichts in plastischen Körpern." Zeitschrift für angewandte Mathematik und Mechanik," Vol. 3, pp. 241—351.
10. JUDD, W. R., ed., 1964, *State of Stress in the Earth's Crust*. Proceedings of the International Conference, held June 13—14, 1963 at Santa Monica, California. New York, N. Y.: American Elsevier Publishing Company, Inc.
11. JUMIKIS, A. R., 1966, *Thermal Soil Mechanics*. New Brunswick, New Jersey: Rutgers University Press, pp. 131—134.
12. JUMIKIS, A. R., 1977, *Thermal Geotechnics,* New Brunswick, New Jersey: Rutgers University Press, pp. 277—300.
13. JUMIKIS, A. R., 1979, "Cryogenic Texture and Strength Aspects of Artificially Frozen Soils," *Engineering Geology* (Amsterdam), Vol. 13, 1979, pp. 125—135.
14. KÁRMAN, TH., 1911, "Festigkeitsversuche unter allseitigem Druck." *Zeitschrift des Vereins deutscher Ingenieure*, Vol. 55, pp. 1749—57.

15. LOVERDO, DE, 1910, "La congélation du sols dans les travaux du Metropolitain de Paris." Monographie sur l'état de l'industrie du Froid en France, published by Association Française du Froid; II. Internationale Kältekongress, Vienna, 1910.
16. MISES, R., VON, 1913, "Mechanik der festen Körper im plastisch-deformablen Zustand." Göttingen: *Nachrichten der Gesellschaft der Wissenschaften zu Göttingen.* Mathemathisch-physikalische Klasse, pp. 582—592.
17. NÁDAI, A., 1950, *Theory of Flow and Fracture of Solids.* New York, N. Y.: McGraw-Hill Book Company, Inc., Vol. 1, 2nd edition.
18. National Academy of Sciences, 1966, "*Rock Mechanics Research.*" Washington, D. C.: NAS-National Research Council Publication 1466.
19. National Convention of the ASCE: *An Overview of the Alaska Highway Gas Pipeline: The World's Largest Project.* Convention held April 24—28, 1978 at Pittsburgh, Pennsylvania. Published in 1978 by the ASCE (130 pages).
20. ÔURA, H. (editor), 1967, *Physics of Snow and Ice. Proceedings of the International Conference on Low Temperature Science,* Vols. 1 and 2. Sapporo, Japan: The Institute of Low Temperature Science, Hokkaido University, 1967.
21. PRANDTL, L., 1921, "Über die Eindringungsfestigkeit (Härte) plastischer Baustoffe und die Festigkeit von Schneiden," Zeitschrift für angewandte Mathematik und Mechanik, Vol. 1, No. 1, February 1921, pp. 15—20.
22. JESBERGER, H. L., editor (1978) of the *Proceedings of the International Symposium on Ground Freezing,* held March 8 to 10, 1978, at the Ruhr-University Bochum, Germany.
23. PROCTOR, R. V. and WHITE, T. L., 1946. *Rock Tunneling with Steel Supports,* with an Introduction to Tunnel Geology by Karl Terzaghi. Youngstown, Ohio: The Commercial Shearing and Stamping Co.
24. SAINT-VENANT, B., DE, 1870, "Sur l'établissiment des equations des mouvements intérieurs opérés dans les corps ductiles au delà des limites où l'élasticité pourrait les ramener à leur premier état." *Comptes Rendus,* Paris, Vol. 70, 1870, pp. 473—480.
25. SANGER, F. J., 1969, *Foundations of Structures in Cold Regions.* Cold Regions Science and Engineering Monograph III — C4. U.S. Army Material Command, Terrestrial Sciences Center, Cold Regions Research and Engineering Laboratory, Hanover, New Hampshire, June, 1969.
26. SIPRE, 1951, *Review of the Properties of Snow, Ice, Permafrost.* Research Establishment, Corps of Engineers, U.S. Army, SIPRE Report, No. 7, July 1951, by the University of Minnesota, Minneapolis, March, 1951.
27. SMITH, G. R., 1962, "Freezing Solidifies Tunnel Shaft Site." *Construction Methods and Equipment.* Vol. 44, No. 10, October, 1962, pp. 104—108.
28. STEWART, G. C., GILDERSLEEVE, W. K., JAMPOLE, S., and CONNOLLY, J. E., 1963, "Freezing Aids Shaft Sinking for a Water Tunnel Under New York Harbor." *Civil Engineering,* Vol. 33, No. 4, April, 1963, pp. 52—54.
29. STINI, J., 1952, "Der Gebirgsdruck und seine Berechnung," *Geologie und Bauwesen,* No. 3, 1952, pp. 165—200.
30. SWINZOW, G. K., 1963, "Tunnelling and Subsurface Installations in Permafrost." *Proceedings Permafrost International Conference,* held 11—15 November, 1963, at Purdue University, Lafayette, Indiana. National Academy of Sciences-National Research Council publication 1287. Washington, D.C., 1963, pp. 519—526.
31. SZÉCHY, K., 1966, *The Art of Tunneling.* Budapest: Akadémiai Kiadó.
32. TRESCA, H., 1864, "Mémoire sur l'écoulement des corps solides soumis à des fortes pressions." *Comptes Rendus,* Académie des Sciences, Paris, July—December, Vol. 59, 1854, pp. 748—758; Vol. 64, 1867, 809.

PART 2
ROCK EXPLORATION

CHAPTER 4

METHODS OF ROCK EXPLORATION

4-1. Need for Rock Exploration

In rock engineering, rock exploration is as important as is soil exploration in soil and foundation engineering.

Civil engineering practice has demonstrated that careful design of an engineering structure is not yet all that is needed for its safety and stability. Before any design and construction are contemplated, a comprehensive and careful exploration should be made of the environment of the structure to be built. The said applies especially to the structure-supporting geological material, viz., rock upon which the structural foundation is to be laid. In this respect, the advisability of engaging the professional services of a competent engineering geologist should not be overlooked. However, the responsibility for the safety and success of the structure-rock system is the responsibility of the design engineer.

The position of the geologist is not to make restrictive requirements for the engineer, but to furnish full information upon the basis of which the engineer can make the best and most economical design.

Such an attitude is certainly commendable, and should make for the best of relationships in the field.

4-2. Exploration Methods

All kinds of engineering projects on and in rock must be based on rock exploration prior to any design being made and construction begun. The exploration program serves for obtaining data needed for design.

A comprehensive, detailed discussion about the various site exploration methods is beyond the scope of this volume. Only a brief recourse on this topic relevant to the development of the material in this book is presented.

Rock exploration of a construction site and its vicinity is usually made along the following seven lines, namely:

1. Geological exploration,
2. Hydrological exploration,
3. Geophysical exploration,
4. Thermal exploration,
5. Evaluation of rock as to its diminution and workability,
6. A study of procurement of rock for construction, and
7. In-situ and laboratory testing of strength properties of rocks.

Which one of the ground exploration methods to pursue and to what extent, depend upon local site conditions, the particular engineering construction project on hand, and the data needed for the design of that structure.

A publication about subsurface exploration and sampling has been prepared by HVORSLEV (50) for the American Society of Civil Engineers.

4-3. Geological Exploration

In any appraisal of rock for a construction project, it is necessary to determine the nature of the geological environment and its critical features in which the rock exists. Performancewise, igneous, sedimentary, and metamorphic rocks differ greatly among themselves. For geotechnical engineering purposes the ground is explored relative to morphology, geology, hydrology, physical properties of soil and rock, chemical properties of soil, rock and groundwater regimen, biological properties of the ground and groundwater, and mineralogic-petrographic properties of soil and rock.

The geological exploration for an underground opening (tunnel, power plant, storage facility), of a dam and reservoir site, and other important engineering structures is usually delegated to a competent and experienced engineering geologist.

Geological exploration is concerned with geological conditions of rocks of the terrain and the construction site, as well as with the various properties of rock. Of interest are geological structures; stratification; tectonic conditions such as faults and systems and nature of rock weaknesses and discontinuities; petrographic nature of rocks and their effect on supporting loads; and rock performance under the action of load, water, and temperature. The weathering process of rock, too, should not be overlooked. Blasting of rock may open fissures and cracks and facilitate influx of water into underground openings.

Design and construction in rock in earthquake-prone regions should conform with local aseismic design building codes.

In order that the geologist's report be of any value for its use by engineers in planning, design and construction, the geological exploration report should be prepared clearly and in easily understandable language. Geology must be made not only available, but also accessible to engineers. On the other hand, to understand the geologist's report, the engineer must have a background in geology and a thorough understanding of the geological principles involved. The application of the geological subsurface information to the design and construction of engineering structures is the province of the civil engineer.

Geological exploration of rocks involves
1. the study of the geologic history of the region
2. an assessment of past events that have modified features and properties of the rocks in situ
3. the evaluation of the effects of the engineering structures on geologic features in the area and on the strength of the rocks in situ
4. the geologic response to changed environmental conditions during the course of time, and
5. investigation of rock outcrops on the earth's surface and the underlying strata.

The most noticeable feature of an outcrop of sedimentary rocks is the bedding which records the layers in the order of deposition of the geologic material with the oldest at the bottom.

Geological exploration also includes mapping of rock exposures, and the study of rock patterns on the ground, as well as from aerial photographs.

In geological exploration work, aerial photographs are useful for
1. regional reconnaissance and exploration
2. assessing the general terrain at and in the vicinity of the construction site
3. studies of landforms, streams and surface drainage patterns (Fig. 4-1); the main patterns are dendritic, trellis, angular, and radial
4. planning of subsurface exploration work
5. mapping of bedrock and rock outcrops
6. locating and tracing of faults and
7. fault zones (see Fig. 4-2).

Figure 4-2 shows a part of the San Andreas fault in California.

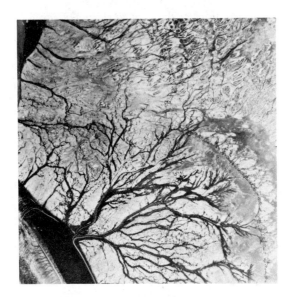

Fig. 4-1
Surface drainage pattern as seen on a vertical aerial photograph

Fig. 4-2
San Andreas fault, California.
Looking in northerly direction, along Elkhorn scarp. The Carizzo Plains are on the left,
Elkhorn Planes on the right.
(Photo by R. E. WALLACE. Photo courtesy of the U.S. Geological Survey.)

METHODS OF ROCK EXPLORATION

From the air, the fault stands out as a giant scar. From the ground level, the jagged stream beds and bold scarps present a clear record of violent movement
8. layout of borings
9. locating of sources of construction materials
10. planning and design of large, important engineering structures, and possibly for other purposes.

An interesting example of the use of aerial photographs and engineering soil maps is the location of geological features suitable for development of a water supply reservoir. Figure 4-3a is an uncontrolled mosaic of vertical aerial photographs of the Round Valley in Hunterdon County, New Jersey, and Fig. 4-3b in the corresponding soil/rock map of the same area (48, 65).

The floor of the Round Valley is underlain by Triassic shale. The valley is surrounded on three sides by a steep-sided horse-shoe-like diabase ridge known as the Cushetunk Mountain. The reservoir was formed by closing the gaps of the valley with three dams (Fig. 4-3b). The rock foundations in the dam locations are gneiss, quartzite and Triassic shale.

The Round Valley reservoir, constructed by the State of New Jersey is an initial step in the long-range water conservation and development program.

To reveal rock structure and material at the site under investigation, core drillings or borings are made and rock cores are extracted.

Subsurface Exploration

For deep subsurface exploration, borings, exploration shafts and exploration tunnels are made. A boring is a borehole drilled into the ground for purpose of extracting soil and/or rock samples for examination and testing. Shafts and tunnels afford a direct exploration of the geologic formations encountered, and observation of their stratification in place.

Depending upon the need, boreholes may be drilled vertically, horizontally, and at inclination.

Generally, there are two kinds of rock core samples procured from the exploration site, namely:
1. the so-called "small" diameter rock core samples, and
2. large diameter rock core or "calyx" samples.

The purpose of core borings is to collect rock core samples of the rock traversed. Cracks and openings in rocks do not appear in the drill cores. The main purpose of a calyx boring is to provide for a shaft. It is an access to the underground for direct inspection of the rock formations in situ as exposed in the wall of the shaft above groundwater table or in dry, dewatered shaft, whichever the case is.

Fig. 4-3a
Uncontrolled airphoto mosaic of Round Valley, Hunterdon County, New Jersey (66).
(Aerial photographs by permission of Aero Service Corporation, Philadelphia, Pa.)

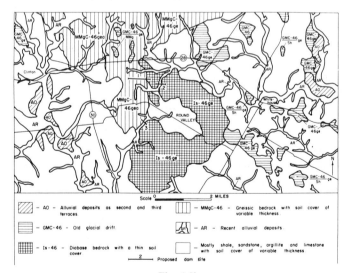

Fig. 4-3b
Engineering soil and rock map of a part of Hunterdon County, New Jersey (48).
(By permission of Rutgers University Press.)

METHODS OF ROCK EXPLORATION

The diameter of small rock cores usually varies from 22.225 mm (7/8 in.) through 50.8 mm (2 in.) to 203.2 mm (8 in.). The length of the rock core cut is limited to the length of the coring barrel, which is usually about 3.5 m (10 feet).

Standard sizes for casings, rods, core barrels and holes are compiled in Table 4-1.

TABLE 4-1
Standard Sizes, in inches, for Casings, Rods, Core Barrels and Holes

Casing Core Barrel	Drill rod	Casing OD	Core barrel bit OD	~Diameter of bore hole	~Diameter of rock core specimen
EX	E	1-13/16	1-7/16	1-1/2	7/8
AX	A	2-1/4	1-27/32	1-7/8	1-3/16
BX	B	2-7/8	2-5/16	2-3/8	1-5/8
NX	N	3-1/2	2-15/16	3	2-1/8

TABLE 4-2
Rock Resistance to Drilling

Rating	Rocks
1	2
Easily drillable rocks:	1. Anhydrite 2. Argillaceous sandstones 3. Gypsum 4. Slates 5. Shales
Moderately difficult drillable rocks:	1. Calcareous sandstones 2. Gneisses 3. Limestones 4. Mica schists
Difficult drillable rocks:	1. Granites 2. Siliceous limestones 3. Siliceous sandstones
Very difficult drillable rocks:	1. Amphibolites 2. Diorites 3. Quartzites 4. Quartz veins

Usually rock cores are cored with NX-size diamond or alloy steel bits to about 10 cm (~ 4 in.) in diameter. Larger bits are used for better and undeformed recovery of samples from softer formations.

Based on their mineral composition, hardness, and texture, the *resistance of rocks to drilling* (and blasting) may be rated as compiled in Table 4-2.

Rock coring in situ is performed in order to
1. obtain rock samples
2. learn about subsurface conditions, stratification systems, and the kind of material encountered
3. disclose the sequence and thickness of the various rock strata traversed by borings
4. study the groundwater regimen
5. identify and classify rocks
6. examine rocks for their defects and weaknesses
7. test rock specimens for their physical and strength properties, and
8. evaluate the rock in situ, along with rock test results as obtained from rock tests in situ, for its suitability as a construction material for the support of structural foundations, hydraulic structures, underground openings, and as a building material.

The boring results should be shown on a boring log. The boring log should also include information about the density of the rocks; their status of fissures and cracks, frequency and width of joints; permeability; fluctuation of the position of the groundwater table; losses of drilling fluid; amount of water absorbed by the rocks during permeability tests; data on petrographic analysis of rocks, and their physical and mechanical properties.

Depth and Spacing of Borings

The depth of borings depends greatly upon the local topographical, geological, and hydrological conditions at the site especially on the magnitude of the hydrostatic pressure head; the kind, form and size of the structure; its purpose; loading intensity and technological importance of the structure as well. Generally, adequate soil and rock exploration should extend to a considerable depth below the proposed elevation of the base of the footing, or dam, or an underground opening. The boring depth should include all stressed zones of soil and/or rock involved in the foundation-soil (rock) system. Also, borings should be made in soil or rock so deep as to leave no doubt about the character of the geologic material below.

Under certain special conditions borings should be carried down deeper than was mentioned above. These would include areas where holes or cavities may be expected, as in mining regions where subsidence occurs; where karst topography is encountered, or when special geological and technical conditions are obtained.

Usually boreholes are spaced from about 30 m up to 100 m (center on center). The depth of borings are governed also by loading intensity and/or the hydrostatic pressure head.

The position and fluctuation of the groundwater table in each borehole should be measured. If artesian water is encountered, the pressure, too, should be recorded.

Site investigation should also be concerned with locating possible springs and seepage, as well as with the permeability problem in rock where it applies.

The effect of spacing and depth of borings on the disclosure of subsurface conditions may be perceived by the following simple illustrative example (Fig. 4-4). If borings are not spaced in sufficient number and deep enough, the soil or rock profile, constructed according to such data, may be very erroneous. The three informatory borings, i, Fig. 4-4a, if spaced far apart, may give the incorrect impression that there is only one layer of soil or rock (a-b-c-d), running almost parallel to the ground surface,

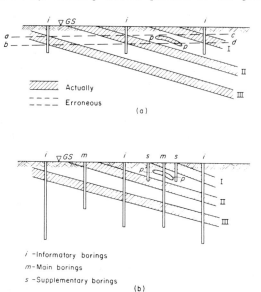

Fig. 4-4
Effect of spacing and depth of borings on disclosure of subsurface conditions

instead of three dipped layers (I, II, III) of the same material. The pocket or lens, (p-p), is not disclosed at all. Additional borings (m), Fig. 4-4b, would disclose the actual conditions of stratification (I, II, III). All the borings should be drilled deeper than shown in Fig. 4-4a.

Study of rock cores enables the construction of graphic profile. The depth at which a recognizable rock stratum is encountered by a series of bore holes gives a clue to rock structure. Core drilling, however, is expensive. A less expensive method of logging is electrical logging. The latter can be utilized by using boreholes drilled by less expensive methods. Groundwater that saturates sedimentary rocks contains dissolved salts. A salt solution is a good electrical conductor. The resistivity of the groundwater at any desired depth is measured by special sensors. Such electrical measurements give an electric log that is widely used in correlation.

Drilling Fluids

Rock coring in situ is usually done by a rotary cutting or drilling tool which rotates to cut or grind the rock. The cutting tool or bit is mounted to the lower end of the drill pipe of a rotary drilling rig.

To cool and lubricate the bit during drilling of the rock and to remove the cuttings from the borehole during drilling operations, a drilling fluid such as water, or clay suspensions, or other kinds of drilling fluid, such as thixotropic fluids, are used.

The drilling fluid is forced down through the drill-pipe and returned to the surface through the concentric, annular space between the casing and the drill pipe, carrying rock cuttings and chips up and out of the borehole.

By definition, thixotropy is the property of a material to undergo an isothermal gel-to-sol-to-gel transformation upon agitation and subsequent rest. Thus, upon mechanical agitation, gels liquefy, and resolidify when agitation is stopped. The word thixotropy is combined by two Greek words: thixis = the touch, the shaking, and tropo = to turn, to change. Thus, thixotropy means "to change by touch."

The thixotropic phenomenon is especially pronounced with sodium (natrium) bentonite suspensions.

An old, simple test for thixotropy, is to shake vigorously a colloidal suspension in a test tube and observing the time necessary for it to gelate to such a consistency that it can no longer flow out from the test tube when inverted (Fig. 4-5). In the left test tube is a bentonite suspension in the form of a gel; whereas in the test tube on the right-hand side in that figure the suspension is in a sol state.

METHODS OF ROCK EXPLORATION 103

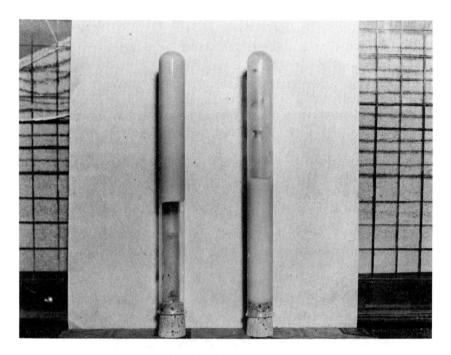

Fig. 4-5
Simple test for thixotropy

The gelation time depends upon the concentration of the sol. WINKLER (115) gives the following gelation times in minutes for sodium montmorillonite as a function of liquid-to-solid-ratio (by volume):

Liquid/solid ratio	Gelation time, minutes
33	1
34	60
36	45

The basic functions of a drilling fluid in rock coring operations are:
1. to transport bit-cuttings of rock up to the ground surface
2. to buoy the drill shaft (pipe or rod)
3. to suspend the rock cutting in the borehole when circulation of fluid is stopped for a while. During such a stopping time the thixotropic drilling fluid gelates, gains strength in shear and thus is able to

support the bit cuttings by the mass of the gel, and
4. to cool the drilling bit.

These properties of a drilling fluid must be retained under various temperature conditions.

The clay used for preparing drilling fluids is usually a commercial bentonite, chiefly a Na-bentonite (montmorillonite with Na⁺ as the exchangeable cation, because these readily become thixotropic).

Bentonite, a geologic rock formation, is one of the well known ultrafine clays, mainly composed of the montmorillonite group of clay minerals. Montmorillonite — Al_2O_3 (Ca, Mg) 0.5 $SiO_2 \cdot nH_2$) — results from the alteration of volcanic ash.

A scanning electron micrograph of montmorillonite is shown in Fig. 4-6.

Montmorillonite is capable of entraining water, so that the clay swells and increases in volume considerably just like glue or gelatin. Bentonite

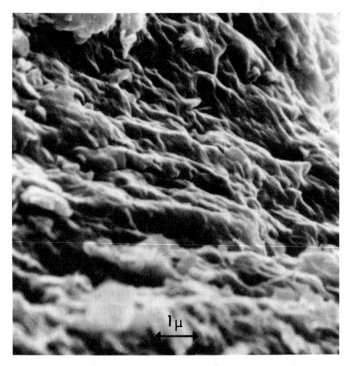

Fig. 4-6
Scanning electron micrograph of a montmorillonite, x 10,000.
(Photo courtesy of Dr. Victor Greenhut, College of Engineering, Rutgers University)

injections are used in sealing cracks in concrete dams, rocks, and in fortifying walls of excavations for building cutoff walls (cores) for dams in weak soils.

Attapulgite — a clay mineral — is also a suspending agent, and gives high viscosity because of interaction of their needle-like particles (about 1 micron in length, and approx. one-hundredth microns across). Viscosity is the resistance to flow. It is that property of semi-fluids, fluids and gases by virtue of which they resist an instantaneous change of shape or arrangement of molecules. In other words, viscosity is any resistance to deformation that involves dissipation of energy by internal friction. Some technical uses of attapulgite are as an oil-well drilling fluid (85).

Sepiolite (Meerschaum or sea foam) — $Mg_8(H_2O)_4(OH_4)Si_{12}O_{30}$ — is a magnesium-rich clay mineral. It is very light, and, when dry, will float on water. Hence the name "sea foam." Sepiolite is a spongy, or fibreous hydrous magnesium silicate, white or cream in color. The clay particles are lath-shaped-thin, long mineral crystals (Fig. 4-7). The lath-shaped particles split off when shear stress is applied to a sepiolite slurry. Although soft and

Fig. 4-7
Scanning electron micrograph of a sepiolite, x 2000.
(Photo courtesy of Dr. VICTOR GREENHUT, College of Engineering, Rutgers University)

easily carved, sepiolite is tough and durable. The laths bundle together to provide the thixotropic properties, rather than the cohesion of thin plates found in layered swelling clays.

Sepiolites have been found useful as dispersoids in oil-well drilling fluids for operations that penetrate high-salt formation waters.

Many rock drillers are using sepiolite clay mud as a drilling fluid in geothermal hot-well drilling operations (23, 38).

In hot well drilling operations, sepiolite clay fluid is used for temperature stabilization, control of loss of fluid, viscosity control, and reduction of cost.

Sepiolite holds water in zeolitic channels along the laths and is relatively unaffected by electrolytes. Regardless of the water used, sepiolite provides viscosity while preventing bottom-hole gelation because of high temperatures. In addition, it remains stable in down-hole condition over 232 °C (= 450 °F), and retains its gel characteristics after heating. Temperature fluctuation in sepiolite drilling fluids is not a problem as is with sodium bentonite. Besides, sepiolite drilling fluid does not thicken excessively, and prevents weighting material to settle.

Zeolites. There are nearly 2000 species of zeolites in sedimentary rocks of volcanic origin. So far, about 15 zeolite minerals have been identified (86). Zeolites are formed by alteration of feldspars by stream and circulating waters, and are often found in cavities of igneous rocks.

The clay minerals of the zeolite group are the most abundant and significant hydrous silicate compounds of aluminum, calcium and sodium. Zeolites have a marked system of interstices, containing water molecules. Zeolites are usually colorless or white. Their hardness vary from $H = 3.5$ to $H = 5.5$, and their specific gravity is from $G = 2.0$ to $G = 2.4$.

Many zeolites can be dehydrated, without collapse of their lattice structure, and are then capable of taking up other compounds, especially gases and vapors. Some zeolites are soluble in water.

Their adsorption, ion-exchange, catalytic and dehydration properties are the basis for dozens of commercial applications in many areas of industrial and agricultural technology such as paper industry, in pozzolanic cements, as lightweight aggregate, as ion-exchangers in pollution-abatement processes, in the separation of oxygen and nitrogen from air and as acid-resistant absorbents in gas drying and purification (85).

Some zeolites dissolve in hydrochloric acid. With zeolites, ion exchange means that zeolites will take calcium or other bases into their composition in the place of sodium, and are often used as water softeners. Because of this property, materials containing significant amounts of zeolites are unsuitable for use as concrete aggregates.

Some zeolites are fibrous and woolly while others are platy and micaceous.

Some of the members of the zeolite family are

analcime $\quad$ Na $(AlSi_2O_6) \cdot$ H_2O
chabazite $\quad$ $Ca_2(Al_4Si_8O_{24}) \cdot 13 H_2O$
erionite $\quad$ $H_2CaK_2Na_2(Al_{20}Si_6O_{17}) \cdot 5 H_2O$

and others.

Figures 4-8, 4-9, and 4-10 show scanning electron micrographs of zeolite species analcime, chabazite, and erionite, respectively. The cubic, pseudocubic and rhombic analcime and chabazite crystals (Figs. 4-8 and 4-9) are from Ischia, Italy and Nevada, U.S.A., respectively.

The crystals range from 5 to 25 microns in size. The sedimentary chabazites, $H = 4$ to $H = 5$; $G = 2.05$ to $G = 2.15$ (Fig. 4-9), have been

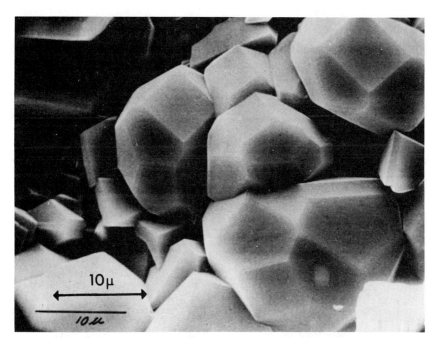

Fig. 4-8
Scanning electron micrograph of an assemblage of "analcime" zeolite crystals from Ischia, Italy. Crystals range from 5 to 25 microns in size. Magnification: 2300x.
(Photo by Dr. W. C. Ormsby, Federal Highway Administration, Washington, D.C. and F. A. Mumpton, State University College, Brockport, New York)

Fig. 4-9
Scanning electron micrograph of zeolite, "chabazite", from Reese River Valley, Nevada.
Magnification 1800 x.
(Photo by Dr. W. C. ORMSBY, Federal Highway Administration, Washington, D.C. and
F. A. MUMPTON, State University College, Brockport, New York)

Fig. 4-10
Scanning electron micrograph of zeolite, "erionite", from Hector, California.
Magnification: 1500 x.
(Photo by Dr. W. C. ORMSBY, Federal Highway Administration, Washington, D.C. and
F. A. MUMPTON, State University College, Brockport, New York)

used in the purification of methane gas from landfills and as an ion-exchange material for the removal of radioactive cesium from low-level nuclear wastes (85).

Figure 4-10 is a scanning electron micrograph of erionite-zeolite from Hector, California. The bundles consist of hundreds of individual needles, each about a half-micron thick.

Chabazite does not effervesce in hydrochloric acid.

Large-Diameter Borings

Site exploration should be supplemented by drilling of several large-diameter boreholes or shafts and/or explorations drifts. They afford visual in situ inspection, photography and sampling of the exposed rock material at the walls, a detailed survey of the exposed joint patterns and a study of possible gouge inclusions in cracks and joints.

Large-diameter borings (76.2 cm to 137.16 cm, or 30 in. to 54 in.) are drilled by the *calyx*- or the *shot core* drilling method.

In rock engineering usage, the term "calyx" has several connotations.

1. A calyx is a special feature of one make of a shot drill: it is a special sludge barrel device for collecting coarse rock cuttings which are falling back from the upward flowing drilling fluid. The term "calyx" derives its name because the sludge barrel resembles the calyx — the outermost set — of a flower. Calyx forms an essential part of a shot core barrel. The advantages of the sludge barrel (calyx) are:
   a) it keeps the drillhole clean of cuttings, and
   b) it reduces cleaning time of the borehole.
2. "Calyx" also indicates a very large borehole ($>$ 76.2 cm or $>$ 30 in.) drilled by the shot drilling method.
3. The term also means the actual method of shot core drilling, called the calyx boring.
4. The term "calyx" is also to indicate large-diameter rock core samples. Cores of 1.80 m in diameter have been drilled to depth of $\sim$ 300 m.

The performance of the calyx- or shot core drilling may be described approximately as follows.

In this method of drilling, chilled-steel shot of suitable size is fed into the wash water in the casing. The shot settles at the bottom of the hole and lodges or embeds into the lower part of the soft steel rock-coring barrel, some outside, and some inside. Others find their way under the edge of the hollow mild-steel drilling bit. The bit rotates and exerts a vertical pressure on the shot. The shot is crushed into sharp and highly abrasive particles, thereby causing them to mill or grind away the rock. Thus, rock cutting is a combination of abrasion and wear.

The diameter of the shot is about 2.54 mm (0.10 in.), and the shot is known by the trade name "Calyxite".

Shot-cut rock cores of necessity have a rough cylindrical mantle surface.

The principle of shot core drilling is shown in Fig. 4-11. Figures 4-12a, b, and c show some large-diameter rock cores.

Fig. 4-12c shows a calyx rock drilling operation of the Robert Moses Power Dam on Barnhart Island, N.Y. (Power Authority of the State of New York, St. Lawrence Power Project), and portrays a calyx rock core which is 90 cm (36 inches) in diameter. Fig. 4-12d shows a drilled rock core 12.5 cm (5 in.) in diameter taken from the same area.

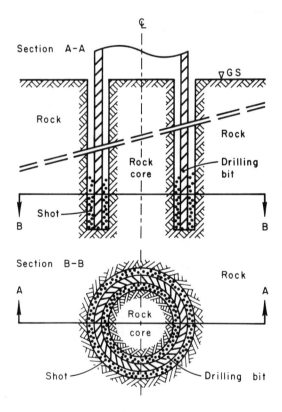

Fig. 4-11
Principle of shot-core drilling

METHODS OF ROCK EXPLORATION 111

Fig. 4-12a
Large diameter rock core obtained by the shot-core drilling method
(Photo courtesy of W. EBERT, Ingersoll-Rand Company, Rock Drill Division, Phillipsburg,
New Jersey)

Fig. 4-12b
Calyx rock cores
(Photo courtesy of W. EBERT, Ingersoll-Rand Company, Rock Drill Division, Phillipsburg,
New Jersey)

Fig. 4-12c
Calyx rock core, 90 cm (36 inches) in diameter, extracted from the ground during the exploration work for the St. Lawrence Power Project.
(Photo courtesy of the Power Authority of the State of New York)

Fig. 4-12d
Drilled rock core, 12.5 cm (5 inches) in diameter, taken from the Robert Moses Power Dam area. St. Lawrence Power Project.
(Photo courtesy of the Power Authority of the State of New York)

Large-scale rock cores, no doubt, afford a better observation of the rock condition in situ than small-scale rock cores. But large diameter core drilling is expensive. However, shortage of finances should not be an obstacle against finding the truth. On the other hand, occasionally practical realization of physical possibilities does not always turn out to be compatible with economics. Engineering limitations, and exploration and construction costs draw a hard line between what is desirable and possible, and what is practical.

A successful instrument for borehole examination to a depth of about 30 m is the borehole periscope (72). Small diameter TV cameras have also been used for examination of borehole walls (22, 116).

Borings, rock sampling and testing provide the engineer with complete information about the subsurface conditions of rock along the lines of each boring, but the materials between these borings inevitably remain unexplored. Unfortunately, Nature is not obliged to comply with any general continuity rule which would allow interpolation of results in between. The results of subsurface exploration still leave a wide margin for interpretation. Also, the findings of the geologist require expert translation into physical and mechanical terms. Besides that, it requires the evaluation of the most unfavorable mechanical possibilities which might occur under the existing geological conditions. It is so because, for reasons of safety, one designs an engineering structure for the worst condition that can be possibly expected. The important thing is to make a correct, realistic assessment of the situation.

To decide whether or not all the known facts allow the geotechnical engineer to advise the suitability of a given site for construction is perhaps the greatest decision that professional men have to make. Foundation work in rock cannot be schematizied, nor rules set for such work. Each case is different and should be handled differently. The numerous examples in practice of successful dams, for example, and their extensive geological preparatory work should bring out this point of view here as an illustration and explanation only.

Small Rock Samples versus Large Ones

The effect of the size, viz., diameter of the rock sample is an important economical and technical factor in rock exploration for construction purposes. The performance of rock in situ depends upon the geological defects it contains, its weaknesses, orientation, attitude and nature of the rock mass. But the small intact rock samples — cores and blocks — forwarded to, or prepared in, the laboratory for testing, usually represent the stronger rock materials than the rock mass in situ from where the samples are obtained.

Fig. 4-13a
Various size rock core specimens prepared for laboratory testing

METHODS OF ROCK EXPLORATION

The reasons for studying small-size rock samples in the laboratory may be thought of as being two-fold, namely:

1. tests on small-scale specimens of intact rock may often render a practical identification of the rock quality. This may help the technical evaluation of the rock mass in situ;
2. laboratory testing on suitable intact rock specimens affords determination of a particular resistance to deformation of the discontinuities or weaknesses of the rock.

In rock exploration and testing work, it is advisable to use as large a diameter rock core as is consistent with economy and technological importance of the structure. The larger and more important the structure, the more extensive and detailed the exploration should be.

Small-Size Rock Core Specimens

Figures 4-13a and 4-13b show various small-size rock core specimens prepared for laboratory testing. The rock test specimens can be prepared by means of a set of equipment such as a heavy-duty, commercial dia-

Fig. 4-13b
Granite core specimen equipped with a strain-gauge rosette

mond-impregnated rock saw with a circular metal blade 50.8 cm (20″) in diameter (Fig. 4-14); an automatic medium-duty rock saw, and a light rock saw; a rock saw for trimming rock cores (Fig. 4-15); a laboratory rock-core drill (Fig. 4-16); a universal hydraulic compression testing machine; strain scanner, and data keypunch console (Fig. 4-17); a rock classification hammer, strain gauges, and other necessary accessories.

Petrographic Analyses

Petrographic analyses aid in the evaluation of performance of rock under the action of load, water and temperature after the structure is built, and give information on rock crystal interlocking, microfracturing in rock, and on rock weathering.

Fig. 4-14
Heavy-duty rock saw

METHODS OF ROCK EXPLORATION 117

Fig. 4-15
Rock saw for trimming rock specimens

Fig. 4-16
Laboratory rock-coring drill

Fig. 4-17
Laboratory rock testing devices.
Left to right:
universal compression testing machine
strain scanner, and
data keypunch console

4-4. Hydrological Exploration

Hydrological exploration should furnish information on precipitation; surface runoff; floods and tide conditions; groundwater-carrying layers (their numbers, depth); position of the groundwater table and its fluctuation; groundwater flow (direction and velocity); location of springs, possibility of influx of water in rock excavations and underground openings. Also, groundwater must be tested for its aggressiveness with regard to soil, rock, and foundation construction materials.

Saturated rock, on the average, has less strength than a dry one.

One must always count upon an influx of water into an underground opening. Influx of water in tunnels has a detrimental effect on the safety of tunnel traffic, especially if cold air at a temperature below 0 °C has access into the tunnel. Water on the roadway pavement may freeze to form ice.

Also, water escaping from shale rock jointing into a freezing atmosphere may cover road pavements and freeze, thus resulting in icing on pavements causing hazardous driving conditions to motorists. Figure 4-18 shows water frozen to ice from shale jointing along a road. Therefore, in investigating seepage through shale, the jointing pattern in such and other jointed rocks should be established or ascertained in order to design and install proper drainage facilities along the road.

Because water seeks the path of least resistance to flow, there should be a maxim observed to provide, in every dewatering system of rock, for the most convenient way possible for water to drain and flow away freely toward an outlet. Or else, because one must always count upon the influx of water, the use of efficient pumping systems must be considered. Also, water flowing out from rock slopes influence rock slope stability. When discharging on a highway pavement during freezing periods, the ice formed may create a traffic hazard.

Movement of underground water brings about the dissolution of certain minerals in rocks (karst regions of limestones).

Migrating groundwater may wash out soft clays and other decomposed materials from joints and faults. Under varying humidity and alternate drying and wetting conditions, some rocks separate on their bedding planes or, in time, disintegrate completely.

Fig. 4-18
Water from shale frozen to ice along a road

In foundation and rock engineering, the chemical content of water is evaluated mainly from the point of view of its aggressiveness relative to structural foundation materials, rocks, and tunnel-lining materials.

Karst Regions

A karst region is one underlain by limestone, or dolomite honeycombed by solution subsidence, cavities, holes, sinks, and tunnels dissolved out or leached away by groundwater that much of the drainage is underground. A karst topography is characterized by streamless valleys, or by streams that often and abruptly disappear underground plunging into underground tunnels and caverns to reappear as great springs elsewhere. A topography with such features is called karst topography after the Karst region in the Dinaric Alps of Yugoslavia on the eastern side of the Adriatic Sea. The term karst is now generally applied to karst regions all over the world.

In the United States of America, karst regions are located in Alabama, Florida, Indiana, Kentucky, Tennessee, Texas, and in Puerto Rico.

Construction in karst regions without thorough underground exploration at the site may result in a disaster by the collapse of the limestone rock formation and of the structure.

With caverns and sinkholes riddled limestone may be stabilized to some extent with concrete rubble and by grouting.

Figure 4-19 shows a sinkhole in limestone in New Jersey discovered just before filling a prepared reservoir site with water.

4-5. Geophysical Exploration

Geophysical exploration is practiced to detect changes in the rock of some physical property—for example, specific gravity, magnetism, or the transmission or reflection of seismic waves.

Magnetometric surveys are most successful in regions where igneous rocks (containing magnetite) project into overlying sedimentary rocks by intrusion, or faulting, or as erosional hills beneath surfaces of unconformity.

Rock exploration in situ can be made quickly and relatively inexpensively by means of seismic methods. In seismic exploration, an artificial earthquake is brought about by exploding a buried charge. From the travel time of seismic waves, which are picked up at the ground surface by means of seismographs, the thickness and condition of various rock strata can be determined. Sometimes seismic methods are used to determine rock moduli.

However, from the rheological point of view, a seismic test is a short-time test during which a rock performs rather elastically if no time is allowed to pass. In sustained long-term performance, rock exhibits a creep phenomenon.

4-6. Thermal Exploration

Thermal and hydrothermal exploration should furnish information about underground temperatures and thermal gradients in rock. This information is needed for design of ventilation and/or air conditioning systems for underground structures during their construction and service.

Fig. 4-19
A sinkhole in limestone

Also, this phase of exploration should pay attention to anticipated influx of gases into underground opening and subsurface structures. Gas exploration results may have a bearing on the capacity and efficiency of the ventilation and refrigeration systems and other kinds of machinery involving heat transfer.

Temperature variations may bring about spalling of a rock. Large variations in temperature induce thermal stresses in rock, especially in granite and other coarse-grained rocks, resulting in "peeling" of the rock surface. Such rocks are usually rejected in summer-winter climates for their use as a riprap material in protecting slopes against erosive wave action.

4-7. Evaluation of Rock as to its Diminution and Workability

Evaluation of rock as to its diminuity and workability is of great importance, and necessary for procurement and effective use of equipment for work in rock. This evaluation also has a bearing on overbreak, in making decisions as to whether or not rock bolting is needed, or if ring supports are to be used for supporting tunnel walls, and the like.

Rock crushing and grinding constitute an integral part of temporary industry.

Rocks differ greatly in their value for building purposes.

They vary markedly in their weathering qualities-resistance to atmospheric agents such as rain, frost, and wind.

Rocks vary in hardness; this factor affects the rate of drilling them and thus the cost.

The discussion about procurement of rock for construction is beyond the scope of this book. For in-situ and laboratory testing of strength properties of rocks refer to Sections 4-26 through 4-38 of this book.

4-8. Properties

Rock is a natural substance. As such, it has structural features which are not encountered in most other engineering materials.

The performance of soil and rock under the action of load, water, temperature, and tectonics of the earth's crust depends upon *physical and mechanical* (strength) *properties* of those materials.

One should distinguish between macro- and microscale properties of rock. Essentially, macroscale rock mass properties are properties of the compound, whereas microscale (small-scale) rock properties are those of the rock substance.

For safe and economical design of a structure on and in rock, adequate knowledge about the technological properties of rock is indispensible. Some of the main technological properties of rocks necessary to consider in the design of concrete dams, rock slopes, and underground structures are:

1. Unit weight,
2. Movements and deformations under load (absolute or relative, or strains),
3. Static and dynamic strength (compressive, tensile, bending, shear, punch, torsion) of dry and saturated rock
4. Angle of internal friction φ
5. Cohesion c
6. YOUNG's modulus of elasticity E
7. POISSON's number $m = 1/\mu$ (μ = POISSON's ratio)
8. Natural or primary rock stress
9. Impact resistance
10. Resistance to mechanical abrasion (wear)
11. Creep
12. Dilatancy
13. Density against water absorption
14. Permeability to water
15. Resistance to weathering
16. Resistance to frost action in soil (freezing and thawing)
17. Resistance to chemical influences
18. Thermal properties (conductivity, heat capacity)
19. Electrical properties, and
20. Other possible properties.

Also, it can be brought out that some of the properties of rock materials may be elastic, plastic, viscous, or a combination thereof.

4-9. Physical Properties

The physical properties of rocks affecting design and construction in rock are:

1. Mineralogical composition, structure, and texture;
2. Specific gravity G
3. Unit weight γ
4. Porosity n
5. Void ratio e

6. Natural moisture content w
7. Saturation moisture content
8. Degree of saturation S
9. Permeability to water k
10. Chemical effects
11. Thermal properties
12. Electrical properties.

Mineralogical composition is the intrinsic property controlling the strength of the rock.

Although there exist more than about 2,000 kinds of known minerals, only about nine of them partake decisively in forming the composition of rocks. They are: quartz, feldspars, micas, hornblende, augite, olivine, calcite, kaolinite, and dolomite. Rocks containing quartz as the binder are the strongest, followed by calcite and ferrous minerals as the cementing agent "glue". Rocks with clayey binder are the weakest.

Rock Structure

According to KRYNINE and JUDD (73), the term "rock structure" applies to well-pronounced macroscopic features of the rock (columnar structure, for example). This term may signify special features of the rock, such as the position, arrangement, and attitude of a system of joints (open or closed); fractures; successive strata of rock formations; holes and/or cavities (small or large), gouge fillings, and the geologic features which determine that attitude such as stratification, faults, folds (anticline and syncline), and igneous intrusions, for example.

Figure 4-20 shows a columnar structure of the intrusive diabase rock of the Palisades along Hudson River, New Jersey.

Figure 4-21 illustrates a dipped layered structure of Triassic shale rock in New Jersey.

Figure 4-22 depicts a folded rock structure of limestone.

Rock Texture

The term "rock texture" refers to the arrangement of its grains or particles on a freshly exposed rock surface, easily seen by the naked eye. Texture is the appearance, megascopic or microscopic, seen on a smooth surface of a mineral aggregate, showing the geometrical aspects of the rock including shape, size, and arrangement.

METHODS OF ROCK EXPLORATION   125

One distinguishes between coarse-textured (coarse-grained) and fine-textured rock. A coarse-textured rock is one in which the large crystals or other kinds of grains can be seen easily by the naked eye.

A fine-textured rock is one whose grains cannot be seen without magnification.

An igneous rock whose coarse crystals are embedded into a fine-textured matrix is called porphyry.

Fig. 4-20
Columnar structure of diabase rock of the Palisades, New Jersey
(Photo by ANDRIS A. JUMIKIS)

Fig. 4-21
Dipped layered structure of Triassic shale rock in New Jersey

Fig. 4-22
Folded rock structure of limestone
(Photo by Andris A. Jumikis)

Obviously, rock structure and texture affect the strength properties of the rock.

From a morphological point of view, rock texture may be grouped into three main groups, namely:

1. Homogeneous,
2. Nonhomogeneous or heterogeneous, and
3. Layered.

Texture of igneous rocks develops mainly as a function of composition and rate of cooling of the magma and/or lava, whichever the case may be. In most igneous rocks, the texture has an overall aspect of a network of interlocking crystals.

The main significance of texture in rock mechanics is its effect on the strength and permeability of the rock.

Some rock textures are shown in Figures 4-23a, b, c, and d. Figure 4-23a shows a homogeneous texture of a sandstone. Figure 4-23b illustrates a heterogeneous layered texture of a gneiss. Figure 4-23c is a heterogeneous texture of a porphyry rock specimen, and Figure 4-23d shows a heterogeneous texture of a Triassic shale specimen.

Specific gravity G of a material is defined as the ratio of the weight of a material to the weight of an equal volume of water. In other words, G is a number expressing how many times a material is heavier (lighter) than an equal amount of volume of water.

True or absolute specific gravity G of a soil and/or rock is the specific gravity of the actual soil and/or rock grains or solids:

$$G = \frac{W_d}{W_{\text{sat in air}} - W_w - W_{\text{sat in water}}} \qquad (4\text{-}1)$$

$$G = \frac{\text{weight of rock (soil) particles in g}}{(\text{volume of soil particles})(1.00)} = \frac{W_s}{V_s \cdot \gamma_w} = \frac{\gamma_s}{\gamma_w} \qquad (4\text{-}2)$$

A rock mass may comprise several different minerals of varying absolute specific gravities, but its true or absolute specific gravity is the average determined from a representative rock specimen.

a) Homogeneous texture of a sandstone specimen

b) Heterogeneous layered texture of a gneiss specimen

c) Heterogeneous texture of a porphyry specimen

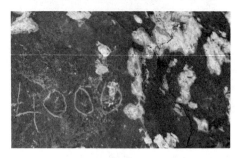
d) Heterogeneous texture of a Triassic shale specimen

Fig. 4-23
Some rock textures

Specific gravity G is used in computing other soil and rock properties; for example, the porosity and void ratio of a rock and soil. Table 4-3 shows specific gravities of some minerals.

TABLE 4-3
Specific Gravity of Some Minerals

No.	Mineral	Specific gravity G
1	2	3
1	Biotite (black mica)	2.70–3.20
2	Calcite	2.71–3.72
3	Dolomite	2.80–3.00
4	Feldspar	2.50–2.80
5	Gypsum	2.20–2.40
6	Hornblende	2.90–3.50
7	Kaolinite	2.50–2.65
8	Montmorillonite	2.00–2.40
9	Muskovite (white mica)	2.76–3.00
10	Quartz	2.65
11	Talc	2.60–2.70

Bulk or apparent specific gravity G_b is the ratio of the dry weight W_d of rock to the weight W_w of volume of water equal to the *total* volume of rock including voids.

For water, its specific gravity is assigned the numerical value of $G = 1.00$.

$$G_b = \frac{W_d}{W_w} \qquad (4\text{-}3)$$

where $W_w = W_{\text{sat in air}} - W_{\text{sat in water}}$

Unit Weight

The unit weight γ of soil and rock above the groundwater table is expressed as the ratio of the total weight of the rock (soil) in air W to the total volume V of the rock, all voids included:

$$\gamma = \frac{W}{V} \qquad (4\text{-}4)$$

Rocks containing heavy minerals have higher unit weight than rocks with lighter minerals. Usually, igneous and metamorphic rocks have greater unit weight than sedimentary rocks. The more porous the rock is, the less is its unit weight.

The dry unit weight of rock (soil) γ_d is computed as:

$$\gamma_d = \frac{W_d}{V} = \frac{G}{1+e} \cdot \gamma_w = \frac{\gamma}{1+w} \qquad (4\text{-}5)$$

where W_d = dry weight of rock specimen;
V = total volume of the specimen;
G = absolute or true specific gravity of soil, or average specific gravity of rock;
e = void ratio;
γ = moist unit weight;

$$\frac{1}{1+e} = \frac{V_s}{V} \qquad (4\text{-}6)$$

V_s = volume of solids;
γ_w = unit weight of water = 1 g$_f$/cm³ = 1 t$_f$/m³;
w = moisture content, by dry weight, in decimal fractions.

The unit weight γ of a moist soil calculates as:

$$\gamma = (1-n) \cdot G\gamma_w + nS\gamma_w \qquad (4\text{-}7)$$

where

$$S = \frac{w}{w_{sat}} = \frac{n_w}{n} = \frac{V_w}{V_v} = \frac{wG}{e} \qquad (4\text{-}8)$$

S = degree of saturation, in decimal fractions
w = moisture content in rock;
w_{sat} = saturation moisture content;
n_w = relative pore volume at moisture content w;
n = porosity of rock in decimal fractions;
V_w = volume of water in rock;
V_v = volume of all voids (pores) in rock;
G = average specific gravity of rock material, and
e = void ratio.

The unit weight of solids by absolute volume is

$$\gamma_s = G \cdot \gamma_w \qquad (4\text{-}9)$$

Saturated unit weight γ_{sat} of soil or rock is:

$$\gamma_{sat} = (1-n)G\gamma_w + n\gamma_w \qquad (4\text{-}10)$$

where

n = porosity of rock or soil,
G = specific gravity,
γ_w = unit weight of water
   = 1 g$_f$/cm³ = 1 t$_f$/m³ = 9.81 kN/m³ = 62.4 lb/ft³

When inundated, the rock and soil particles are subject to buoyancy. The buoyant or submerged unit weight γ_{sub} is expressed as:

$$\gamma_{sub} = (1 - n)(G - 1)\gamma_w = \gamma_{sat} - \gamma_w \qquad (4\text{-}11)$$

Dry unit weights of some American rocks are compiled after GRIFFITH (43) in Table 4-4:

1 lb/ft³ = 0.157092 kNm³ = 0.0160184 t$_f$/m³ = 16.01846 kg$_f$/m³

TABLE 4-4

Dry Unit Weight and Porosity Values of Some American Rocks
(After J. H. GRIFFITH, Ref. 43)

Type of rock	Bulk or apparent specific gravity	Dry unit weight t$_f$/m³	Porosity n %
1	2	3	4
Igneous:			
Basalt	2.21 – 2.77	2.21 – 2.77	0.22 – 22.06
Diabase	2.82 – 2.95	2.82 – 2.95	0.17 – 1.00
Gabbro	2.72 – 3.00	2.72 – 3.00	0.00 – 3.57
Granite	2.53 – 2.62	2.53 – 2.62	1.02 – 2.87
Sedimentary:			
Dolomite	2.67 – 2.72	2.67 – 2.72	0.27 – 4.10
Limestone	2.67 – 2.72	2.67 – 2.72	0.27 – 4.10
Sandstone	1.91 – 2.58	1.91 – 2.58	1.62 – 26.40
Shale	2.00 – 2.40	2.00 – 2.40	20.00 – 50.00
Metamorphic:			
Gneiss	2.61 – 3.12	2.61 – 3.12	0.32 – 1.16
Marble	2.51 – 2.86	2.51 – 2.86	0.65 – 0.81
Quartzite	2.61 – 2.67	2.61 – 2.67	0.40 – 0.65
Schist	2.60 – 2.85	2.60 – 2.85	10.00 – 30.00
Slate	2.71 – 2.78	2.71 – 2.78	1.84 – 3.61

Porosity of Rock

Voids, i.e. pores and fractures in rock, are very important forms of non-uniformity of structure and texture of rocks.

Pores in rock are little, interconnected voids having connection to the air also.

The quality of rock porosity is characterized by the term "porosity". The degree of porosity depends upon the kind and structure of the rock.

Pores have a considerable effect on heat, gas, and water conduction and on the behavior of rock relative to its moisture content.

The presence of voids or pore spaces in rocks affects negatively its mechanical—viz., strength—properties. A small amount of porosity in the form of cracks has a large effect upon the deformation of rock. All polycrystalline materials, among them also rocks, are relatively porous depending upon their type, mineral composition, and mode of formation. Porosity is the result of internal stresses developed by only modest changes in stress and temperature; or as a result of imperfect sintering, exsolution of gas; or, as in compact crystalline rock, the opening of small fissures by internal stresses (20). Upon quick cooling of the magma or lava, whichever the case is, fine crystals form in the igneous rock, resulting in a relatively non-porous rock; whereas upon slow cooling, large crystals are formed. When gases are released upon cooling, very porous rock results.

The porosity of a sedimentary rock depends upon the cementing material ("glue") of the aggregate present, and the granulometric contents of the rock. Usually, the largest porosities are observed in sedimentary rocks.

The amount of voids in a rock and soil mass can be expressed in terms of porosity n, and as void ratio e.

Porosity n is the amount of voids based on total volume V of rock:

$$n = V_v/V = \frac{e}{1+e} = \frac{V - (W_s/G\gamma_w)}{V} \tag{4-12}$$

where V_v = volume of voids,

$$e = \frac{n}{1-n} = \text{void ratio},$$

W_s = dry weight of rock solids,
G = specific gravity of soil or rock,
γ_w = unit weight of water, and
$$V = V_s + V_v = V_s + V_a + V_w \tag{4-13}$$
= total volume

V_s = volume of solids in the rock specimen,
V_a = volume of air or gas in the rock specimen,
V_w = volume of water in the rock specimen.

Porosity values of some American igneous, sedimentary, and metamorphic rocks are compiled in Table 4-4.

In general, a cubic meter of *solid rock* excavation will make more than a cubic meter of fill or embankment, because there will remain void spaces between the rock fragments as placed in the fill. A *"swell"* of approximately 25 % can be expected, depending upon the size and gradation of the rock fragments and the applied compactive effort.

Porosity values are most important in the study of the physical and mechanical properties of rocks. Porosity also affects the electrical properties of rocks, and porosity values aid in interpreting electric logs.

A small amount of voids, viz., small porosity in the form of cracks has a great effect on the deformation of rock. Closing and opening of cracks under the influence of stress makes the rock stress-strain relationship nonlinear. Sliding between crack surfaces brings about hysteresis in the stress-strain graphs.

Void ratio e is the ratio of the volume of voids V_v to the volume of solids V_s of the rock, expressed in decimal fractions. It is a dimensionless number which simply shows how many times there are more voids than solids in the rock:

$$e = \frac{V_v}{V_s} = \frac{n}{1-n} \qquad (4\text{-}14)$$

The volume of solids V_s calculates as:

$$V_s = \frac{W_s}{G \cdot \gamma_w} \qquad (4\text{-}15)$$

Water in Rock

In rock engineering, one must always count upon the influx of water into underground openings. Water always seeks the path of least resistance to flow.

Fractures and other rock defects are the easiest passages of ingress for groundwater in the rock. Hence, the extent and the geometry of the fracture and jointing system in rock of various lithology, topography, orientation, relation to surface streams and drainage pattern, and depth below ground surface should be helpful in interpreting the hydraulic regimen of, and permeability conditions in the rock.

Rock pores such as cavities, fissures, and fractures may contain a certain amount of water that may be free or bound (water film, for example).

As to the water film, it is present in the earth's crust in great quantity. This interstitial water cannot be drained mechanically or pumped away. The water film moves thermo-osmotically under the action of a thermal gradient (or sometimes under an electrical gradient) in rock through the smallest, thinnest voids such as fissures and seams. In due time, this category of water brings about alteration, weathering, and, thus, fragmentation of the rock. Hence, these alterations affect the stress and hydraulic field within the rock mass. Therefore, the study of moisture migration in fissured and fine porous rocks requires the consideration of the thermo-osmotic phenomenon, which phenomenon, in its turn, requires knowledge of the electrical parameters of the rock — dielectric constant, electrical conductance and resistivity as a function of structure, texture, porosity, and moisture content of the rock materials (62, 64, 67).

Moisture content w of rock is the weight of water W_w present in a rock expressed in percentage by oven-dry weight of the rock:

$$w = \frac{W_w}{W_s} \cdot 100 = \frac{W - W_s}{W_s} \cdot 100 \qquad (4\text{-}16)$$

The dry weight of the solids of rock calculates as:

$$W_s = \frac{W}{1 + (w/100)} \qquad (4\text{-}17)$$

where W = total weight of rock specimen, including moisture.

Most rock contains moisture of less than 1 % to more than 35 % in porous rocks (such as sandstone, for example). In many mines and other underground works, the rock is nearly saturated. Results are leaky tunnels and water seeping and dripping into the opening and out from fractures, faults, and joints (Fig. 5-5).

Alternate wetting and drying causes some rocks to expand and contract, thus affecting their properties. Also, riprap stones may become destroyed by alternate wetting and drying.

Moisture adversely affects electrical and electronic equipment, installations, and optical devices.

TABLE 4-5
Some Soil Physical Constants

Properties to be Determined by Test	Quantities to be Calculated for a Soil in its Undisturbed State	
	Description	Equation
1	2	3
1. γ = unit weight of soil	1. Unit weight of soil skeleton or dry unit weight	(a) $\gamma_d = \dfrac{\gamma}{1+w}$ (b) $\gamma_d = \dfrac{G}{1+e}\gamma_w$
2. w = moisture content by dry weight	2. Porosity	$n = 1 - \dfrac{\gamma_d}{G\gamma_w}$
	3. Void ratio	$e = \dfrac{n}{1-n} = \dfrac{G\gamma_w - \gamma_d}{\gamma_d}$
3. G = specific gravity of soil particles	4. Relative volume of voids	$n = \dfrac{e}{1+e}$
	5. Relative volume of solids	$n_s = \dfrac{1}{1+e}$
4. γ_w = unit weight of water	6. Relative volume of water in soil	$n_m = nS$
	7. Soil moisture content by dry weight	$W_w = w\gamma_d$
	8. Soil moisture content by volume	$V_w = \dfrac{W_w}{\gamma_w}$
	9. Moisture content upon full saturation	$w_{sat} = \left(\dfrac{1}{\gamma_d} - \dfrac{1}{G\gamma_w}\right)\gamma_w$
	10. Degree of saturation	$S = \dfrac{w}{w_{sat}} = \dfrac{w\gamma}{n(1+w)\gamma_w}$
	11. Void ratio upon full saturation	$e = wG$
	12. Volume of air in a unit volume of soil	$n_a = (1-S)n$
	13. Saturated unit weight of soil	$\gamma_{sat} = \dfrac{G+e}{1+e}\gamma_w$
	14. Submerged (buoyant) unit weight	$\gamma_{sub} = \dfrac{G-1}{1+e}\gamma_w$

Degree of saturation S is determined as:

$$S = \frac{V_w}{V_v} = \frac{w}{w_{sat}} = \frac{n_w}{n} = \frac{w \cdot \gamma}{n(1+w)\gamma_w} = \frac{wG}{e} = \frac{wG(1-n)}{n} \quad (4\text{-}18)$$

The functional relationships between various physical properties of soil or rock are summarized in Table 4-5. The volumetric and gravimetric relationships of soil or rock are illustrated graphically by means of a phase diagram as shown in Fig. 4-24. The phase diagram illustrates absolute and relative volumetric and gravimetric proportions of solids, water, and air in a unit of volume of soil or rock.

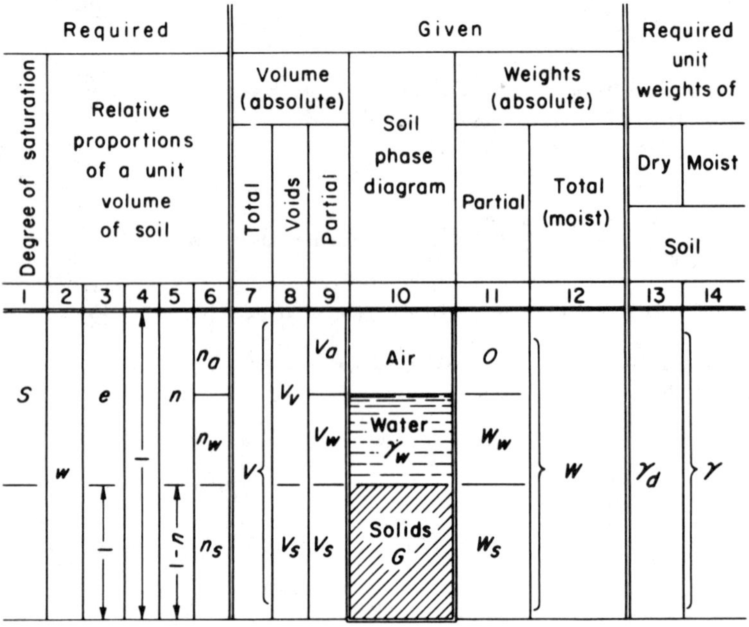

Fig. 4-24
Phase diagram illustrating absolute and relative volumetric and gravimetric proportions of solids, water, and air in a unit mass of soil

Permeability

Rock porosity and permeability to water are important physical properties in rock engineering.

Permeability is defined as the property of a porous material that permits the passage or seepage of fluids, such as water and/or gas, through its interconnecting voids.

METHODS OF ROCK EXPLORATION

The resistance to flow depends upon the type of the rock (soil), the geometry of the voids of rock (size and shape of the voids), and the surface tension of water (temperature and viscosity effects).

Although theoretically all rocks (soils) are more or less porous, in practice the term "permeable" is applied to rocks (soils) that are porous enough to permit flow of water through it. Conversely, rocks that are permeated with great difficulty are termed "impermeable".

1. Gravitational or free water drains from the voids of the rock by gravity.
2. Water film does not drain freely.
3. Water also may resist drainage by gravity if it is held in the soil by surface tension forces (capillarity).
4. Drainage lowers the position of the groundwater table.

Water seeps through soil at velocities usually well below the critical value at which turbulence appears (Fig. 4-25). The total discharge Q through a gross cross-sectional area A during time t is given by DARCY's law $v = ki$ (29), namely:

$$Q = vAt = kiAt \ [m^3] \quad (4-19)$$

$$v = kit = \text{seepage velocity} \quad (4\text{-}19a)$$

where

$k = v/i = \tan \alpha =$ coefficient of permeability of the rock (soil) (it expresses the degree of permeability of rock or soil to water); and

$i = h/L =$ slope of the pressure gradient, or the pressure head h_l lost in a unit of distance.

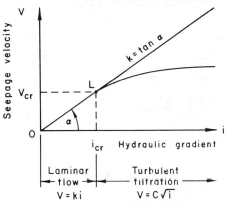

Fig. 4-25
Laminar region of DARCY's law from 0 to i_{cr}

The coefficient of permeability is a function of rock type, pore size, entrapped air in the pores, temperature in the rock, and viscosity of water.

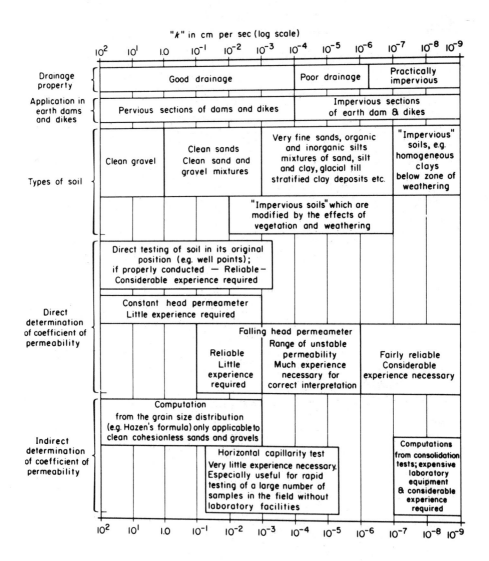

Fig. 4-26
Coefficient of permeability
(After CASAGRANDE and FADUM, Ref. 25)

Beyond point L on Fig. 4-25, where $i > i_{cr}$, the filtration is turbulent, having a seepage velocity $v > v_{cr}$. In the turbulent region the seepage velocity can be approximately expressed as $v = C\sqrt{i}$, where C = CHEZY's coefficient characterizing the seepage medium (soil, or rock), the nature of the walls of the pores and/or fissures, the viscosity of water, and the turbulence of flow. The transition from laminar to turbulent flow occurs at velocities from ~ 0.3 cm/s to ~ 0.5 cm/s. Under natural conditions a mixed flow regimen can exist.

The coefficient of permeability k of soils can be determined in the laboratory (60, 66) as well as in situ.

The equation 4-19a expressing DARCY's law of permeability pertains explicitly to flow of water through filter sand. DARCY's law is valid for *saturated* flow if the seepage velocity is small, the flow is laminar, and the medium through which seepage takes place has a continuous porous structure such as sand, for example.

Notice that the coefficient of permeability k represents the overall permeability of the entire water-bearing stratum in the vicinity of the pumping test well, and is influenced by the diameter of the pumping well and pumping operations. Therefore, k is the *"system's permeability."* Some system's permeability values k are compiled in Table 4-6a.

TABLE 4-6a

Average Coefficients of Permeability k of Soil

Soil Type	k m/s
1	2
Coarse gravel	5×10^{-3} to 1×10^{-2}
Coarse gravel with lenses and pockets of sand	4×10^{-3} to 5×10^{-3}
Medium gravel	3.5×10^{-2}
Fine gravel	3×10^{-2}
Coarse river sand	2×10^{-3} to 8.8×10^{-3}
Sand, 4—8 mm	3.5×10^{-2}
2—4 mm	2.5×10^{-2} to 3×10^{-2}
Find sand & clay	8×10^{-4} to 3×10^{-3}
Fine, clean, sharp sand + some clay	5×10^{-4} to 1×10^{-3}
Dune sand	2×10^{-4}
Very fine sand	1×10^{-4}
Silty fine sand	1×10^{-3} to 1×10^{-4}
Loess, $e = 1.3$	1×10^{-3}
Loess, $e = 0.55$	2×10^{-7}
Clays	2×10^{-7} to 1×10^{-10}

It is customary to indicate k-values at a temperature of 20°C. For a table of dynamic viscosity values of water, η in g/(cm · s) and for dynamic viscosity correction tables for water, $\eta_T/\eta_{20°C}$, refer to Appendices 7 and 8 in this book (60).

The ratio of dynamic viscosity η to density γ is called the kinematic viscosity $\nu : \nu = \eta/\gamma$ (cm²/s, or stoke).

For a systematic treatment and discussion on gravity and artesian wells, perfect and imperfect wells, and discharge and recharge wells, refer to the author's books on Soil Mechanics, and Mechanics of Soils (Refs. 66 and 60a).

Conditions allowing DARCY's law ($v = ki$) to be applied to permeability problems in rocks rarely exist in the rock mass in situ. In dam engineering, one should distinguish between the perviousness of the rock substance and that of a rock mass. For example, a calcareous rock specimen may test out as being practically impervious, whereas the rock mass in situ from where the specimen was sampled may have a high permeability because of the existing fissures and cavities in that rock.

In contradistinction to soils, in fissured rock groundwater spreads out and distributes only through interconnecting fissures, joints and voids. In some *isolated* fissure systems and voids there may be entrapped water present, whereas others may be dry.

Because of the rock defects, viz., irregularity in the amount of fissures and voids and their distribution, permeability of rocks is termed nonlinear or nonuniform. Nonuniform permeability in rock may also be caused by contraction and expansion of rock fissures.

Because of anisotropic conditions in rocks, it is impossible to establish a permeability such as for soils shown in Fig. 4-26. The permeability of in situ soils and rocks are usually determined by means of pumping- and/or water-pressure tests (60a; see Lugeon test). The actual performance of these tests are not standardized as yet. Again, the permeability tests in situ depend upon the kind of the structure to be built, the kind and nature of rock encountered at the construction site, the hydraulic regimen prevailing at the site, and the specific site involved.

For purpose of one's orientation, the approximate order of magnitude of permeability k and porosity of some rocks are compiled in Table 4-6b.

Leakage in rock endangers the safety of underground openings.

In practice, the solution of drainage is frequently complicated because of the difficulty of providing satisfactory outlets in rock for the collection of water.

METHODS OF ROCK EXPLORATION

TABLE 4-6b

Approx. Coefficient of Permeability k of Rocks at 15 °C, and Porosity n

In situ rock	Coefficient of permeability k cm/s	Porosity n %	References
1	2	3	4
Igneous rocks			
Basalt	10^{-4} to 10^{-5}	1 to 3	
Diabase	10^{-5} to 10^{-7}	0.1 to 0.5	
Gabbro	10^{-5} to 10^{-7}	0.1 to 0.5	
Granite	10^{-3} to 10^{-5}	1 to 4	
Syenite			
Sedimentary rocks			
Dolomite	4.6×10^{-9} to 1.2×10^{-8}	—	
Limestone	10^{-2} to 10^{-4}	5 to 15	
Sandstone	10^{-2} to 10^{-4}	4 to 20	
Shale	10^{-3} to 10^{-4}	5 to 20	
Metamorphic rocks			
Gneiss	10^{-3} to 10^{-4}	—	
Marble	10^{-4} to 10^{-5}	2 to 4	
Quartzite	10^{-5} to 10^{-7}	0.2 to 0.6	
Schist	10^{-4} to 3×10^{-4}	—	
Slate	10^{-4} to 10^{-7}	0.1 to 1	

Relative to groundwater flow through rocks, the flow takes place through the fissures, jointing, and other defects of rock. Here, one must deal with fissure water in basalt, diabase, granite, gneiss, quartzite, limestone, sandstone and shale, for example.

Generally, the groundwater movement through fissures takes place according to the same laws as through the voids of soil, but with some difference. For example, in a porous medium like soil, water spreads out and distributes through the interconnecting pores in all directions, whereas in a fissured rock, water spreads and moves only along the directions of the fissures and joints.

While pore size in soil varies within relatively narrow limits, the size of fissures and joints may vary within very wide limits — from hair cracks to several meters in width.

Because of this irregularity, permeability of fissured rocks is said to be nonuniform. Also, expansion and contraction of fissures cause nonuniform and turbulent flow. Fissures also change size by erosion processes, and during the flow processes calcites and silicates may be washed out at one place and deposited at another place.

The correct evaluation of distribution of fissures and fissure water is possible only if based on:

1. Careful observation of size of fissures and their frequency distribution,

2. Research on transmissibility of water through rock; and

3. Rational systematization and classification of rock fissures and underground fissure water.

It should be distinguished between

1. *Tectonic fissures*—formed upon the formation of the structure of the rock;

2. *Lithogenetic fissures*—formed upon the formation of rock; and

3. *Weathering fissures*—brought about by weathering processes of rocks by water, wind, glaciers, and temperature.

Influx of water in underground openings and foundation pits requires its removal by pumping and/or drainage facilities.

The influx of water into the Bologna-Florence tunnel was up to 1200 liters per second, and into the St. Gotthard tunnel up to about 230 liters per second.

Influx of water in foundation pits and underground openings in rock as well as in soil frequently presents problems that are difficult to cope with. Also, if the rock near the toe of a slope is already stressed to near its point of failure, the added fissure and joint-water pressure may be large enough to bring about the failure of the slope.

The water seeping into the rock also pulls air with it into the depth, pushing air bubbles and air cushions ahead through the voids, thus promoting the air exchange and circulation within the rock.

Sometimes underground works affect the regimen of the surface courses: springs may become dry, and streams and lakes may disappear.

Stratigraphy is the basis for exploring hydrological conditions in rock at a site. Hydrological studies involve stream drainage pattern locations of springs. Also, sound knowledge about the physical properties of the rocks, but especially sedimentary rocks, are essential.

Chemical Action

Fissures, joints, and fault zones can be invaded by surface waters or by hot solutions from below, or by both. Hence, chemical alterations in such zones is likely to take place.

The chemical action of rainwater which enters the rock from the surface is referred to as the *chemical weathering*. Hot water (magmatic water) entering the voids in rock from great depth is said to bring about *hydrothermal alteration*. Hydrothermal agents and carbon dioxide are the main factors altering the feldspars of igneous and metamorphic rocks into kaolin and other clays.

Hardness of water — viz., the contents of various dissolved mineral salts, especially those of calcium and magnesium in groundwater—is of some interest to rock engineering. On the average, water is considered to be hard whose degree of hardness is pH < 7, containing free carbon dioxide, calcium chloride, chlorate, magnesium salts, and sulfates.

A soft water is one containing little dissolved electrolytes. It occurs in igneous rocks and glacial formations. Generally, the softness of water increases with the increase in coldness of the climate.

In foundation and rock engineering, the chemical effects of water are evaluated mainly from the point of view of the aggressiveness of water with respect to foundation materials, rocks in underground openings, tunnel-lining materials, and tunnel rock-bolt anchorages.

Depending upon the mineralogical composition of rocks, movement of underground water in rocks may result in dissolution of certain soluble mineral components and in the formation of karsts. For example, rocks containing pyrite FeS_2 and marcasite FeS_2 (white iron pyrite), may form sulfuric acid which is a strong solvent of calcite and magnesium salts.

In limestone, openings of solution channels and cavities are brought about by the decomposition of calcium carbonate produced by water containing dissolved carbon dioxide.

Thus, chemical actions may bring about changes in the mechanical properties of the rock.

In some instances, the groundwater is relatively acid. Humus and sulfur-containing water are detrimental to concrete even at relatively low concentrations.

Thus, the chemical alterations of rocks brought about by the solution depends upon the chemical composition of the solution and the type of rock, as well as upon the depth and temperature at which the changes take place.

Therefore, because of chemical decay, presence of air, and the dissolution of certain rock-forming minerals by water and other agents in rock, the chemical nature of the groundwater must be investigated carefully.

Carbonic acid is a relatively weak acid, but it attacks concrete, and dissolves limestone and dolomite by removing calcium and magnesium from these rocks.

Sodium chloride (NaCl) is found in seawater and in many rocks. Sodium chloride solutions are not too dangerous as compared with solutions of magnesium sulfate ($MgSO_4$).

The salt termed sodium sulfate (Na_2SO_4) occurs in solution in seawater and in some mineral waters.

Calcium sulfate gypsum ($CaSO_4$) is very aggressive, and magnesium sulfate ($MgSO_4$) is even more detrimental than gypsum. Hence, rock formations containing gypsum, rock salt, and several other soluble minerals are subject to a rapid process of alteration, especially when changes in groundwater level fluctuation take place.

The factor of the possibility of swelling pressures of clay minerals to occur within the joints of a rock mass because of change in their moisture content is a very important factor in the stability of underground openings and rock slopes.

Gases

Rocks and building stones exposed to atmosphere can be detrimentally affected by atmospheric gases. Also, in-situ rocks can be seriously affected by underground gases.

The occurence of gases in underground openings depends upon the geological formations of the rock. Harmful gases may be encountered mainly in weathered, fissured, and jointed igneous rocks.

One of the most harmful gases is carbon dioxide (CO_2). It usually occurs in igneous rocks in the proximity of coal formations and other deposits of organic nature. In fault zones, the gas flows upward from the magma through faults, fissures, and other suitable passages.

The presence of carbon dioxide in underground works requires great safety precaution.

Usually carbon monoxide (CO) occurs in the proximity of coal deposits, as a result of methane and/or coal dust explosions.

Methane or marsh gas (CH_4) is encountered in the vicinity of coal and oil fields. It is also brought about by the decay of organic matter. Prevented from escape into the atmosphere, the gas can migrate long distances through the fissures of the rock.

Methane influx from a shale formation into the Great Apennine tunnel (Bologna-Florence) resulted in a great fire (107).

Sulfur dioxide (SO_2) and sulfur trioxide (SO_3) are harmful gases, also. When combined with atmospheric moisture, sulfuric trioxide forms sulfuric acid.

Sulfuric gases acting on calcite result in sulfate. As a result, rock scaling will take place. Therefore, under such conditions, limestone and marble do not resist weathering too well.

The inflammability of hydrogen (H_2) always involves the hazards of explosion and fire.

Nitrogen (N) is not inflammable. This gas is encountered in young eruptive rocks, or as the product of decay of organic matter of floral origin.

The aforegoing discussion about gases brings to the fore that it is important to explore gas exfiltrations, gas outbursts, and rock temperatures. Gas and temperature are significant factors for the safety and health of workmen and miners.

Hydrochloric Acid

Hydrochloric acid (HCl) is an aqueous solution of hydrogen chloride HCl that is a strong corrosive liquid acid. The acid is very soluble in water.

Hydrochloric acid can be used as a simple identification test for detecting the presence of calcite (calcium carbonate) in a rock. Upon application of a few drops of hydrochloric acid on limestone the latter will effervesce, i.e., it will bubble, hiss, and foam as gas escapes. Even a small amount of calcite will bring about a strong effervescence.

Acid Precipitation

Acid precipitation from the sky in the form of acid rain and snow is a skyborne peril. It is a newly recognized and increasingly harmful kind of pollution that is invisible and insidious. Actually acid rain contains weak solutions of sulfuric and nitric acids. Rising high into sky and borne hundreds of miles by wind, the chemicals of floating industrial emissions and automobile exhausts mix and react with water vapor in the air to form sulfuric and nitric acids. The acids then falling to earth in the form of rain and snow can cause damage from monuments to living organisms. The acids from the sky can also corrode stone statues, limestone buildings and metal rooftops. For example, Athen's Parthenon and Rome's Colosseum have deteriorated severely during the past two decades because of suspected acid rain as a prime source.

"Rock Cancer"

"Rock cancer" eats away the Taj Mahal. About 20 000 workers, stonemasons, sculptors and jewellers worked 22 years to build the Indian mausoleum of Taj Mahal. Now, after 330 years the "rock cancer" eats away the marble structure (supposedly by the gases emitted from a nearby refinery).

4-10. Electrical Properties

Most rocks are dielectrics, hence subject to dielectric constant measurement. This and other electrical properties depend upon the type of rock, its structure and texture, porosity, and moisture content (67).

Electrical properties of rocks are of interest for a number of practical purposes.

Electrical properties such as dielectric constant and electro-conductance are of great value in geo-electrical prospecting of groundwater resources in soil as well as in rock; in well-logging; in interpretation of electrical logs of soil and/or rock; in electrical sounding and profiling of rock; and in electrical prospecting for minerals.

The resistivity of rocks varies from 10^{-2} to 10^{16} ohm · cm.

The order of magnitude of some electrical properties of rocks and water is shown in Table 4-7.

TABLE 4-7

Some Electrical Properties of Rocks and Water

Rocks and Water	Specific resistance, Ωm	Dielectric Constant
1	2	3
Rock-forming minerals	10^{10}—10^{14}	4— 8
Igneous rocks	10^3 —10^7	7—14
Sedimentary rocks, dry	10^3 —10^9	7—14
Sedimentary rocks, wet	10^1 —10^4	7—14
Fresh water	10^1 —10^3	80
Salt water	10^{-1}—10^0	80

Radioactive Properties

Some minerals such as autuntite, carnotite, monazite, torbernite, uraninite (an uranium mineral pitchblende containing radium), zircon, and others are radioactive, spontaneously emitting particles capable of exiting a Geiger counter or of turning photographic film black. The dangers of radioactivity require special handling (lead shielding, decontamination) when large quantities of such materials are being worked with or stored (man-made hazards).

4-11. Thermal Properties of Rocks

Heat is a transient form of energy, a thermal potential for translocation soil or rock moisture along a thermal gradient (62, 64).
ground openings (vertical shafts, power plants, shelters, storage spaces, and mines, for example). Also, it is known that an increase in temperature lowers rock strength and increases ductility. Therefore, knowledge about the thermal conditions in rock and of the geothermal gradient is of great technical and economical importance. This obtains especially with respect to observing health conditions and heat endurance limits for workmen; to designing efficient underground ventilation and air-conditioning systems; to choosing an appropriate method of construction; or to designing soil and rock artificial freezing and thawing facilities.

The thickness of the overburden cover of geological materials above an underground opening also influences the temperature in the opening.

Changes in temperature in rock may bring about rock exfoliation.

Heat

Heat is a transient form of energy, a thermal potential for translocation soil or rock moisture along a thermal gradient (53, 57).

Heat has a profound effect upon matter. Heat in the ground tends to expand and contract rock, water, gas, and air. Upon the addition of heat, solids and liquids increase in volume in most cases. Upon removal of heat, they contract. Heat has only one measurable property-temperature. Temperature is the measure of the intensity of heat energy. One of the characteristics of heat is its tendency to move.

When different parts of a rock are at different temperatures, heat flows from points at higher temperature to points at lower temperature. Because of nonuniform warming of the various geological strata of the earth's crust, heat exchange takes place constantly between these layers.

Change in temperature with depth within a stratum of rock and/or soil takes place in conformance with natural laws that are more complex than under usual, ordinary conditions.

The natural course of the air temperature follows the periodic oscillations of the atmospheric conditions, viz., climate.

The daily or diurnal temperature variations in soil can be measured to approximately 1 m below the ground surface. The monthly temperature variations in soil can be detected to a depth of about 7 to 10 m. The annual temperature variations extend down to about 20 m below the ground surface.

It is known that the temperature in rock formations of the earth's crust at shallow depths below the ground surface is a function of the temperature of the atmosphere above the ground surface. At a depth of about 25 to 30 m, the temperature is constant year round; it is about equal to the mean annual air temperature. However, deeper than about 30 m, the temperature starts to increase proportionally with depth.

The rate of increase in temperature with depth is expressed by a *geothermal step*. It is defined as the vertical distance along which there is a temperature increase of 1 °C. The inverse of the geothermal step is the *geothermal gradient*. It expresses the temperature increase for every 1-m depth.

The geothermal gradient is influenced by the width of water passages in the rock, distribution of the passages, temperature of the groundwater, extent and duration for filling the voids, velocity of flow of the fissure water, thermal conductivity of the rock, temperature of the trapped gas in rock, and other possible factors.

Tunnels driven through mountains made up of porous rocks cool considerably upon receiving meltwaters from snow and ice, or cold precipitation waters at temperatures somewhat above 0 °C.

In some mines, the rock temperature ranges from about −18 °C to +60 °C.

The temperature of the rock prevailing in the interior of the mountain depends upon the following factors (1):

1. Position of the geoisotherms under the mountain ranges (geothermal step),
2. Soil temperature of the ground surface above the tunnel,
3. Thermal properties of the rock,

4. Hydrological conditions above and below the ground surface, and
5. Position and elevation of the tunnel.

Table 4-8 shows overburden thickness and temperature in some tunnels (1, 107).

TABLE 4-8

Overburden Thickness and Temperature in Some Alpine Tunnels

Tunnel	Length m	Overburden thickness m	Max. temperature °C	Geothermal step m/°C	Rock
1	2	3	4	5	6
Albula	5,886	750	11–52	49	Granite
Apennine	18,500	2,000	64	—	—
Arlberg	10,250	715	18.5–34	38.6	Gneiss with mica granite, slate
Gotthard	14,998	1,752	40.4	47	
Karawanken	7,976	916	15.0	144	—
Lötschberg	14,605	1,673	34	45	Granite and slate
Mont Cenis	12,236	1,610	29.5	58,4	Sandstone and limestone
Simplon	19,729	2,135	55.4	37	—
Tauern	8,551	1,567	23.9	49	Granite, gneiss, and mica schist

Some of the more important thermal properties of rocks are:

1. heat capacity,
2. latent heat of fusion,
3. thermal conductivity,
4. thermal diffusivity or temperature conductivity, and
5. thermal expansion and contraction.

Heat Capacity

Among the important thermal properties of a material is its heat capacity. Mass heat capacity c_m of a substance is the actual amount of heat energy Q necessary to change the temperature of a unit mass, say 1 kg, of the substance by one degree:

$$c_m = Q/\Delta T \quad [\text{Cal}/(\text{kg})\,(^\circ\text{C})] \qquad (4\text{-}20)$$

where c_m = unit of heat capacity for the mass m of a substance,
$\Delta T = T_1 - T_0$ = rise in temperature,
T_0 = initial temperature, and
T_1 = final temperature ($T_1 > T_0$).
Cal = large or kilo-calorie = 1000 cal (small or gram-calories)
kg = kilogram

Numerically (but not dimensionally), the magnitude of the heat capacity is equal to the dimensionless value of specific heat.

Specific Heat

The specific heat of a substance is the dimensionless ratio of the heat capacity of a substance to the heat capacity of water.

In heat transfer problems in soil and rock engineering, the concept of volumetric heat capacity c_v of a substance is defined as the amount of heat necessary to change the temperature of a unit of the substance by one degree.

The relationship between the mass heat capacity c_m and volumetric heat capacity is:

$$c_v = c_m \cdot \gamma_d \quad [\text{Cal}/(\text{m}^3)\,(^\circ\text{C})] \qquad (4\text{-}21)$$

where γ_d = dry unit weight of material, in kg/m³.

One observes that the coefficient of the volumetric heat capacity of soil or rock depends upon its unit weight and its moisture content.

At freezing-point temperatures, the values of mass heat capacity are (57):

Water: $c_{mw} = 1.0$ [Cal/(kg) (°C)]
Ice: $c_{mi} = 0.5$ [Cal/(kg) (°C)]
Rock and dry soil
  mineral particles: $c_{mr} = 0.2$ [Cal/(kg) (°C)]

The relationship between the volumetric heat capacity c_w and dry unit weight for various soil moisture contents w in percentages are:

METHODS OF ROCK EXPLORATION

For unfrozen soil or rock:

$$c_{vu} = \gamma_d \left[c_{mr} + \frac{(c_{mw})(w)}{100} \right] \quad [Cal/(m^3)(°C)] \qquad (4\text{-}22)$$

For frozen soil or rock:

$$c_{vf} = \gamma_d \left[c_{mr} + \frac{(c_{mi})(w)}{100} \right] \quad [Cal/(m^3)(°C)] \qquad (4\text{-}23)$$

In Table 4-9, mass heat capacity values of some minerals and rocks are compiled (27, 64, 103).

TABLE 4-9

Mass Heat Capacity of Minerals and Rocks

Substance	Temperature °C	Mass heat capacity c_m Cal/(kg)(°C)
1	2	3
Calcspar, CaCO₃	0– 50	0.188
Calcspar, CaCO₃	0–100	0.2005
Hornblende	20– 98	0.195
Mica (Mg)	20– 98	0.2061
Mica (K)	20– 98	0.2080
Basalt (fine, black)	12–100	0.1996
Dolomite	20– 98	0.222
Gneiss	17– 99	0.196
Granite	12–100	0.192
Limestone	15–100	0.216
Marble	0–100	0.210
Quartz sand	20– 98	0.191
Sandstone	–	0.22

Latent Heat of Fusion

When water changes its phase from liquid to solid, heat is liberated (latent heat of fusion). When ice melts, it absorbs heat from the air in contact with it. The amount of heat energy necessary to make this change isothermally, i.e. with no temperature change, is termed the latent heat of fusion, L, in Cal/kg. Because 1 kg of water liberates (or absorbs) $L = 80$ Cal/kg of latent heat, the total latent heat of fusion, Q_L, of $w^0/_0$ water per 1 m³ of soil or rock material is:

$$Q_L = (\gamma_d)(w)(L) \quad \left[\frac{Cal}{m^3}\right] \qquad (4\text{-}24)$$

where γ_d = dry unit weight of the material; and
w = moisture content of soil or rock by dry weight, in decimal fraction.

Notice that the amount of latent heat of fusion depends upon the dry unit weight of the rock (soil) material, and its moisture content w.

Thermal Conductivity

The total amount of heat Q transferred in the steady state by conduction through unit area perpendicular to flow ($A = 1.0$) during a time interval t under a temperature gradient of $(T_1 - T_2)/x$ is calculated as:

$$Q = K A t (T_1 - T_2)/x \quad [\text{Cal}] \qquad (4\text{-}25)$$

where
K = coefficient of thermal conductivity of rock, in Cal/(m · h · °C); here
Cal = kilo-calorie (large calorie)
$T_1 - T_2$ = temperature difference in °C between two points in the material x distance apart, and
$(T_1 - T_2)/x$ = temperature gradient in °C/cm.

Thermal conductivity K is not necessarily a constant but, in fact, is a function of temperature for all phases, and in liquids and gases depends also upon pressure. When the thermal conductivity is constant, a linear thermal gradient exists.

The coefficient of thermal conductivity is very sensitive to rock or soil type and its moisture content.

Thermal Diffusivity

The term thermal diffusivity is also known by the term temperature conductivity. When dealing with variable heat flow, in thermal calculations one uses the quantity thermal diffusivity a. It is connected with the thermal conductivity K as:

$$\alpha = \frac{K}{(c_{mr})(\gamma_d)} = \frac{K}{c_v} \left[\frac{m^2}{h} \right] \qquad (4\text{-}26)$$

where c_{mr} = mass heat capacity of rock,
c_v = volumetric heat capacity, and
γ_d = dry unit weight of rock (kg$_f$/m³).

Thus, thermal diffusivity is a measure of the rate at which a change in temperature spreads through a body.

Equation (4-26) shows that at high moisture content, the thermal diffusivity a decreases, because the product $(c_{mr})(\gamma_d)$ in the denominator increases more rapidly than the thermal conductivity K.

Thermal conductivity and diffusivity values of some geological materials are compiled in Table 4-10 (27, 43, 64, 103).

TABLE 4-10

Thermal Conductivity K and Diffusivity α Values
of Some Geological Materials

(Refs. 27, 43, 64, 103)

Material	K Cal/(m · h · °C)	α m²/h
1	2	3
Clays	0.21	0.0035
Sands (dry)	0.23	0.00072
Granite	2.08	0.0027
Limestones	1.58	0.0022
Sandstones	1.4	0.0022
Shales	0.8 – 1.25	0.0020

4-12. Coefficient of Thermal Expansion and Contraction

This topic should be of interest to geotechnical engineers because temperature has a marked effect on materials and engineering design. Temperature changes induce thermal stresses and strains in construction elements. Besides, one should be cognizant that upon freezing of soil and/or rock water, ice expands, while soil and rock mineral particles generally contract.

The ratio of the increase in length (linear coefficient), area (superficial), or volume (cubical) of a body for a given rise in temperature (usually from 0 °C to 1 °C) to the original length, area, or volume is known as the coefficient of thermal expansion. The three coefficients are approxiamately in the ratio of 1:2:3.

If the temperature of a body is warmed (or cooled) from a temperature T_1 to T_2 ($\Delta T = T_2 - T_1$), the change ΔL (expansion or contraction) of the initial length L of a specimen of the body calculates as:

$$\Delta L = \alpha_t \cdot L \, (T_2 - T_1) = \alpha_t \cdot L \cdot \Delta T \qquad (4\text{-}27)$$

Its new length L_t is approximately equal to:
$$L_t = L + \Delta L = L\,(1 + \alpha_t \cdot \Delta T) \qquad (4\text{-}28)$$

Its new volume V_t is:
$$V_t = V\,(1 + \beta \cdot \Delta T) \qquad (4\text{-}29)$$

where V = initial volume,
α_t = coefficient of linear thermal expansion, expressed in units of cm per cm per degree of temperature change as
$$[(\text{cm/cm})/^\circ\text{C}] = [1/^\circ\text{C}]$$
β = coefficient of volumetric (cubical) thermal expansion $[1/^\circ\text{C}]$.

Here,
$$\alpha_t = \frac{1}{L} \cdot \frac{dL}{dT} \quad [1/^\circ\text{C}] \qquad (4\text{-}30)$$

or
$$\alpha_t = \frac{\Delta L}{L} \cdot \frac{1}{\Delta T} \qquad (4\text{-}31)$$

and
$$\beta = \frac{1}{V} \cdot \frac{dV}{dT} \quad [1/^\circ\text{C}] \qquad (4\text{-}32)$$

or
$$\beta = \frac{\Delta V}{V} \cdot \frac{1}{\Delta T} \qquad (4\text{-}33)$$

The coefficient α_t of linear thermal expansion (contraction) for solid bodies is the relative change in length dL/L per degree C increase (decrease) in temperature (ratio of dL per degree to the length L at $0\,^\circ\text{C}$).

The coefficient β of volumetric thermal expansion (contraction) for solids is the relative change in volume dV/V per degree increase (decrease) in temperature (ratio of dV per degree to the volume of $0\,^\circ\text{C}$).

Water has a volumetric (cubical) expansion coefficient of (46):
$$V_t = V\,[1 + a\,(\Delta T)^2 + b\,(\Delta T) + c\,(\Delta T)^3 + \ldots] \qquad (4\text{-}34)$$

where $a = -6.43 \times 10^{-5}$,
$b = 8.50 \times 10^{-6}$, and
$c = 6.79 \times 10^{-8}$.

METHODS OF ROCK EXPLORATION

For homogenenous bodies such as most of the sound rocks that expand equally in all directions, the coefficient β of volumetric thermal expansion is three times the linear one:

$$\beta = 3\alpha_t \tag{4-35}$$

One notices that thermal expansion is proportional to the rise (fall) in temperature.

Some coefficients of linear thermal expansion α_t of some geological materials are compiled in Table 4-11 (42).

TABLE 4-11

Average Coefficients α of Thermal Expansion of Some American Rocks at Room Temperature to 100 °C

(after GRIFFITH, Ref. 42)

Name of Rock	α 1/° C
Igneous	
Granite series	$34 \times 10^{-7} - 66 \times 10^{-7}$
Basalt series	$22 \times 10^{-7} - 35 \times 10^{-7}$
Diabase	$31 \times 10^{-7} - 35 \times 10^{-7}$
Gabbro	$20 \times 10^{-7} - 30 \times 10^{-7}$
Sedimentary	
Limestones and dolomites	$24 \times 10^{-7} - 68 \times 10^{-7}$
Sandstones	$36 \times 10^{-7} - 65 \times 10^{-7}$
Metamorphic	
Gneisses	$34 \times 10^{-7} - 44 \times 10^{-7}$
Marbles	$34 \times 10^{-7} - 51 \times 10^{-7}$
Quartzites	$60 \times 10^{-7} - 61 \times 10^{-7}$
Schists (crystalline)	$34 \times 10^{-7} - 43 \times 10^{-7}$
Slates	$45 \times 10^{-7} - 49 \times 10^{-7}$

It should be noticed that the coefficients (α_t) of layered soils and rocks are anisotropic relative to thermal expansion (contraction) because of the anisotropy of the layered soil (rock) systems. Also, with increase of all-sided confining pressure, the linear and volumetric thermal coefficients of *expansion* decrease.

The relationship between thermal expansion and various other factors such as composition of soils and rocks, density, moisture content, and others are practically unexplored as yet.

Mechanical Properties of Rocks

4-13. General Properties

Generally, the mechanical properties of a material characterize its reaction to the effect of the force field of its environment. Particularly, the mechanical properties of rocks depend upon:

1. The nature of the rock substance,
2. The stratigraphy of the rock in situ,
3. Rock defects, and
4. Testing methodology.

The most important mechanical properties of rocks that must be thoroughly investigated when designing foundations, hydraulic structures, and underground openings relative to failure of a geologic material are:

1. Hardness,
2. Durability,
3. Permeability to water,
4. Elasticity,
5. Plasticity,
6. Deformability, and
7. Strength.

The mechanical properties of a rock material are elasticity, plasticity, viscous deformations, and combinations thereof.

However, for safe, effective work in rock, the mechanical properties of most concern are the strength properties relative to the competency of rocks in underground opening and open-cut excavations. These rock properties vary with depth of the rock formation.

Hardness

Hardness (written H) of a mineral and rock is the resistance to abrasion. Every mineral and every rock have a hardness or range of hardness that ultimately depends on the strength of chemical bonds.

To rate hardness of minerals and rocks, the empirical MOHS' hardness scale is used (Table 4-12). The hardness may be determined by scratching one mineral with another. Table 4-12 shows an arbitrary numerical scale or rating of relative hardness based on common minerals. The higher rating numbers represent harder substances. Each material in the MOHS' scale of hardness scratches the previous one by the following one.

TABLE 4-12

Mohs' Scale of Hardness of Minerals

Number of relative hardness scale or rating H	Mineral	Chemical composition	Remarks
1	2	3	4
1	Talc	$Mg_3Si_4O_{10}(OH)_2$	Softest; can be scratched by fingernail.
2	Gypsum	$CaSO_4 \cdot 2H_2O$	Can be scratched by fingernail.
3	Calcite	$CaCO_3$	A copper penny or a brass pin can scratch calcite.
4	Fluorite	CaF_2	Fluorspar. May be scratched easily by a steel point.
5	Apatite	$Ca_5F(PO_4)_3$	Any of the calcium phosphate minerals. Can be scratched by a knife. A window glass may be rated as $H = 5.5$ on the hardness scale.
6	Orthoclase (feldspars)	$KAlSi_3O_8$	Can be scratched by a knife blade of a good-quality steel; a hardened steel file may be rated as $H = 6.5$
7	Quartz	SiO_2	Scratches steel, glass, and all of the minerals whose $H < 7$.
8	Topaz	$Al_2SiO_4(F, OH)_2$	Great hardness. A valuable jewelry stone.
9	Corundum	Al_2O_3	Harder than any other natural mineral except diamond. An important industrial abrasive and refractory. Has many gem varieties, among them ruby and sapphire.
10	Diamond	C	The hardest substance known; not all diamonds are of the same hardness, however.

Sometimes hardness is used as a strength criterion in rocks.

The strength of sandstone rock increases with increasing amount of quartz.

The hardness of the minerals contained in a composite rock reflects to a certain extent the mechanical properties of a rock.

For example, based on his experimental results, PRICE (92) showed the effect of quartz content in sandstone with a calcite matrix, and in siltstone with a clay matrix, on the uniaxial compressive strength of these materials, namely: the strength of the rock materials tested increased considerably with increasing quartz content.

In general, excavation of solid, hard rock for its removal, which requires blasting, is more difficult and therefore more expensive than excavation of soil materials.

4-14. Durability

Durability of rock is a relative term. This property depends upon the nature of the rock environment—such as the climate and atmosphere, for example—and the amount of exposure of rock or building stone in the structure (viz., weathering).

Also, the rate of solution action on limestone, dolomite, and cement grout; resistance of a rock to frost action (freezing and thawing); rate of weathering; and porosity of rock may be correlated with durability.

According to the National Bureau of Standards [ESHBACH, (31)], the estimates for the life of rock, viz., building stone, are:

Ohio sandstone, best variety	1 year to many centuries
Limestone	20 to 40 years
Marble, coarse	40 years
Marble, fine	50 to 200 years
Gneiss	50 years to many centuries
Granite	75 to 200 years

Generally, sound sandstone is durable, and is rated as having good fire resistance.

Therefore, by means of petrographic and other appropriate analyses, and by freeze-thaw tests, the resistance of rocks to weathering and chemical aggressiveness by groundwater and precipitation should be ascertained.

4-15. Elasticity

Elasticity is a universal property of an ideal material. Every solid material deforms under the action of a load, viz., stress. To every kind of stress there is a corresponding strain. If the stress is not too large, the strained ideal material will recover its unique, natural state, the original shape and size to which the material returns when the external loads are removed. All stresses, strains, and displacements are measured from this natural state; their values are taken as zero in this state. The property of recovering from strain of a material is called *elasticity*. If a material recovers completely, it is called perfectly elastic. If the material does not recover completely, the strain that remains when the stress is removed is called *permanent set,* and the material is said to be in an elastic state.

A relatively small but easily measurable part of the total deformation of solid bodies under load is said to be of an elastic nature.

Relative to small reversible distortions or deformations, ordinary solids have *isotropic elasticity* (quartz and diamond, for example).

Depending upon how closely a rock approximates the ideal material, the concept of elasticity applies also to rocks. Thus, one gets the notion that in practice, the property of elasticity of rocks depends upon their continuity, homogeneity, and isotropy.

Because of the manifold factors involved in rock strength, in rock engineering practice when designing in rock and performing stability analyses of rocks, certain idealizations and assumptions as to the nature of rock as a construction material are usually made. The following are some of the most important and most frequently made idealizations:

1. The rock is assumed to be a continuous, homogeneous, linear-elastic, isotropic material. The group of materials whose mechanical property does not depend upon the direction is said to be isotropic.

2. This material obeys Hooke's law of proportionality between stress and strain; that is, the strains are linear functions of stresses.

3. The deformations (strains) of the loaded rock material are so small that they may be neglected in setting up equilibrium conditions.

These assumptions are supported by measurements of the observed strains within the elastic range in metals and rocks, with some exceptions of porous solids and other materials, to a sufficient degree for most practical applications.

In certain rock mechanics analyses, one thus usually adheres to the fundamental theory of elasticity, to the basic elasticity constants such as Young's modulus of elasticity E, and Poisson's ratio $\mu = 1/m$, where $m = 1/\mu$ is the Poisson's number.

4-16. Plasticity

Plasticity of a solid material is its property to be continuously and permanently deformed, that is, a property to change shape in any direction without rupture under a stress exceeding the yield value of the material. Thus, plasticity of a material is characterized by the existence of a yield point beyond which permanent strains appear. In other words, the plastic deformation of a material is the permanent deformation after complete unloading of the material, assuming the unloading to be elastic. In the plastic state, permanent deformation of a material may occur without fracture. The term "fracture" implies the appearance of distinct surfaces of separation in the material.

The exact definition of plastic deformation, valid in all cases, refers to strain rates, of course. Plastic flow of a solid is the phenomenon of increasing strain at constant stress.

Physically, plastic deformation is anisotropic. Any initial isotropy which may have been present is usually destroyed by plastic deformation.

The conditions prevailing in the deeper strata of rock, such as long duration of small differences in principal stresses, elevated temperatures, and high average pressure, are all contributing to plastic deformations of rocks.

As shown in Chapter 8, plastic flow tends to relieve the high stress concentrations that would ordinarily develop in perfectly elastic materials.

Today, the plastic state of matter is of considerable interest to many branches of science and of engineering, among them rock mechanics and soil mechanics. The changes in minerals and rocks brought about by plastic deformation are in many respects analogous to certain phenomena observed in the changes in the structure of metals.

In studies of the plastic state of stress in soils and rocks, use is made of the so-called slip surfaces (refer to Sections 8-6 to 8-8). The slip surfaces are curved according to the general mathematical, viz., physical, curve of a logarithmic spiral (57-61), namely:

$$r = r_i \cdot e^{\pm \omega \cdot \tan \varphi} \qquad (4\text{-}36)$$

where r = general, variable radius vector;
r_i = initial radius vector;
e = base of the system of natural logarithms;
ω = amplitude (angle between r_i = const and variable r);
φ = angle of internal friction of soil or rock, whichever the case is;
$\tan \varphi$ = coefficient of friction.

Here r and ω are polar coordinates.

Equation (4-36) represents a family of two orthogonal, equilateral logarithmic spirals, viz., slip lines (Fig. 8-7).

Criteria of Rock Failure

4-17. General Notes about Failure Criterion

An important problem in rock mechanics is to ascertain the mechanical conditions which cause rock to deform permanently, or to fracture. If the mechanical, viz., stress condition in a rock is known, there arises immediately the question about the criterion of rock failure or fracture (rupture) and/or lasting plastic deformation without increase in stress. The term "fracture" is here used in the sense of brittle fracture or failure. This implies a complete loss of cohesion across the surface of failure.

The criteria of rock failure which are most useful in rock mechanics have not been derived merely from simple mathematical assumptions. Rather, the criteria are the expressions of physical hypotheses. Obviously, in designing in rock, and remembering that all physical laws have been derived from experiment, it is desirable to have a criterion of rock failure based on experiment. Such a theory should specifiy how the strength of the rock is affected by the state of stress in rock, time effects, temperature, and other factors involved. Unfortunately, no complete, plausible theory of failure has been put to the fore for a complex polycrystalline material, such as rock, for example. As a result, during the course of time, various empirical assumptions about the criteria for failure have been made.

In appraising the danger of a failure occurring in rock, a distinction should be made among some of the most commonly used failure criteria. Some of the better known criteria of failure that are based on experiment and are considered to be reasonably acceptable in soil mechanics as well as in rock mechanics are (28, 40, 41, 80, 111):

1. The maximum tensile stress;
2. TRESCA's criterion, or that of the maximum shear stress;
3. COULOMB's criterion of failure;
4. MOHR's criterion; and
5. GRIFFITH's criterion of brittle failure in tension.

Because of their different physical background, it is well, here, to keep COULOMB's and MOHR's criteria separate.

Here, the term criterion is now advisedly used, because in a polycrystalline rock material where very little is known about the mechanisms of deformation and failure, it is impossible to devise a theory which will comprehend and fit all rocks in all states of stress, under various confinement, at various temperatures, and at all times.

4-18. Various Criteria

Maximum Tensile Stress Criterion for Rock

By this criterion, the rock material is assumed to fail by brittle fracture in tension when the applied least principal stress $-\sigma_3$ to the rock is equal to its uniaxial tensile strength $\sigma_{t\ ult}$:

$$\sigma_3 = -\sigma_{t\ ult} \qquad (4\text{-}37)$$

Maximum Shear Stress Criterion

TRESCA's criterion of failure is valid for isotropic, ductile materials. The criterion may be stated as a function of principal stresses σ_1 and σ_3.

According to this criterion, the material is assumed to fail when the maximum shear stress τ_{max} is equal to the shear strength s of the material (111):

$$s = \tau_{max} = \frac{\sigma_1 - \sigma_3}{2} \qquad (4\text{-}38)$$

where σ_1 and σ_3 are the major and minor principal stresses, respectively. The intermediate principal stress σ_2 plays no part in this criterion. Hence, the decisive factor in TRESCA's criterion for failure is the maximum shear stress in the material.

One notices from Eq. (4-38) that TRESCA's failure criterion is a special case of COULOMB's criterion (refer to Section that follows).

Coulomb's Criterion of Failure

Empirically, the shear strength s of a cohesive soil or rock varies with the normal stress σ_n on the rupture plane according to the classical COULOMB's law written as (28):

$$s = \tau = \sigma_n \cdot \tan \varphi + c \qquad (4\text{-}39)$$

where τ = shear stress,
 φ = shearing strength parameter or angle of friction (a test parameter), and
 c = test parameter known as the cohesion.

Geometrically, Eq. (4-39) represents a straight line, $t\text{-}t$ (Fig. 4-27), known as the COULOMB's shear strength line.

METHODS OF ROCK EXPLORATION 163

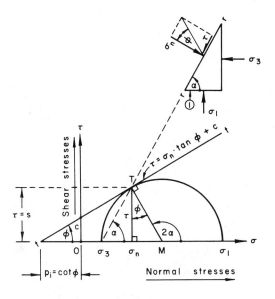

Fig. 4-27
COULOMB-MOHR failure criterion (general case)
$r-r$: rupture or failure plane
$t-t$: COULOMB's shear strength line, tangent to circle
$\sigma_1 - \sigma_3$: diameter of MOHR's stress circle
$\tau = s$: shear strength

From geometry as in Fig. 4-27, the normal stress σ_n on the rupture or shear plane r-r is calculated as

$$\sigma_n = \frac{\sigma_1 + \sigma_3}{2} + \frac{\sigma_1 - \sigma_3}{2} \cos 2\alpha \qquad (4\text{-}40)$$

wherein σ_1 and σ_3 are the major and minor principal stresses, respectively, and α is the angle of rupture.

The shear stress τ in the rupture plane is:

$$\tau = \frac{\sigma_1 - \sigma_3}{2} \cdot \sin 2\alpha \qquad (4\text{-}41)$$

When $c = 0$, the shear strength at failure is, according to Eq. (4-39):

$$s = \tau = \sigma_n \cdot \tan \varphi \qquad (4\text{-}42)$$

a straight line passing through the origin of the normal stress/shear stress coordinates (Fig. 4-28).

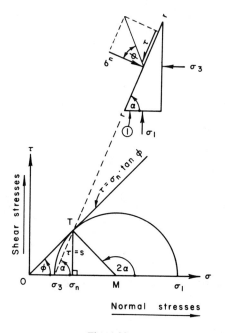

Fig. 4-28
COULOMB-MOHR failure criterion when $c = 0$

When $\varphi = 0$, then Eq. (4-39) transforms into:

$$s = \tau = c = \frac{\sigma_1 - \sigma_3}{2} = \text{const} \qquad (4\text{-}43)$$

(Fig. 4-29), i.e., the shear strength is constant and independent of normal stress. One sees that TRESCA's criterion is a special case of COULOMB's criterion for failure.

Furthermore, when $\sigma_3 = 0$ and $\varphi = 0$, then:

$$s = \tau = c = \frac{\sigma_1}{2} \qquad (4\text{-}44)$$

i.e., the shear strength of a pure cohesive material is equal to one-half its compressive strength σ_1.

Accepting MOHR's criterion, the discussion of which follows immediately in the next section, one notices in Fig. 4-27 that COULOMB's criterion is equivalent to MOHR's linear envelope to MOHR's stress circles.

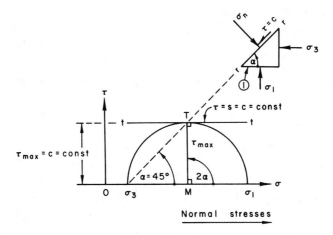

Fig. 4-29
COULOMB-MOHR-TRESCA failure criterion for a pure cohesive material ($\varphi = 0$)

COULOMB's shear strength criterion may also be written from geometry as in Fig. 4-27 in terms of principal stresses as:

$$\sin \varphi = \frac{\sigma_1 - \sigma_3}{\sigma_1 + \sigma_3 + 2 p_i} \tag{4-45}$$

where

$$p_i = c \cdot \cot \varphi \tag{4-46}$$

is the initial stress of the test specimen.

When $c = 0, p_i = 0$, and

$$\sin \varphi = \frac{\sigma_1 - \sigma_3}{\sigma_1 + \sigma_3} \tag{4-47}$$

Referring to Fig. 4-27, it can be reasoned out that as the confining pressure σ_3 is increased, the normal stress σ_n also increases on the incipient plane of shear (rupture). Thus, the necessary shear stress $\tau(s)$ to bring about failure also increases. From geometry, as in Fig. 4-27 or Fig. 4-28:

$$2\alpha = 90° + \varphi \tag{4-48}$$

and

$$\alpha = 45° + \varphi/2 \tag{4-49}$$

meaning that, theoretically, failure in shear takes place on rupture planes at an angle of rupture $\alpha = 45° + \varphi/2$ to the major principal plane ①.

Theoretically, in the case of a pure cohesive material, Fig. 4-29 shows that the rupture plane intersects the line of action of the axial load at $a = 45°$. In reality, however, the angle of rupture a varies from one type of rock to another.

Although COULOMB's straight-line criterion is used extensively to predict failure in rocks, it must be said that this criterion does not represent exactly the curvilinear envelope to MOHR's stress circles of most rocks.

Mohr's Criterion of Failure

MOHR's criterion merely assumes the existence of a shear failure envelope (Fig. 4-30), which itself may be a straight line (t-t), or a curvilinear one (e-e). The envelope in MOHR's criterion is the governing condition for the occurrence of rupture surface brought about by the shear stress upon exhaustion of the shear strength of the material. Analogous to soil mechanics, MOHR's failure criterion can also be used in rock mechanics.

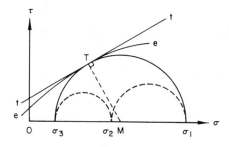

Fig. 4-30
Straight-line (t—t) and curvilinear envelopes in MOHR's criterion of failure

If the stress-circle envelope is a straight line, then MOHR's and COULOMB's criteria are identical (80).

MOHR's failure criterion says that a compressed material fractures when the shear stress τ on the incipient rupture plane increases to a value which depends upon the magnitude of the normal stress σ_n on the rupture (shear) plane, i.e., $\tau = f(\sigma_n)$. This relationship has been established experimentally.

On MOHR's stress circle diagram as in Fig. 4-30, it can be seen that failure of rock (soil) takes place when the $(\sigma_1-\sigma_3)$-diameter stress circle just tangents the shear strength envelope e-e. Also, one notices in this diagram that in the three-dimensional representation of MOHR's failure criterion, the intermediate principal stress σ_2 (if any) does not affect failure. Only the major (σ_1) and minor (σ_3) principal stresses play an active part in this criterion. One should also be cognizant that in MOHR's stress-circle diagram, the failure envelope (*e-e*) is not represented by any mathematical formula but is obtained experimentally: the envelope is drawn to tangent the various stress circles of a set of experiments (Fig. 4-31).

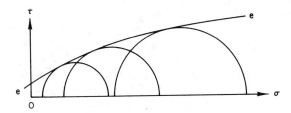

Fig. 4-31
MOHR's curvilinear strength envelope *e-e*

MOHR's failure criterion specifies on a macroscopic scale not only the state of stress at failure, but also the direction of the failure plane. However, COULOMB's and MOHR's criteria do not account for the cause of failure of the rock material on an internal or microscopic basis.

Although MOHR's criterion does not explain completely and satisfactorily all the observed variations and deviations in rock compressive and shear strength, at the present time it enjoys its widest acceptance in rock mechanics.

MOHR's failure criterion can also be used with reasonable validity in studies of shear strength in rock faults, joints, and other kinds of discontinuities (Fig. 4-32).

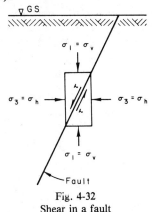

Fig. 4-32
Shear in a fault

Griffith's Criterion of Tensile Failure

It is known that crystalline substances contain microfractures. In his criterion of failure, GRIFFITH (40, 41) assumes the existence of thin, flat, narrow, elliptical uniform microcracks in the material. These cracks bring about stress concentration at the tips or ends of GRIFFITH's microcracks. According to GRIFFITH's criterion, fracture of material is caused by these stress concentrations, causing the crack to propagate and ultimately contributing to macroscopic failure of the material. Also, the GRIFFITH criterion postulates that fracture is initiated in a brittle material by failure in tension around the tips of the microcracks, and that the crack retains its form until the moment of failure. GRIFFITH's hypothesis has been corroborated by experimental work on glass. Insufficient studies have been made on rocks to determine whether GRIFFITH's criterion would predict compression failure by knowing the tensile strength of the rock.

As to rock, GRIFFITH's theory does not account for the closure of the crack before failure. If the cracks do close, it can be expected that frictional forces will be acting between the closed surfaces. In such a case, GRIFFITH's criterion becomes a special case of COULOMB's criterion.

The actual mechanism of failure in rock is still a subject of hypothesis. Despite its theoretical elegance, there can be little general application of GRIFFITH's failure criterion to rocks as yet, mainly because the varying rock properties require an empirical approach as represented by the COULOMB or MOHR failure criteria.

Also, joints in rock are not equivalent to thin elliptical openings, because joints in situ more commonly abut a network of other joints. Besides, the lengths of the cracks in a rock mass are not small with respect to the stressed rock volume, and the spacings of the joints are frequently close enough to affect the stress concentrations around any flaw. The stress concentrations around a GRIFFITH elliptical crack are computed based on the theory of elasticity. Therefore, the mechanism of failure should be time-independent, and hence would not account for variation in strength with stress or strain rate.

In the words of FARMER (37), "the GRIFFITH's criterion applied to rock is at best a rather complicated way of arriving at a logical solution."

Deformation of Rocks

4-19. Definition

The manner in which a rock or rock mass deforms upon action of loads on them, or because of changes in internal stresses brought about by all kinds of excavations in rock, is an important consideration in rock mechanics.

The term "deformation of rock" means any change in the original form or volume of a rock specimen; or change by externally applied loads on in-situ rock; or by tectonic forces (compressive and/or shear forces). In nature, the common modes of deformation may be folding, faulting (shear), and solid flow.

4-20. Mutual Dependence of Stress and Strain

In studies of rock deformations and rock strength, as well as in theory and design of structures in rock, the concepts of stress and strain are basic: every stress brings about strain and displaces individual rock particles. The concepts of stress and strain are also important in studying stress fields.

Upon subjecting a test specimen of rock to compression, it deforms: to a certain particular stress there occurs a corresponding definite deformation, viz., strain. The rock test specimen in the testing machine resists the deformation. The specimen props firmly against the externally applied load with an induced stress that corresponds to the externally applied load. Thus, the simultaneously induced stress in the rock and its deformation depend upon each other.

Without deformation there is no stress possible, and without stress no deformation can be brought about. Also, every stress imparting a strain on the rock displaces individual rock particles.

4-21. Stress-Strain Diagrams

For any axial symmetrical load applied to the cylindrical rock core specimen, the axial and lateral deformations are measured and the corresponding strains ε_1 and ε_3 are calculated. The plot of stresses and corresponding strains of the tested rock results in a stress-strain diagram.

Now, in the discussion that follows, and for the sake of one's indoctrination in the understanding of the stress-strain relationships of various materials, let us avail ourselves to the study of several stress-strain diagrams that follow.

In Figs. 4-33a, b, and c are shown three idealized materials characterized by the modes according to which they deform. Figure 4-33a illustrates elastic deformation of an elastic material represented by a straight-line stress-strain (force-deformation) diagram. Figure 4-33b shows viscous deformation of a viscous substance represented by a straight-line shear-force versus rate-of-shear diagram. Figure 4-33c represents plastic deformation of a perfectly plastic substance represented by a straight-line stress-strain diagram.

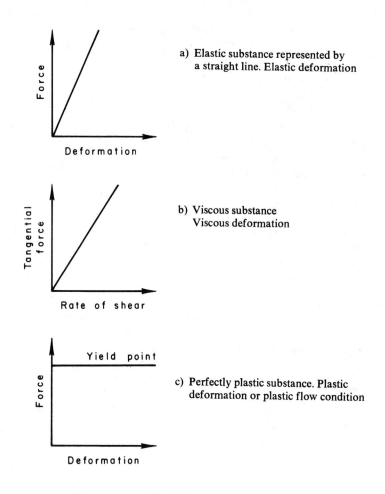

a) Elastic substance represented by a straight line. Elastic deformation

b) Viscous substance Viscous deformation

c) Perfectly plastic substance. Plastic deformation or plastic flow condition

Fig. 4-33
Three idealized substances a), b) and c), characterized somewhat by the modes according to which they deform

Work and Strain Energy

If a specimen of an elastic material is loaded uniaxially by a centric load in compression or tension, the load or force P brings about an absolute axial deformation ΔL — a shortening or elongation — of the specimen. This means that the force P performs mechanical work W on the specimen. The deformation ΔL is the displacement of the force P. When a specimen of an elastic material is axially shortened or elongated, part of the work done on the specimen (rock or soil) is transformed into heat energy and part is transformed into permanent deformation of the specimen.

The nature of the plot of displacement ΔL versus force P is shown in Fig. 4-34 by the force-displacement line $\overline{OF}$.

Theoretically, the physical triangular area, $\triangle$ AOF, under the force-displacement line (or curve) OF represents the axial work done by the force P on the elastic specimen:

$$\text{axial work} = W = \triangle \text{AOF} = (1/2)(P)(\Delta L) \quad [N \cdot m] \quad (4\text{-}50)$$

If plotting a stress-strain (σ-ϵ) curve, $\overline{OF}$ (Fig. 4-35), where $\sigma = P/A$; A = cross-sectional area of the specimen $\perp$ to force P; $\epsilon = \Delta L/L$ = relative deformation or strain, and L = initial length of the specimen, the physical triangular area $\triangle$ AOF beneath the stress-strain curve (or line) $\overline{OF}$ of the elastic material times its test volume V, is defined as strain energy S_e:

$$\text{strain energy} = S_e = (1/2)(\sigma)(\epsilon)(V) \quad [N \cdot m] \quad (4\text{-}50a)$$

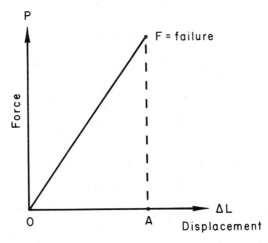

Fig. 4-34
Work = $\triangle$ AOF = $(1/2)(P)(\Delta L)$

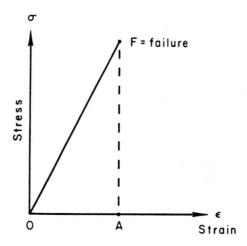

Fig. 4-35
Strain energy = $(\Delta AOF)(V) = (1/2)(\sigma)(\epsilon)(V)$

Factors Affecting Deformation

Rock deformation and strength are affected by the so-called deformability factors. These are the geological details such as:

1. Various rock defects;
2. Rock petrographic structure (rock matrix);
3. Geometric orientation and attitude of the rock formation (dip and strike);
4. Degree of weathering or alteration of the rock;
5. Elastic, plastic, and rheological properties of rock;
6. Anisotropy of rock;
7. Direction and magnitude of acting loads on rock;
8. Degree of compression and/or decompression of rock;
9. Fissures and/or hair cracks brought about by blasting and/or excavation of, or drilling in, the rock;
10. Seismic factors; and
11. State of stress (internal stress) within the rock mass.

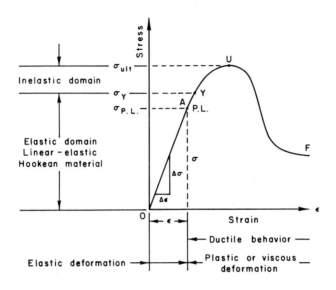

Fig. 4-36
Stress-strain diagram

Hence, because of the anisotropy of the rock, rock deformations cannot be characterized by a single parameter. Also, deformations depend upon the kind of load imposed on rocks—i.e., whether static, dynamic, or cyclic—and their magnitude and duration (creep, for example). Besides, plasticity, too, affects deformability and magnitude of deformations of rocks considerably.

Whereas elastic deformations entail changes in the volume of rock, plastic deformations (viz., the relative movement or sliding of the ruptured or sheared-off parts of the rock past each other) take place usually at a constant volume.

Slow deformation of rock at a constant load (solid flow) is a rheological phenomenon known by the term "creep." It is a feature of rock like creep in concrete. Creep contributes to genuine pressure in a rock mass.

Figure 4-36 shows an idealized stress-strain relationship diagram for a ductile material. In this diagram, it can be observed that up to a certain magnitude of the externally applied stress σ_Y (yield point Y) on the rock, the corresponding strain ε is constant. In other words, the stress σ (or $\Delta\sigma$) is proportional to strain ε (or $\Delta\varepsilon$) by HOOKE's law *"ut tensio sic vis"*:

$$\sigma = E \cdot \varepsilon \qquad (4\text{-}51)$$

or

$$\Delta\sigma = E \cdot \Delta\varepsilon \qquad (4\text{-}51a)$$

where $E = \sigma/\varepsilon = $ Hooke's coefficient of proportionality, known also as Young's modulus of elasticity, defined for uniaxial stress; the modulus of elasticity E represents the stiffness of the material.

When $\varepsilon = \Delta L/L = \sigma/E = 1.0$, then $E = \sigma$. Here $L =$ original, axial length of test specimen, and $\Delta L =$ absolute linear axial deformation of the same test specimen.

It should be recalled that for most materials, the proportional limit (point P.L.) practically coincides with the yield point (point Y). For this reason and for the sake of brevity, only the terms yield stress σ_Y and yield point Y will be used in the discussion that follows.

The other elasticity constant of the elastic material used in its strength analysis is Poisson's ratio μ:

$$\mu = 1/m = \varepsilon_3/\varepsilon_1 \qquad (4\text{-}51)$$

where $m = 1/\mu$ is Poisson's number, ε_3 is the lateral or transverse strain, and ε_1 is the longitudinal strain of the material tested.

The so-called elasticity constants E and μ may be obtained experimentally not merely from uniaxial stress-strain diagrams, but also from biaxial and triaxial stress-strain diagrams as well.

Referring to Hooke's law Eq. (4-51), one notices that between the stress and strain a linear relationship exists. This means that up to point Y in Fig. 4-36, the strains are elastic strains, and the material is said to be linear-elastic. A material is called linear-elastic if the relationship $\sigma = E \cdot \varepsilon$ holds accurately.

It should be kept in mind that the E-value is different for various materials.

Point Y, at which the transition from elastic to ductile behaviour of the material takes place, is termed the *yield point,* and the corresponding stress σ_Y is called the *yield stress*. Plasticity is characterized by the existence of a yield point beyond which permanent strains appear.

Upon continuing the loading beyond the yield point Y up to the ultimate stress σ_{ult} to point U on the stress-strain curve, failure of the material takes place, and with increasing strain beyond U the stress drops, as shown by the part UF of the curve. In other words, the material deforms without limit under this stress unless constrained. Thus, Figure 4-36 represents a stress-strain curve for a perfectly elasto-plastic material.

In this stress-strain diagram, the following three domains can be distinguished, namely:

1. The elastic domain,
2. The partly elastic-plastic domain, and
3. The plastic domain,

the domains as in 2 and 3 above being called the *inelastic* domain. Deformation in the inelastic domain is generally termed plastic or viscous deformation.

A material is called *purely elastic* if the deformation is recovered when the stress is removed.

A material is called *purely plastic* if the deformation does not disappear when the stress is removed and if the stress determines the amount, but not the rate, of deformation.

A material is called *purely viscous* if the stress acting upon it determines the rate of deformation.

Theoretically, as mentioned already earlier in this Chapter, within the domain of elasticity of an ideal elastic material, unloading or removing of stress recovers the corresponding strain completely. That is, there should be no permanent, irreversible deformation in the material.

Refer now to Fig. 4-37. If upon loading beyond the yield point Y the stress-strain curve continues to rise within the inelastic domain from Y to F above yield point, the material is said to be *strain-hardened*.

If the material is unloaded from σ_B to $\sigma_D = 0$, then only part of the strain is recovered: an *irreversible,* permanent, relative deformation or strain (permanent set) of the magnitude ϵ_{ir} results (Fig. 4-37), known also by the term plasticity deformation:

$$\varepsilon_{ir} = \varepsilon_{total} - \varepsilon_{el} \qquad (4\text{-}52)$$

Upon reloading the material with the stress σ_B, the stress-strain curve takes the course CF. Thus it can be seen that strain-hardening increases the yield stress from σ_A to σ_B but does not affect the elastic constant of the material.

The ratio e_{el} of the elastic strain ε_{el} to the total strain ε_{total} of the material is termed the *degree of elasticity* of the rock:

$$e_{el} = \varepsilon_{el}/\varepsilon_{total} \qquad (4\text{-}53)$$

The ratio e_{pl} of the plasticity strain $\varepsilon_{pl} = \varepsilon_{ir}$ to the total strain ε_{total} of the material is termed the *degree of plasticity* of the material:

$$e_{pl} = \varepsilon_{ir}/\varepsilon_{total} \qquad (4\text{-}54)$$

Unloading of rock in underground structures and rock slopes is an important stress release phenomenon around every underground opening, every hole, and every open cut. This phenomenon is based on plastic performance of the rock, and lies within the plastic domain of the stress-strain diagram.

In repeated loading-unloading-loading tests, the ratio E_c of stress σ to the total strain $\varepsilon_{total} = \varepsilon_{el} + \varepsilon_{ir}$ is called the *compression modulus* or *modulus of deformation:*

$$E_c = \frac{\sigma}{\varepsilon_{total}} = \frac{\sigma}{\varepsilon_{el} + \varepsilon_{ir}} \qquad (4\text{-}55)$$

This modulus is thus based on the total measured strains, i.e., elastic plus inelastic (irreversible or plastic) strains, ε_{el} and ε_{ir}, respectively.

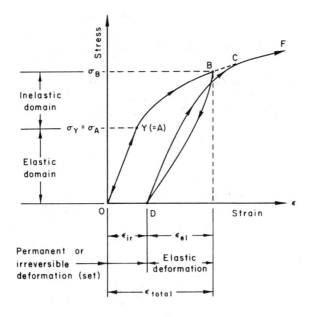

Fig. 4-37
Stress-strain diagram of strain hardening characteristic of a ductile material

METHODS OF ROCK EXPLORATION 177

The true modulus of elasticity E from such loading test results is then expressed on the elastic strain basis as:

$$E = \frac{\sigma}{\varepsilon_{\text{total}} - \varepsilon_{\text{ir}}} \quad (4\text{-}56)$$

Depending upon the amount of jointing in a rock mass, viz., elastic and plastic strains, the deformation modulus E_c may be quite different from the true, elastic modulus of elasticity E.

4-22. Stress-Strain Diagrams for Rocks

As to the deformations of rock, it has been learned that usually many kinds of rock fail, or fracture at the proportional limit P.L. of elasticity or somewhat beyond it very near to the yield stress. Such materials are referred to as brittle. Rocks are characterized as failing in brittle fracture if they fail with no previous plastic deformation. Rocks are referred to as ductile if they deform appreciably (plastically) before failure. After previous plastic deformation, the rock fails by ductile rupture. For ductile materials, there is no brittle fracture.

Under normal temperatures and pressures, rocks usually tend to exhibit a brittle kind of rupture of failure mechanism.

The term "fracture" is used here in the sense of brittle fracture or failure; this implies a complete loss of cohesion across a surface.

Because most of the rock materials are brittle, the plasticity domain of such rocks and thus their plastic deformation and degree of plasticity are relatively very small.

For most rocks, the stress-strain diagrams take an approximately linear course, like that of a perfectly elastic solid where stress is proportional to strain and where there is no yield point, ending abruptly in failure (at point F on the diagram as in Fig. 4-38a). If the relationship $\sigma = E \cdot \epsilon$ for such a material holds strictly, the material is referred to as linear-elastic. Generally, igneous rocks and sedimentary rocks under ordinary compression loads deform very little before they break. However, at high temperatures and/or pressures, rocks deform plastically. References 45 and 71 contain many papers dealing with such experimental data.

The brittle-ductile transition is of great interest in geology and geophysics in connection with the behavior of rock materials in the lower earth's crust. It is not of great interest in rock mechanics, because the necessary large pressures and high temperatures rarely are encountered in engineering practice.

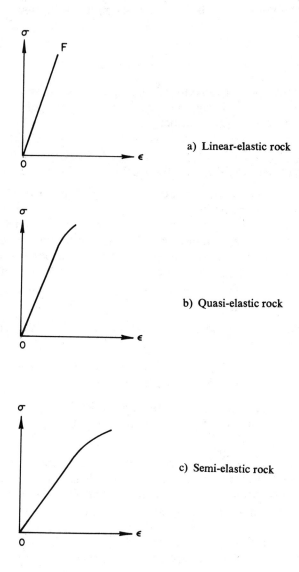

Fig. 4-38
Various kinds of stress-strain diagrams for rocks

METHODS OF ROCK EXPLORATION 179

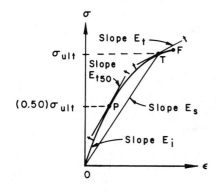

d) Generalized stress-strain diagram for rock. Perfectly elastic material showing 50 % — tangent modulus E_{t50}, and secant modulus E_s (OT)

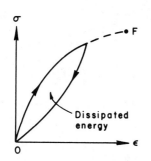

e) Elastic material with a hysteresis loop formed upon a single loading and unloading cycle

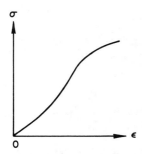

f) Nonelastic material

Fig. 4-38
Various kinds of stress-strain diagrams for rocks

In Figure 4-38b is shown a σ-ϵ-diagram for what is termed by FARMER (37) a *quasi-elastic* rock. These rocks approximate a brittle elastic material with a nearly linear σ-ϵ-relationship to the point of failure. Examples of such rocks are the sound, dense, fine-grained, extrusive igneous rocks and fine-grained metamorphic rocks.

Figure 4-38c shows a σ-ϵ-diagram in which the slope decreases as the stress increases. Termed by FARMER *semi-elastic,* these rocks represent the less elastic, coarser-grained igneous rocks and fine-grained, densified, low-porosity sediments with a reasonable amount of cohesion.

Figure 4-38d illustrates a stress-strain diagram for a perfectly elastic material. Because of its nature, the diagram shows three kinds of modulus of elasticity, namely:

1. E_i — the initial tangent modulus at zero load
2. E_t — the tangent modulus at a particular point T on the stress-strain diagram for a specified stress (also shown is the 50% tangent at point P), and
3. E_s — the secant modulus for a particular point T.

For rock, normally the initial tangent modulus E_i is referred to because it is the most accurately obtained under test conditions.

Figure 4-38e shows a stress-strain diagram for an elastic material with a hysteresis loop formed by a single loading and unloading cycle.

Figure 4-38f depicts a variable stress-strain diagram for nonelastic material such as a less cohesive and weak sedimentary rock with large void spaces.

A different kind of testing methodology applied to the same kind of rock will result in different stress-strain diagrams, of course.

Creep

No material is perfect. All materials combine in some proportion the characteristics of elasticity, plasticity, and flow.

Plastic flow of rock is of particular interest in rock engineering in the design of underground openings at great depth below ground surface. When a material deforms slowly in a continuous way, this kind of permanent deformation is known as creep—a deformation phenomenon when strain increases linearly with time.

Creep—a complex response of strain to stress—is the time-dependent movement (plastic displacement) of rock under a sustained load. In other words, creep is the phenomenon of increase in strain during the course of time under constant stress. The amount of creep depends upon the stress level: at a high stress, creep accelerates, and failure is attained quickly.

It appears that no general equation exists that adequately defines the rheological properties of rocks. The various rheological formulas in the technical literature do not give any consideration for accelerated strain. For the same reason, the behavior of most of the real rock materials cannot be represented by rheological models. However, rheological models in rock mechanics increase our awareness of the variety of strain-time patterns that might exist in rocks.

Probably the most accurate creep data on rock to be used in design are obtained experimentally.

Figure 4-39 illustrates elastic and creep strains ϵ at constant stress σ as a function of time. If strain increases linearly with time, the deformation is known as creep. Here, in this experiment, essentially a constant stress (σ = const) is applied to a rock specimen, and the corresponding strain ϵ is a function of time; i.e., $\epsilon = f(t)$.

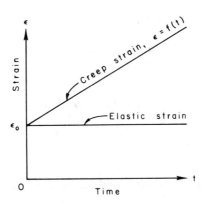

Fig. 4-39
Elastic and creep strains at constant stress

The creep test, generally, is one of more qualitative than quantitative nature, because it describes behavioral characteristics of rock rather than furnishes numerical values.

Figure 4-40 illustrates time-dependent deformation at constant stress. In time-dependent deformation such as in rock creep, three major phases

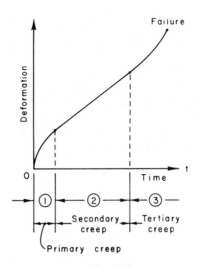

Fig. 4-40
Time-dependent deformation at constant stress

of strain responses to a constant stress can be visualized, namely:

1. Primary or transient creep,
2. Secondary or steady-state creep (pseudo-viscous flow) at a medium high stress), and
3. Tertiary or accelerated phase of creep—the phase of failure—occurring at high pressure.

In the primary phase, strain increases at a decreasing rate. In the second phase, the rate of strain is constant. During the tertiary phase of creep, strain increases at an increasing rate until the specimen fails or ruptures. This accelerated rate of strain in the tertiary phase occurs when the specimen becomes weaker and increasingly fractured.

4-23. Appearance of Deformed Rock Specimens

Some typical appearances of deformed rock specimens of brittle fracture to ductile flow by compression are shown schematically after GRIGGS and HANDIN (1960) in Fig. 4-41.

KÁRMAN's tests (69) on marbles and sandstones have shown, though, that the same material may be brittle or plastic, depending upon the magnitude of the confining pressure. Thus, testing conditions such as, for example, increase in either confining pressure σ_3 or in temperature, increase the ductility of the rock specimen and bring about the material to transform toward uniform flow (87).

METHODS OF ROCK EXPLORATION 183

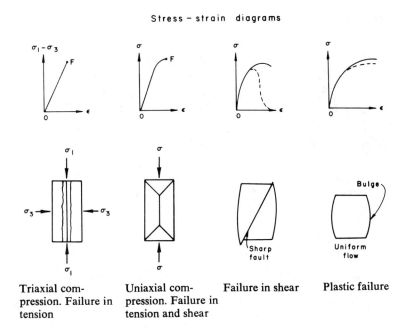

Fig. 4-41
Typical appearance of deformed rock specimens of brittle fracture to ductile flow

Because of the varied appearance of the stress-strain diagrams, it is impossible to conclude from such diagrams alone the nature of such rock deformation.

Depending upon the constraint offered by the load-bearing platens of the compression testing machine, as well as upon the quality of the parallel end surfaces (smooth or rough), rock core specimens tested for their unconfined compression strength fail either in tension or in shear (Fig. 4-42). The greater the friction between the platens and the rock test specimen end surfaces, the more likely the specimen will fail in shear.

Fig. 4-42a shows an irregular, longitudinal splitting of a rock core specimen resulting from an unconfined compression test.

Fig. 4-42b shows a rock failure in shear. Its characteristic is shear displacement along the surface of rupture, viz., shear plane corresponding to a geological fault.

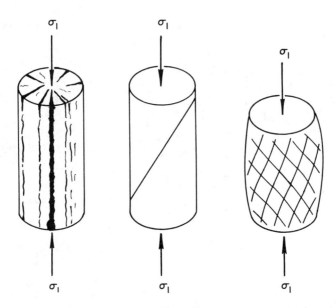

Fig. 4-42
Some modes of failure of rock in unconfined compression

Fig. 4-42c depicts a multiple shear failure; the plastic deformation is accompanied by a network of shear planes.

Fig. 4-43 shows the appearance of some deformed rock cores subjected to unconfined compression test. Specimen a), a pegmatite gneiss, failed by axial splitting. Specimen b), also a pegmatite gneiss, failed from the characteristic cones formed on the shear planes. Specimen c), a Vermont marble, and specimen d), a Triassic Brunswick shale, failed in shear.

From the discussion on σ-ε relationships, it becomes apparent that generally rock deformation properties cannot be studied on the basis of merely simplifying assumptions, as is usually done for metals. Nor can rock deformations be described by a single parameter. This is so because of heterogeneity and anisotropy of rocks, and the testing methodology used. Also, rock plasticity greatly affects rock deformation and stress reversibility; viz., full recovery of strain after unloading the rock is not ideally possible even in the elastic range of testing.

METHODS OF ROCK EXPLORATION

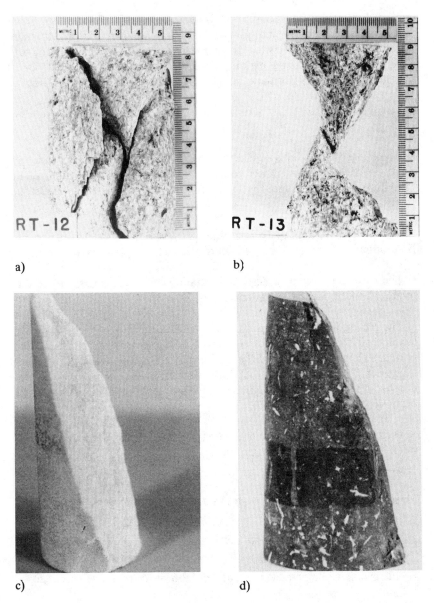

Fig. 4-43
Deformations of some rock core specimens:
a) and b) Pegmatite gneiss
c) Vermont marble
d) Triassic Brunswick shale

4-24. Modulus of Elasticity of Rocks

Knowledge about the modulus of elasticity of rocks is an important factor in evaluating rock deformation under various loading conditions. It is also necessary in studies of seismics in rocks. The modulus of elasticity of rock varies from one geologic region to another because of the existence of various types of rock formation. For this reason alone, different values of the modules of elasticity of different rocks can be anticipated. Hence, there exists a need for the determination of the numerical values of the modulus of elasticity of rocks.

Generally, the modulus of elasticity of rock is affected by rock type, porosity, particle size, and water content. The variation in the values of the modulus of elasticity is more pronounced in clayey shales than in sandstone, for example.

The modulus of elasticity is greater perpendicular to stratification and fissure than that determined parallel to the bedding planes and fissures.

The various modes of genesis or formation of rocks also affect the magnitude of the modulus of elasticity.

The reason for high modulus of elasticity values of some rocks probably lies in their structures themselves, as well as in their mineral chemical compounds. The known values of the modulus of elasticity of various rocks may suggest that eruptive igneous rocks of basic nature—like basalts, for example—have a larger modulus value than acidic rocks—such as granite, for example.

Technically, the modulus of elasticity of a rock can be increased considerably by means of rock grouting.

Elasticity constants E and μ of rocks are determined by:
1. static tests
2. longitudinal resonance tests
3. electrical resistance (strain gauge) tests
4. ultrasonic velocity measurements on laboratory rock test specimens, and
5. dynamic tests, in which velocities of propagated energy waves in situ are measured. Measurements of the primary wave (P-wave) and secondary wave (S-wave) velocities thus afford the indirect determination of E and μ of geologic materials in situ.

TABLE 4-13
Elasticity Constants E and μ of Some Rocks

Rock	Young's modulus of elasticity E		Poisson's ratio	References
	kg_f/cm^2	N/m^2	μ	
	Multiply by 10^5	Multiply by 10^{10}	Average values	
1	2	3	4	5
Igneous rocks				
Basalt	2.0 – 10.0	1.96 – 9.81	0.14 – 0.25	37, 63, 68, 107
		4.85 – 11.15	0.22 – 0.25	18
Diabase	3.0 – 9.0	2.94 – 8.83	0.125 – 0.25	42, 63
		–	0.333	68
		2.20 – 11.40	0.103 – 0.184	68
		8.00 – 10.75	–	18
Gabbro	6.0 – 11.0	5.88 – 10.78	0.125 – 0.25	37, 63, 68
		5.84 – 8.71	0.154 – 0.48	18
Granite	2.6 – 7.0	2.55 – 6.86	0.125 – 0.25	37, 63
		–	0.155 – 0.338	18
		–	0.150 – 0.240	112, 68
		2.13 – 7.05	–	18
Syenite	6.0 – 8.0	5.88 – 7.85	0.25	37
		6.29 – 8.63	0.17 – 0.319	18
			0.15 – 0.34	68
Sedimentary rocks				
Dolomite	2.0 – 8.4	1.96 – 8.24	0.08 – 0.20	37, 63
		7.10 – 9.30	0.08 – 0.20	68, 107
		–	0.32 – 0.37	18
Limestone	1.0 – 8.0	0.98 – 7.85	0.10 – 0.20	63, 107
		–	0.16 – 0.23	68
		–	0.33 –	63
		0.80 – 2.10	0.14 – 0.30	18
Sandstone	0.5 – 8.6	0.49 – 8.43	0.066 – 0.125	112
		–	0.230 – 0.300	56
		–	0.17	63, 68
		–	0.07	112
		–	0.62	56
Shale (clay)	0.8 – 3.0	0.78 – 2.94	0.11 – 0.54	37, 68
		0.98 – 2.35	0.10	56
		1.20 – 4.40	0.23 – 0.30	107
		–	0.04 – 0.12	18
Sandstone	4.5 – 5.2	4.41 – 5.10	0.21 – 0.24	63

TABLE 4-13 (Continued)
Elasticity Constants E and μ of Some Rocks

Rock	Young's modulus of elasticity E		Poisson's ratio	References
	kg_f/cm^2	N/m^2	μ	
	Multiply by 10^5	Multiply by 10^{10}	Average values	
1	2	3	4	5
Metamorphic rocks				
Gneiss	2.0– 6.0	1.96– 5.88	0.091–0.25	37, 63
	2.5– 6.0	2.45– 5.88	0.11	112, 107
		1.42– 7.00	0.03 –0.15	56
		–	0.09 –0.20	107
Marble	6.0– 9.0	5.88– 8.83	0.25 –0.38	112, 68
		– 8.50	0.25	
		4.93– 8.70	0.16 –0.27	18
		2.8 –10.00	0.11 –0.20	63
Quartzite	2.6–10.2	2.55– 8.70	0.23	112
		2.80– 8.70	0.11 –0.20	63
		9.75	0.15	18
Schist	4.1– 7.2	4.0 – 7.05	0.01 –0.20	18
		–	0.08 –0.20	112, 68
		–	0.10 –0.17	63
Slate			0.06 –0.44	

For the purpose of one's orientation, some values of the modulus of elasticity E and Poisson's ratio μ for various kinds of rocks as found in the technical literature are compiled in Table 4-13 (18, 27, 37, 55, 56, 63, 68, 90, 103, 107, 112, 113, 114).

The variation in the values of the elasticity constants of rocks may be partly attributed to the nonhomogeneity and anisotropy of rocks and partly to rock testing methodology.

4-25. Poisson's Ratio

Poisson's ratio $\mu = 1/m$ or Poisson's number $m = 1/\mu$ is another important quantity in the theory of elasticity. Poisson's ratio μ is defined as the ratio of induced lateral strain ε_x to the longitudinal strain ε_z in the direction of applied axial load:

$$\mu = \frac{\varepsilon_x}{\varepsilon_z} = \frac{1}{m} \tag{4-57}$$

where $m = 1/\mu =$ Poisson's number.

For small deformations, a material whose Poisson's ratio is $\mu = \frac{1}{2}$ is referred to as incompressible. Poisson's ratio for rocks varies according to the nature of deformation. Its value is usually relatively small. For sound and hard rocks within the elastic domain, Poisson's ratio is of the order of magnitude of about 0.15.

It increases near failure to about $\mu = 0.30$, and at constant deformation to approximately $\mu = 0.50$.

Elasticity constants of anisotropic crystals may be found in the *American Institute of Physics Handbook* (18).

Relative to rock materials in situ, it should also be said that the so-called elasticity constants such as the modulus of elasticity E and Poisson's ratio μ are not constants at all.

In reality, rather, these constants are variable quantities. This becomes clear if one remembers that rock is a nonhomogeneous, anisotropic material in different directions (rock defects). Therefore the modulus of elasticity, too, is different in different directions. These factors, as well as the effect of rock grain interlocks and wedging all affect the magnitude of Poisson's ratio, also.

To conclude the discussion on deformation of rocks, it should be said that because rock is not a perfectly linear-elastic material, certain allowances and intelligent choices must be made for deviations from perfect, ideal elasticity based on the engineer's personal experience, thus reconciling theory with reality and rendering a rational approach to design in rock.

4-26. Strength Properties of Rocks

For engineering design in rock, rock performance involves the determination of appropriate elastic constants and strength properties of the rock.

The ability of a material to resist externally applied forces is called its strength. In ordinary engineering practice, strength may be regarded as the force per unit area necessary to bring about rupture at given environmental conditions.

As applied to rock, the term "strength" is somewhat relative. It can be defined only when all the strength-governing factors such as rock environment; size of rock specimen; kind, intensity and duration of load; all-around lateral confining pressure; temperature; pore-water pressure; and failure criteria are known. The strength of small specimens of sound rock depends entirely on the strength of the rock matrix.

Needless to say, the strength properties of a rock are also governed by the qualitative and quantative mineral composition of the rock. Besides hardness of minerals and durability of rock, there are other factors, too, affecting the mechanical properties of rock. These are:

1. Type of rock;
2. Locality and environment of the rock;
3. Internal strength of individual mineral particles;
4. Strength of mutual bonding of the mineral grains of the rock;
5. Orientation of the mineral crystals and grains relative to loading and subsequent lateral deformation and/or sliding (this condition is especially pronounced with shale and slaty rocks easily cleavable and internally slidable minerals [mica]);
6. Various rock defects such as joints, cracks, haircracks, fissures, voids, gaps, and pores of all sorts;
7. Degree of rock saturation with water;
8. Coarseness of rock particle sizes and their strength;
9. Rock elasticity;
10. Plasticity;
11. Initial stress of rock in situ;
12. Testing methodology (16);
13. Rate of testing; and
14. Time.

One distinguishes between rock
1. Static strength, and
2. Dynamic strength.

The strength of a rock may be established either experimentally by means of laboratory testing of intact rock specimens, and/or by rock-testing in situ (4-15, 32-37, 44, 47). Rock strength tests in situ are performed because a rock may contain various kinds of rock defects and planes of slippage (74).

Static Strength

Rock tests are made for the purpose of obtaining numerical values of their physical and strength properties to be used for design in rock.

The strength of a rock is determined in a variety of ways. For elastic materials, strength is described by means of the elastic constants E and μ, which are characteristic of each material.

Elastic constants are determined by static methods of test, or by means of dynamic methods; in the latter, velocities of propagated wave energy are measured.

Also, strength tests on rocks may be performed in the laboratory, and/or in situ (4-15, 88-92, 93, 96, 105-109, 113, 114).

Important static, laboratory-tested strength properties of rocks are:

1. Compressive strength:
   a. Uniaxial or unconfined compressive strength, and
   b. Triaxial compressive strength.
2. Direct shear strength (involves coefficient of friction f),
3. Tensile strength,
4. Bending or flexural strength, and
5. Punching strength.
6. Sometimes, also, simple torsion tests on rock are made.
7. Thermal stresses and strains,
8. Plastic properties, and
9. Creep.

4-27. Static Laboratory Compressive Strength

There exist unconfined and confined methods of compression tests.

The laterally *unconfined*, uniaxial compressive strength test is generally known to be one of the most rigorous strength tests of a material. It is also the commonest method for studying the mechanical properties of rocks, and is relatively inexpensive to perform.

The unconfined compression test is performed on cylindrical, or prismatic, or cube rock specimens (Fig. 4-44) (13) by compressing or loading them to failure. The commonly used height-to-diameter ratio of cylindrical rock specimens is $h/d = 2.5$ to 3.0. Upon failure, the rock specimen usually fractures by axial, brittle splitting, or fails in shear, depending upon the degree of the end contraints at the ends of the rock specimen offered by

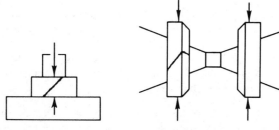

i) Loading of rock fragments ii) Loading of rock pillars in an underground opening

a) Natural loading in situ

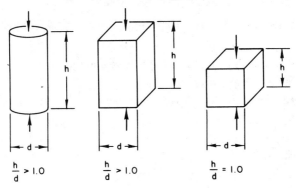

$\frac{h}{d} > 1.0$ $\frac{h}{d} > 1.0$ $\frac{h}{d} = 1.0$

b) Unconfined compression testing of rock cylinders, prisms, and cubes in the laboratory

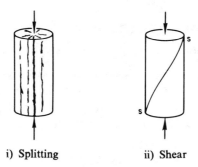

i) Splitting   ii) Shear

c) Modes of failure

Fig. 4-44
Rock test specimens for unconfined compression tests

METHODS OF ROCK EXPLORATION

the platens of the testing machine and the surface quality of the parallel ends of the rock specimen receiving the load (Fig. 4-44c).

Upon failure of the rock specimen at the ultimate compressive stress σ_1, when the shear strength in its potential shear plane s-s is exhausted, the sheared-off parts slide past each other across the inclined s-s plane; i.e., the rock fractures or fails in shear.

For any axial stress applied to the rock core specimen, the axial and lateral strains are measured either by strain gauges fitted on the test specimen, or by measuring displacements. These data are used for determining the POISSON's ratio of the tested rock specimen. Also, deformations are recorded for plotting of stress-strain diagrams to determine the modulus of elasticity E of the tested rock specimen.

The unconfined compression test results also render indirectly the unconfined shear strength $s = \tau$, and the test parameters φ and c (if any), commonly known as the angle of friction and cohesion, respectively, as a function of rupture angle α (Fig. 4-45). Friction on the rupture surface supports a considerable amount of axial stress. Cohesion may be regarded as the tangential strength (no-load strength).

The unconfined shear strength s of a rock is expressed by COULOMB's shear strength equation (Fig. 4-45):

$$s = \tau = \sigma_n \cdot \tan \varphi + c \qquad (4\text{-}58)$$

When $c = 0$,

$$s = \tau = \sigma_n \cdot \tan \varphi \qquad (4\text{-}59)$$

where

$$\sigma_n = \frac{\sigma_1}{2}(1 + \cos 2\cdot\alpha) \qquad (4\text{-}60)$$

is the effective normal stress on the shear plane. From geometry ot MOHR's stress diagram,

$$2 \cdot \alpha = \frac{\pi}{2} + \varphi \qquad (4\text{-}61)$$

and

$$\varphi = 2 \cdot \alpha - \frac{\pi}{2} \qquad (4\text{-}62)$$

The study of friction is of utmost importance in rock mechanics.

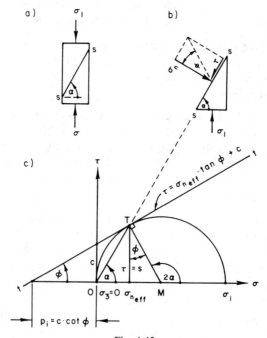

Fig. 4-45
Unconfined compression test

a) Failure of rock specimen
b) Free-body diagram of rock element
c) MOHR's stress diagram and COULOMB's shear strength envelope (tangent $t-t$)

The test parameter φ, viz., $\tan\varphi$, the coefficient of friction, calculates from MOHR's stress diagram as

$$\tan\varphi = \frac{\tau}{\sigma_n + p_i} = \frac{\frac{\sigma_1}{2} - \sigma_n}{\tau} = f \qquad (4\text{-}63)$$

where

$$p_i = c \cdot \cot\varphi \qquad (4\text{-}64)$$

is the initial stress in the rock, and

$$f = \tan\varphi \qquad (4\text{-}65)$$

is the coefficient of friction. From here,

$$\varphi = \arctan(f) \qquad (4\text{-}66)$$

The cohesion c calculates from MOHR's stress diagram as

$$c = \tau - \sigma_n \cdot \tan \varphi \tag{4-67}$$

or

$$c = \tau \cdot \frac{p_i}{p_i + \sigma_n} \tag{4-68}$$

The scatter in unconfined compression test results may be taken as

1. a qualitative indication about the rock in situ
2. a variation in the heterogeneity of rock strength from one domain to another one at the site
3. an indication of the probable anisotropy of the full-scale mass of rock, and
4. an indication of the microfracturing of the rock brought about by tectonic activities at the site.

Thus the unconfined compression test is one of practical utility.

The *triaxial* compressive strength is tested by subjecting a rock specimen to a lateral, all-round confining pressure σ_3 and loading the specimen axially to failure (σ_1). This failure stress σ_1 is referred to as the triaxial compression strength of the rock specimen. The description of the general procedure of performing the triaxial compression test may be found in the book on ASTM standards (15). See also Ref. 99.

Also in the triaxial compression test, the rock specimen may fail either by axial, brittle splitting, or in shear by the characteristic cones formed on the shear surfaces.

Also, from the triaxial compression test data it is possible to evaluate the effective normal stress ($\sigma_{n\,eff}$) and the shear stress (τ) on the rupture plane, as well as the test parameters such as the angle of friction φ and cohesion c (if any) as a function of the angle of rupture a of the rock specimen. Thus, the triaxial compression test, too, gives indirectly the shear strength of the rock.

The shear strength s in this test is (Fig. 4-46):

$$s = \tau = \sigma_{n_{eff}} \cdot \tan \varphi + c \tag{4-69}$$

The φ and c values from a triaxial test are to be determined from the appropriate geometry of MOHR's stress diagram, as in Fig. 4-46.

Triaxial testing has the disadvantage that the geometry of the test specimen changes after any slip deformation, so that continued experiments of long duration are not very possible. However, for small displacements this test is considered to be adequate. It is the method most commonly used for determining the stress parameters φ and c.

Projecting forces acting on the free-body rock element for equilibrium obtain the normal (σ_n) and shear stress (τ) components in terms of the major (σ_1) and minor (σ_3) principal stresses, and as a function of α:

$$\sigma_n = \frac{\sigma_1 + \sigma_3}{2} + \frac{\sigma_1 - \sigma_3}{2} \cdot \cos 2\alpha \qquad (4\text{-}70)$$

$$\tau = \frac{\sigma_1 - \sigma_3}{2} \cdot \sin 2\alpha \qquad (4\text{-}71)$$

When $\sigma_3 = 0$ (as in the unconfined compression test), then

$$\sigma_n = \frac{\sigma_1}{2} + \frac{\sigma_1}{2} \cdot \cos 2\alpha = \frac{\sigma_1}{2} \cdot (1 + \cos 2\alpha) \qquad (4\text{-}72)$$

Static compression tests on *cubes* render rock compressive strength only (12). Like the cylindrical specimens, so also the cubes fail either by brittle splitting, or in shear (Fig. 4-47).

Static tests which are performed during a short period of time and at relatively low loading rates (e.g., 10 psi/s = 68.94 (kN/m²)/s to 100 psi/s = 689.4 (kN/m²)/s) are referred to as *time-independent* as contrasted to *time-dependent* tests. Time-dependent tests pertain to creep tests under long sustained loading, for example.

The end faces of the test specimens must be strictly parallel to each other, and perpendicular to the generatrix of the rock cylinder, or prism, or cube.

The faces must be grounded to close tolerances. The following standards have been suggested by the International Society for Rock Mechanics' Committee on Laboratory Tests of Rocks, Document No. 1, of 1972 (51):

1. In strength tests of rock, the use of capping materials or end-surface treatments other than machining is not permitted. This is to be interpreted that the silicon grease method to eliminate the end friction effects of the rock specimens while testing is not permitted either.
2. The ends of the specimen shall be flat to 0.02 mm.
3. The ends of the rock specimen shall be perpendicular to the axis of the specimen within 0.001 radian (3.5 minutes).
4. The sides of the specimen shall be smooth and free of abrupt irregularities and straight to within 0.3 mm over the full length of the specimen.

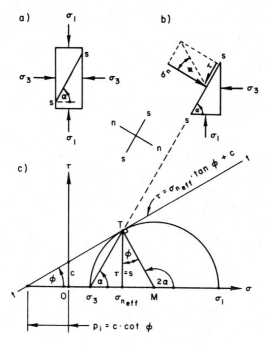

Fig. 4-46
Triaxial compression test
a) Failure of rock specimen in triaxial loading
b) Free-body diagram of rock element
c) Mohr's stress diagram and Coulomb's shear strength envelope ($t - t$)

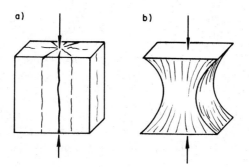

Fig. 4-47
Compression tests on rock cubes
a) Brittle splitting
b) Shear

Usually rock strength test results scatter within a wide range. This brings up the question as to how many specimens are necessary for such tests.

If identical compression tests are performed on n rock specimens, then the compressive strength σ is expressed as:

$$\sigma = \sigma_{\text{ave}} \pm \Delta\sigma \qquad (4\text{-}73)$$

where

$$\sigma_{\text{ave}} = \frac{\Sigma(\sigma_1)}{n} \qquad (4\text{-}74)$$

is the average strength of n specimens, $\sigma_1 =$ ultimate compressive strength of an individual specimen, and

$$\Delta\sigma = \sqrt{\frac{(\sigma_1 - \sigma_{\text{ave}})^2}{n-1}} \qquad (4\text{-}75)$$

is the standard deviation computed by the method of least-squares.

Generally, the compressive strength of rock is a function of the following:

1. Specimen size,
2. Shape,
3. Surface quality of bearing platens and rock specimen end surfaces,
4. Confining pressure,
5. h/d-ratio, where $h =$ height of core specimen and $d =$ its diameter,
6. Rate of loading,
7. Porosity of rock, and
8. Moisture content.

As the confining pressure σ_3 is increased, the strength of the rock increases; also, the amount of the permanent set, or irreversible deformation, before fracture increases.

According to JAEGER (55), this behavior is characteristic of many rocks, even to confining pressures of the order of magnitude of 50 kpsi = 344.75 MN/m², that is, throughout the range of importance in practical rock mechanics.

If the rock specimens having the same diameter are of different length (height), the various ultimate compression-strength data from such specimens are adjusted, for the sake of uniformity of comparison, according to the *ASTM Standard Method of Test for Compressive Strength of Natural Stone* (12), to the strength of a rock specimen having a diameter-to-height ratio of $b/h = 1.00$, as follows:

$$\sigma_{cu} = \frac{\sigma_{ult}}{[0.778 + (0.222) \cdot (b/h)]} \qquad (4\text{-}76)$$

where σ_{cu} = computed ultimate compressive strength of an equivalent cubical (or cylindrical) specimen whose ratio is $b/h = 1.00$, termed here the *adjusted compressive strength;* and

σ_{ult} = ultimate compressive strength of the test specimen whose height h is greater than its diameter $d = b$, i.e., whose b/h-ratio is less than 1; all in consistent units of measurement.

Here b is the width of the cube or prism. In the case of a cylindrical specimen, $b = d$, where d is the diameter of the rock specimen.

In Eq. (4-76), if $b/h = 1.00$, then $\sigma_{cu} = \sigma_{ult}$. Equation (4-76) also shows that the b/d-ratio, too, influences the compression strength test results. For example, for a constant-diameter rock specimen ($d = b$ = const), the strength decreases with increase in height of the specimen.

Generally, small b/h-ratio rock specimens fail because of elastic instability.

According to Moos and Quervain (81), water in rock pores reduces the magnitude of the internal friction of rocks.

Medium b/h-ratio ($\sim$ 0.3 to 0.4) rock specimens are elastically stable. In such specimens the stress distribution is fairly uniform.

Compressive strength of rock also decreases with increase in rock porosity, viz., volume of voids. This is so because at the location of voids there is a lack of bond contact between rock particles. However, at this point, it should be said that one cannot estimate the decrease in strength merely by the volume of the voids: the shape of the voids (narrow, rounded, or otherwise); the magnitude of the internal surface area of the grains because of the voids and gaps, each of which may be of different size; as well as the arrangement of the voids in the rock (interconnected or disconnected)—all these factors affect the strength of the rock.

Frequently one encounters coarse-textured granites which are very porous. Such rocks have considerably less strength than some fine-textured igneous rocks. In other words, the strength of a fine-grained rock is greater than that of a coarse-grained rock.

Basalt is one of the most competent of common rocks. It has a microfine texture, and consists of microcrystals held together by strong mechanical bonding.

Highly foliated schistose gneisses usually exhibit a wide range of strength.

Water in rock, too, affects the compressive strength and the elastic properties of rock. Generally, strength decreases with increase in moisture. Saturated rocks have less strength than dry ones. OBERT and DUVALL (90) are of the opinion that "a part of this decrease in strength with increase in moisture content may be due to an inability of the pore water to migrate freely (within the time limit of the test), and hence a pore (water) pressure may develop as the load on the specimen is increased."

The ratio of dry to wet strength is called the *softening factor*.

The author's experimental studies about New Jersey Triassic shale (68) revealed the great and relatively consistent decrease in unconfined compressive strength of the shale specimens upon soaking them for two weeks in water. The average value of soaked specimens was approximately one-third of the average compressive strength of oven-dry specimens. The softening factor S_f of dry to wet strength in these tests is:

$$S_f = \frac{\sigma_{\text{ult dry ave}}}{\sigma_{\text{ult wet ave}}} \approx 3.36 \tag{4-77}$$

Pore-water pressure in rock is a factor affecting rock strength. If the rock contains an interconnected system of water-filled pores, the fracture, viz., rock strength, is also governed by pore-water pressure, viz., effective stresses. The pore-water pressure reduces the principal stresses. In such a case, the effective principal stresses $\sigma_{1\,\text{eff}}$ and $\sigma_{3\,\text{eff}}$ are expressed as:

$$\sigma_{1\,\text{eff}} = \sigma_{1\,\text{total}} - u \tag{4-78}$$

$$\sigma_{3\,\text{eff}} = \sigma_{3\,\text{total}} - u \tag{4-79}$$

and the effective normal stress as:

$$\sigma_{n\,\text{eff}} = \sigma_n - u \tag{4-80}$$

where u = neutral stress known also by the term "pore-water pressure."

In terms of the effective stress, COULOMB's shear strength equation, Eq. (4-58), is written as:

$$s = (\sigma_n - u) \cdot \tan \varphi + c \tag{4-81}$$

The effect of pore-water pressure on failure is shown in Fig. 4-48.

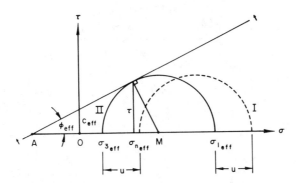

Fig. 4-48
Effect of pore-water pressure u on principal and normal stresses

Circle I is Mohr's stress diagram for actual, total stresses. Circle II is Mohr's stress diagram for effective stresses. Tangent t-t to the effective stress circle is Coulomb's strength envelope to the effective stress circle. As the pore-water pressure u is increased, stress circle I shifts to the left by an amount of u until it tangents the strength line to bring about failure. Notice that the stress difference $\sigma_1 - \sigma_3$, and the slope of the tangent, or envelope, remain unchanged.

Some compressive-strength values of rocks are compiled in Table 4-14 from various sources (37, 107).

4-28. Tensile Strength of Rock

According to the ASTM D 653-67 Standard Definitions of Terms and Symbols Relating to Soil and Rock Mechanics (10), the tensile strength of a material is defined as "the maximum tensile stress which a material is capable of developing."

Practically, the tensile strength is considered to be the maximum stress developed by a given specimen of a material in a tension test performed to rupture under specified conditions (14).

In rock mechanics, knowledge about tensile strength of rocks is important in analyzing rock strength and stability of roofs and domes of underground openings in the tensile zone of the rock (Sections 7-4 through 7-6, and 11-5), in mining for minerals, in preparing rock drilling and blasting programs, and possibly in other endeavors in rock engineering.

Tensile stresses may be induced on the underside of a rock slab or beam subjected to bending, and on the leeward side of a buckling rock pillar

TABLE 4-14

Compressive Strength σ_c, Tensile Strength σ_t, and Shear Strength $\tau = s$ of Various Kinds of Rocks

Rock	Dry, Unconfined Compressive Strength, σ_c		Tensile Strength, σ_t		Shear Strength, ($\tau = s$)		References
	kg$_f$/cm²	MN/m²	kg$_f$/cm²	MN/m²	kg$_f$/cm²	MN/m²	
(1)	(2)	(3)	(4)	(5)	(6)	(7)	(8)
Igneous rocks							
Basalt	800–4200	78–412	60–120	5.9–11.8	50–130	4.9–12.7	107
	1500–3000	147–294	100–300	9.8–29.4	200–600	19.6–49.0	37
Diabase	1200–2500	118–245	60–130	5.9–12.7	60–100	5.9–9.8	107
Gabbro	1500–2000	147–196	50–80	4.9–7.8	40–85	3.9–8.3	107
	1800–3000	177–294	150–300	14.7–29.4	–	–	37
Granite	1200–2800	118–275	40–80	3.9–7.8	50–100	4.9–9.8	107
	1000–2500	98–245	70–250	6.9–24.5	140–500	13.7–49.0	37
Sedimentary rocks							
Dolomite	150–1200	14.7–118	25–60	2.5–5.9	25–70	2.5–6.9	107
	800–2500	78–245	150–250	14.7–24.5	–	–	37
Limestone	40–2000	3.9–196	10–70	1.0–6.9	15–70	1.5–6.9	
	300–2500	29.4–245	50–250	4.9–24.5	100–500	9.8–49.0	107
Sandstone	600–1000	49.0–98	20	19.6	30	2.9	37
	200–1700	19.6–167	40–250	3.9–24.5	–	–	107
Shale							37
(clay)	220–1635	21.6–160	–	–	30–110	2.9–10.8	
	100–1000	9.8–98	20–100	2.0–9.8	30–300	2.9–29.4	107
Sandstone	500	49	–	–	–	–	37
Methamorphic rocks							
Gneiss	800–2500	78–245	40–70	3.9–6.9	30–70		107
	800–2000	78–196	80–200	7.8–19.6	–	–	37
Marble	500–1800	49–177	50–80	4.9–7.8	35–80		107
	1000–2000	98–196	70–200	6.9–19.6	150–300		37
Quartz-ite	870–3600	85–353	30–50	2.9–4.9	–	–	107
	1500–3000	147–294	50–200	4.9–19.6	200–600	19.6–58.8	37
Slate	250–800	24.5–78	–	–	–	–	107
	1000–2000	98–196	70–200	6.9–19.6	–	–	107

107. Széchy, K., 1966, *The Art of Tunneling*. Budapest: Akadémiai Kiadó Budapest, p. 76.

37. Farmer, J. W., 1968 *Engineering Properties of Rocks*. London: E. and F. N. Spon, Ltd., p. 57.

(Figs. 4-49a and b). A piece of rock may also fail in tension, as shown in Fig. 4-49c.

It is generally known that, because of the discontinuities a rock may contain, rock in its natural state is relatively weak in tension. The tensile strength σ_t of a rock is much less than its compressive strength σ_c. The tensile strength of rock is only about 10% that of its compressive strength, i.e.,

$$\sigma_t \approx (0.10) \cdot \sigma_c \tag{4-82}$$

As compared with compressive strength, rock tensile strength has not been studied extensively enough. One possible explanation for this may be the difficulty in preparing a tensile test specimen of rock; fixing it properly in the chucks of the testing machine; and centering exactly the axial, tensile load. Thus, tensile strength tests on rocks are difficult to perform.

To determine the tensile strength σ_t, the rock specimen is subjected to maximum axial tensile force P attained during the test to failure. During the loading process, the strains are measured. Then the corresponding stresses and strains are computed and a stress-strain diagram is plotted. The average tensile strength σ_t is computed as:

$$\sigma_t = \frac{P}{A_o} \tag{4-83}$$

where A_o = initial, unnecked, cross-sectional area of the rock test specimen.

In this test, failure occurs on the weakest cross-sectional plane (such as a bedding plane), or a parting, or a gouged or weakly cemented joint of the rock specimen.

So far, systematic studies on rock *flexural (bending) strength* have been performed very seldom. Flexural strength tests are performed for purposes of determining YOUNG's modulus of elasticity E, and tensile strength of the material.

Flexural strength tests on rock are performed on a test beam made of rock. The simply supported beam at its ends is loaded with a concentrated load P at its midpoint to failure. In such a loading, the underside fibers of the beam are under tension.

From strength of materials, the maximum bending (flexural) tensile stress

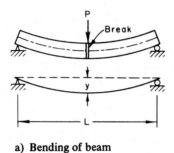

a) Bending of beam

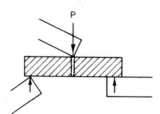

b) Buckling of column   c) Loading in tension

d) Failure of rock in bending

Fig. 4-49
Some modes of deformation of materials in tension

σ_t at the extreme fiber on the underside of a rock beam is considered to be the tensile strength σ_t of the rock:

$$\sigma_t = -\frac{M}{I/c} = -\frac{M}{Z} \qquad (4\text{-}84)$$

$$= -\frac{P \cdot \left(\frac{L}{2}\right)}{I/c} = -\frac{P \cdot c \cdot L}{2 \cdot I} \qquad (4\text{-}85)$$

where M = bending moment,
    I = central moment of inertia of cross-section of the beam,
    c = distance from neutral axis to the farthermost fiber,
    $Z = I/c$ = section modulus of the beam, and
    $L/2$ = one-half of the span of the beam.

The quantity σ_t is also known as the modulus of rupture.

It is generally known that the tensile strength of rocks obtained from flexure tests on rock beams is considerably larger than its uniaxial tensile strength.

The average modulus of elasticity E_{ave} for a central deflection y of the beam is:

$$E_{\text{ave}} = \frac{P \cdot L^3}{6 \cdot I \cdot y} \qquad (4\text{-}86)$$

For a rectangular cross-sectional beam, $I = b \cdot h^3/12$. For a circular cross-sectional beam, $I = \pi R^4/4$.

In order to take into account the difference in YOUNG's modulus of elasticity in tension and compression, DUCKWORTH (30) put forward for rectangular, ceramic beams the following equation for determining tensile strength σ_t:

$$\sigma_t = -\frac{3 M (\varepsilon_1 + \varepsilon_2)}{b \cdot t^2 \cdot \varepsilon_1} \qquad (4\text{-}87)$$

Here, M = bending moment,
    ε_1 = tensile bending strain of the outer fiber,
    ε_2 = compression bending strain of the inner fiber,
    b = width of the gauge section, and
    t = thickness of the gauge section.

Another, indirect method of testing rock tensile strength is the so-called *Brazilian test* (34, 37, 54, 55, 113). In this test, a cylindrical rock specimen, lying on its side, is loaded diametrically with a compression load P, as sketched in Fig. 4-50, to bring about a uniformly distributed tensile stress over the vertical, central, diametrical plane. Ideally, failure in tension should take place as a clean break. The tensile strength σ_t of the rock in this test is:

$$\sigma_t = \frac{2 \cdot P}{\pi \cdot d \cdot h} \tag{4-88}$$

where d = diameter of specimen, and
 h = height (or length) of the specimen.

For one's orientation, some tensile strength and bending strength values for various kinds of rocks are compiled from various sources in Table 4-14.

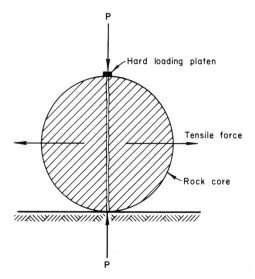

Fig. 4-50
Indirect (Brazilian) tensile strength test

4-29. Shear Strength of Rock

There are many design problems in rock mechanics where knowledge about the shear strength of a rock mass is needed. Among them are:
1. Stability problems of underground openings,
2. Assessment of the degree of stability of rock slopes, and
3. Evaluation of stability of a structure against sliding on its base laid on rock.

METHODS OF ROCK EXPLORATION 207

Some modes of shear failure in rock are sketched in Fig. 4-51.

The shear strength parameters to use in stability calculations of a rock mass and/or structure, in or on rock, point to the kind of shear strength test to be performed (68, 91, 93, 96, 106).

The shear strength of rock is its maximum resistance s to deformation by continuous shear displacement upon the action of a shear (tangential) stress τ.

Upon exceeding the shear resistance, or strength s, failure of rock in shear takes place.

The shear strength s of a rock is the sum of:

1. The surface frictional resistance to translation in the sliding surface,
2. The interlocking effect between the individual rock grains, and
3. Cohesion, if any, in the sliding surface of the rock.
   (Cohesion, it may be said, is the no-load shear strength of the rock.)

When motion, viz., displacement in shear, is just impending, there exists a state of static equilibrium, i.e., the shearing stress τ is equal in magnitude and oppositely directed to the shear strength s:

$$\tau = s \qquad (4\text{-}89)$$

One should be cognizant that the applied stress and the strength stress in these tests are average, uniform stresses, and that they are *statistical averages* that do not represent the actual stresses on the microscopic parts of the rock. The stress is, as it were, "ragged" within microscopic irregularity within each kind of rock.

Methods of Shear Testing of Rocks in the Laboratory

All shear tests on rocks may be grouped in the following groups:

A. Direct shear strength tests with normal stress on the shear plane absent,
B. Direct shear strength tests with normal stress on the shear plane present, and
C. Torsion tests.

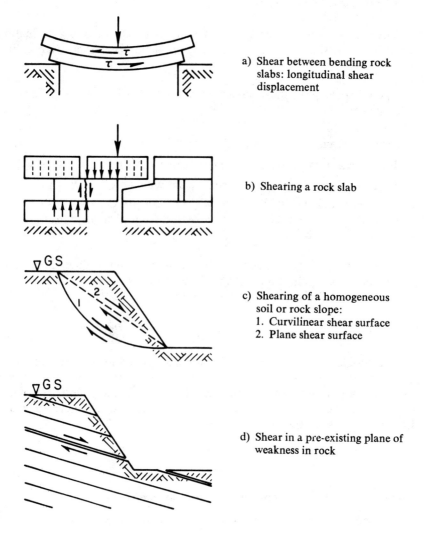

Fig. 4-51
Some modes of shear failure in rock

METHODS OF ROCK EXPLORATION

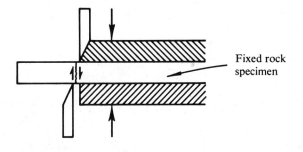

a) Single shear

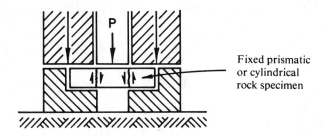

b) Double shear

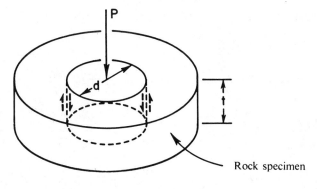

c) Punch shear

Fig. 4-52
Some principles of direct shear

A. Direct Shear Strength Tests with Normal Stress on the Shear Plane Absent

Because in this kind of test the normal stress on the shear plane is zero ($\sigma_n = 0$), the test is also termed the "unconfined" direct shear test. To this kind of test, there belong the following three kinds of direct shear tests:

1. Direct shear test in single shear,
2. Direct shear test in double shear, and
3. Punch shear.

The principles of these direct shear tests are illustrated in Fig. 4-52.

1. Single-Shear Test

The direct shear strength in single shear may be determined by placing a rock specimen between knife edges and exerting through them colinear pressure P on the specimen (Fig. 4-52a). The specimen fails along a predetermined, forced shear plane. The shear strength s of the rock is

$$s = \frac{P}{A} \tag{4-90}$$

where A = cross-sectional area of the shear plane of the tested rock specimen.

2. Double-Shear Test

This kind of uniaxial shear test is sketched in Fig. 4-52b. The prismatic or cylindrical rock specimen is fixed firmly into the specimen-holding jig. A shear force P is applied to the specimen through a die, bringing about shear failure in double shear. The shear strength s calculates as:

$$s = \frac{P}{2 \cdot A} \tag{4-91}$$

where A = cross-sectional area of the rock specimen.

3. Punch Shear

For the punch shear test (Figs. 4-52 and 4-53), flat rock disks are fitted into the testing jig (Fig. 4-53). The test is commenced by placing a rock specimen at the bottom of the casing of a piston-shaped cylindrical jig having a puncher head which fits into a circular shear hole of another hollow cylinder. Then the puncher is placed in position and the entire device is placed in the compression-testing machine and loaded. The punching shear strength s of the rock specimen is:

$$s = \frac{P}{A} = \frac{P}{\pi \cdot d \cdot t} \qquad (4\text{-}92)$$

where P = ultimate punching load,
$\quad\quad\;\; A$ = circumferential shear area of punched-out circle of the thickness t of the disks, and
$\quad\quad\;\; d$ = diameter of the puncher.

Fig. 4-53
Device for testing rock disks in punch shear

Some Triassic-shale disks after failure in punch shear resulting from the author's own experimental studies are shown in Fig. 4-54 (68).

Shale disk specimens tested out with the following average punching shear strength values:

$\quad\quad\quad\quad$ Dry: $\quad s_{\text{ave}} = 337$ kg/cm²
$\quad\quad\quad\quad$ Wet: $\quad s_{\text{ave}} = 114$ kg/cm²

a) Shale disks before test

b) Shale disks after test

Fig. 4-54
Shale disks for punch shear tests
(Author's study, Ref. 68)

The dry/wet factor was:

$$S_t = \frac{337}{114} = 2.96$$

Hence the dry specimens were approximately trice as strong in punch shear as the wet ones. After the punching tests, the deformed shale disks showed radial deformation patterns.

The advantage of the disk-punching method of shear strength is that it is easy to perform. This kind of test may be considered as a realistic representation of punching through a hard shale underlaid by a softer shale stratum.

A disadvantage of this kind of test is that the shear failure process cannot be observed. Also, in the punching shear test, the real stress distribution along the shear surface is very complex, and the shale test results scatter widely, deviating from the average from ± 28 to $\pm 37\%$. Hence only the average punching shear strength data for the shale could be obtained.

B. Direct Shear Strength Tests with Normal Stress on the Shear Plane Present

Among these direct shear strength tests, one distinguishes between:

1. The so-called "box" shear strength tests on rock, and
2. Shear strength tests made on rock cubes.

The shear deformation in these tests take place along a predetermined, forced shear plane.

1. Direct "Box" Shear Strength Test on Rock

The so-called "box" shear strength test involves the fitting, snugly, of a rock specimen into a direct shear-testing box (Figs. 4-55a and b). The box consists of two frames, an upper and a lower. The rock specimen is being loaded with a vertical, normal load N, viz., normal stress $\sigma_n = N/A$, where A = area of shear plane. During the test, the normal load N is kept constant; a horizontal shear (tangential) force T, viz., shear stress $\tau = T/A$, is applied to the upper frame of the shear box; the shear force increments of T are recorded; and the corresponding horizontal displacements are measured. At failure, the shearing stress τ is equal to the shear strength s of the rock.

The plot of $T = f(N)$, or $\tau = f(\sigma_n)$, results in COULOMB's straight-line shear strength relationship (Fig. 4-55c).

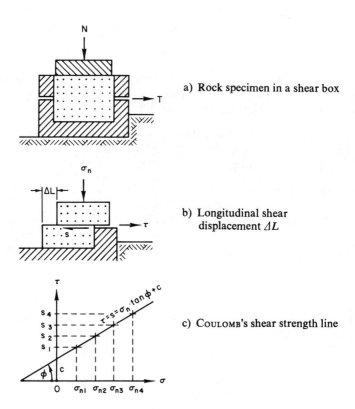

Fig. 4-55
Direct "box" shear strength test

The straight-line relationship between the horizontal shear force T and vertical normal load N is given as:

$$T = N \cdot \tan \varphi + c \cdot A \qquad (4\text{-}93)$$

or

$$\tau = s = \sigma_n \cdot \tan \varphi + c \qquad (4\text{-}94)$$

where the shear strength test parameters are:

$\tan \varphi$ = coefficient of friction of the shear surface,
φ = angle of friction,
c = cohesion (if any), and
A = area of shear.

METHODS OF ROCK EXPLORATION 215

Cohesion, it may be said, is the no-load shear strength.

If $c = 0$, then

$$\tau = s = \sigma_n \cdot \tan \varphi \qquad (4\text{-}95)$$

The $\tan \varphi = f$ value depends upon the quality of the sliding, viz., contact surface between the mutually displacing upper and lower parts of the rock specimen deforming in shear, i.e., whether smooth or rough, level or inclined, dry or wet, quickly or slowly sheared.

If the rock specimen is tested along a forced inclined plane (Fig. 4-56a), the shear strength of the specimen is written as:

$$s = \sigma_n \cdot \tan(\varphi + i) \qquad (4\text{-}96)$$

where i = angle of slope with the horizontal of the forced inclined shear plane.

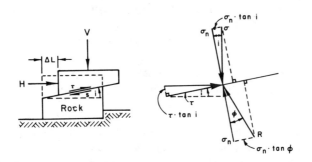

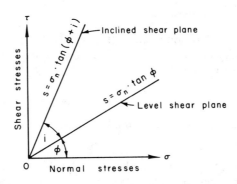

Fig. 4-56
Direct shear test along a forced inclined plane

For such a case, the corresponding plot of $\tau = f(\sigma_n)$ is shown in Fig. 4-56b.

If the direct shear test is performed in water under submerged conditions, then the total hydrostatic uplift or neutral force U reduces the total normal force. Therefore the frictional resistance to sliding is also reduced. Hence, the shear force $\overline{T}$ for impending sliding is:

$$\overline{T} = (N - U) \cdot \tan \varphi \qquad (4\text{-}97)$$

Expressed in terms of effective stresses,

$$\overline{\tau} = (\sigma_n - u) \cdot \tan \varphi \qquad (4\text{-}97\text{a})$$

$$= \sigma_{n_{\text{eff}}} \cdot \tan \varphi \qquad (4\text{-}97\text{b})$$

$$= \overline{\sigma}_n \cdot \tan \varphi \qquad (4\text{-}97\text{c})$$

where $\overline{\tau}$ = shear stress on effective stress basis,
σ_n = normal stress, and
$\overline{\sigma}_n$ = $\sigma_n - u = \sigma_{n_{\text{eff}}}$ = effective normal stress.

The magnitude of the total normal load N and the total neutral pressure U acting on the failure surface must be determined from rock exploration work in the field.

When working in the laboratory with a controlled-strain, direct shear testing device, or when working in situ, the test data for each single run of the test result in a shear stress-displacement diagram, as shown in Fig. 4-57a.

The angle φ_r of residual shearing resistance (Fig. 4-57b) is frequently used in stability calculations of jointed rock masses against their sliding along planes of weakness (93).

2. Direct Shear Strength Tests of Rock Cubes

This test is performed on prepared rock cubes. The loading of the cubes in this direct shear test is accomplished by means of a universal compression testing machine and a special device or jig (rock cube specimen holder, Fig. 4-58a) (68). The rock cube is placed in the holder diagonally at an angle of $\alpha = 45°$ with respect to the horizontal. Other jigs are designed for

METHODS OF ROCK EXPLORATION

other angles with the horizontal (Fig. 4-58b). The bottom parts of these jigs are mounted with two loops of spherical bearings as a movable displacement base.

In the 45-degree jig, the loading device transfers the vertical load P to the rock specimen, forcing the rock cube to shear and fail in shear along a predetermined, forced-shear plane *s-s* (Fig. 4-59).

The normal stress σ_n is here calculated as:

$$\sigma_n = P \cdot \sin 45°/A = N/A \qquad (4\text{-}98)$$

and the shear stress τ as:

$$\tau = P \cdot \cos 45°/A = T/A \qquad (4\text{-}99)$$

(see Fig. 4-59), where N and T are forces on the shear plane and A is the area of the shear plane.

This kind of direct shear test is designed to bring out the shear strength of the rock cubes parallel and perpendicular to the laminarity of the rock. Thus this test is a forced direct shear test.

Figure 4-60 shows some shale specimens after failure in the 45-degree direct shear test.

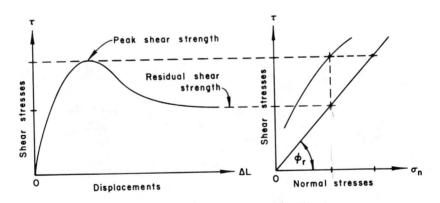

a) Shear strength as a function of displacement ΔL at constant normal stress σ_n

b) Angle φ_r of residual shear resistance

Fig. 4-57
Shear stress-displacement diagram and angle of residual shear resistance φ_r

a) Rock cube holder for a 45-degree forced-failure plane with the horizontal

b) Rock cube holders for various angles of forced-failure plane with the horizontal

Fig. 4-58
Devices for testing rock cubes in direct shear
(Author's study, Ref. 68)

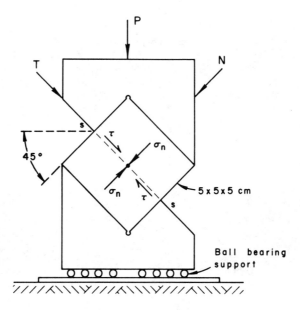

Fig. 4-59
Schematic rendering of a rock cube specimen holder
(Author's study, Ref. 68)

With jig inside inclinations other than 45 degrees, the rock can be forced to shear at different angles with the horizontal.

C. Torsion Tests

In the torsion tests, a solid-shaft cylindrical rock specimen, provided with square, prismatic ends for easy fixing into the chucks of the torsion-testing machine, is subjected to torsion or twisting. Upon torsion, the shear stress in the circular-shaft cross-section increases. It increases linearly from zero at the center of the circular cross-sectional area of the rock specimen to a maximum value τ_{max} at the outermost fiber of the shaft, where $r = r$ (Fig. 4-61):

$$\tau_{max} = \frac{(16) \cdot T}{\pi \cdot d^3} \qquad (4\text{-}100)$$

where $T = P \cdot (2 \cdot r) =$ torsion moment (applied force couple),
  $P =$ force,
  $d = 2 \cdot r =$ diameter of the solid-shaft rock specimen.

Fig. 4-60
Shale cube specimens after failure in direct shear test
(Author's study, Ref. 68)
Top: wet specimens
Bottom: dry specimens

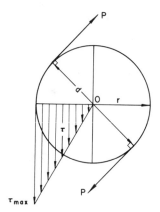

Fig. 4-61
Distribution of shear stresses in a cross-section of a circular,
solid-shaft test specimen

For a rectangular shaft, a satisfactory formula for computing the maximum shear stress is (102):

$$\tau_{max} = \frac{T}{a \cdot b^2} \cdot \left(3 + 1.8 \cdot \frac{b}{a}\right) \tag{4-101}$$

where a = long side of the rectangular section, and
 b = its short side.

For one's orientation, some shear strength values of various rocks are compiled from various sources in Table 4-14.

Friction

Resistance to sliding, viz., shear parallel to a geological discontinuity, is almost entirely frictional in character. Therefore, in rock mechanics and foundation engineering, the study of friction is of paramount importance (91, 113).

Friction occurs:

1. Between opposing surfaces of minute cracks in rock;
2. Between individual rock and soil particles; and
3. On joint or fault surfaces, small as well as on very large ones.

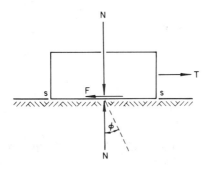

Fig. 4-62
Block at static equilibrium

So far, few studies have been devoted to friction on minerals and rocks.

From statics, at equilibrium (Fig. 4-62), the force of shearing resistance F is expressed as:

$$F = N \cdot f = T \tag{4-102}$$

where N = normal force of the block (including its own weight) on the contact surface s-s, whose area is A;

$f = \tan \varphi$ = coefficient of static friction; here f depends upon the quality of the contact surfaces, such as new or old surfaces, smooth or rough, for example;

T = externally applied tangential or shear force parallel to the surface of contact s-s necessary to initiate a sliding movement;

$F = N \cdot f$ = resistance to shear force T.

The normal force N presses the surfaces in contact together.

Upon dividing of the F-equation by the surface area A of contact, obtain the shear stress τ to bring about sliding:

$$\tau = f \cdot \sigma_n \qquad (4\text{-}103)$$

Herein σ_n = normal stress on the contact surface s-s.

When sliding is in progress at a constant speed, the resisting force to sliding is also proportional to the load, but is independent of the area of the contact surface. In the case of sliding, the constant of proportionality is f'; this coefficient f' is called the coefficient of dynamic friction. The magnitude of the coefficient of dynamic friction is less than that of static friction, i. e.,

$$f' < f \qquad (4\text{-}104)$$

The coefficient f' varies with the velocity of the sliding motion.

The values of coefficients of friction are determined in the laboratory by means of compression tests, as well as in situ.

If in the shearing plane s-s there is neutral pressure u present, then the principal and normal stresses are to be decreased by u to perform calculations with effective stresses.

In Table 4-15 are compiled coefficients of friction f as found in the various sources of literature by various researchers using different methodology and different types of equipment (18, 37, 55, 103). The lower values of φ pertain to tests on wet rock surfaces.

4-30. Dynamic Properties of Rocks

In the laboratory, dynamic properties of rocks are determined on cubes by means of an impact device. One such dynamic property of rock is the *specific impact work* measured per 1 cm³ of rock.

Frequently, the so-called *toughness index* I_T of rock is determined:

METHODS OF ROCK EXPLORATION 223

TABLE 4-15

Angles of Friction φ and Coefficients of Friction f

Rock	Angle of Friction φ, degrees	Coefficient of Static Friction $f = \tan \varphi$	References
Igneous rocks			18, 37, 55, 103
Basalt	48–50	1.11–1.19	
Diabase	50–55	1.19–1.43	
Gabbro	10–31	0.18–0.60	
	11.3–31	0.20–0.60	
Granite	56–58	1.48–1.60	
	45–60	1.00–1.73	
Sedimentary rocks			
Dolomite	22	0.40	
Limestone	35–50	0.70–1.20	
Sandstone	27–34.2	0.51–0.68	
	26.6–35	0.50–0.70	
Shale	15–30	0.27–0.58	
Metamorphic rocks			
Gneiss	31–35	0.60–0.70	
Marble	32–37	0.62–0.75	
	35–50.2	0.70–1.20	
Quartzite	25.6–60	0.48–1.73	
	50–60	1.20–1.73	
Schist	62.25	1.90	90

$$I_\text{T} = \frac{\text{specific impact work}}{\text{compressive strength}} \left[\frac{(N \cdot \text{cm}) \cdot \text{cm}^{-3}}{N \cdot \text{cm}^{-2}} \right] \quad (4\text{-}105)$$

In these tests, a falling weight of 50 kg$_\text{f}$ $\approx$ 490 N is usually used. At first, 1 cm is chosen as the height of fall. In consecutive tests, the height of fall is increased by 1 cm.

However, the results obtained from these tests scatter very much: deviations of 50 to 100 % from the average are almost a normal occurrence. Therefore, perhaps only series of a large number of tests may somehow render statistical, comparable results. A disadvantage of these multinumber tests is their high cost. For this reason, such tests are very seldom performed.

Recently, studies on elastic properties of rocks (obsidian, in this case) have been attempted by the method of ultrasonic interferometry (77).

4-31. Summary on Laboratory Testing of Rocks

Rock tests performed in the laboratory are an important part of the engineering discipline of rock mechanics. The information needed for design in rock is obtained by direct testing, either on laboratory specimens or in the field. Much information can be obtained from laboratory tests of rock specimens with known orientation.

Laboratory testing of rocks, model studies, and indoor laboratory experimental work have the advantage over field work in that:

1. Loads, pressure, time, temperature, thermal gradients, and water can be controlled easily, whereas these and other factors for most rocks in their natural state in situ can be evaluated only crudely.
2. Laboratory testing equipment is convenient to use.
3. Laboratory tests are easy to perform; therefore, great precision is possible.
4. A laboratory-size rock specimen can be imposed easily upon an unconfined or triaxial load, and a sonic or shock wave propagated through it.
5. Laboratory tests enable one to obtain good and relatively inexpensive compressive strength data and shear strength parameters of intact rock.
6. Laboratory tests complement the test data as obtained from in-situ tests of rock.

Laboratory testing of rocks has also some disadvantages over in-situ testing, namely:

1. Disturbance is imparted on rock samples upon their sampling while separating them from the rock massif in situ.
2. There is possible disturbance, to some degree, of the structure of rock samples by shocks during their shipment from the sampling site to the laboratory.
3. To some extent, there is disturbance of rock properties during preparation of test specimens (sawing, coring, lapping, for example).
4. The strength properties of an intact laboratory rock specimen may differ greatly from the strength of the in-situ rock from which the specimen was sampled. It should be realized that the strength of a rock mass in situ, besides its own inherent intact strength, depends also to a high degree upon its various geological defects (viz., directional anisotropy). The effect of the defects of the prototype rock cannot be comprehended by laboratory tests on intact rock specimens free of fissures. Therefore, laboratory rock specimen test results do not represent the in-situ properties of the overall rock mass.

5. Laboratory test results do not shed any light about the internal stress conditions, primary or secondary, in the in-situ rock.

However, correct evaluation of rock strength properties obtained from appropriate laboratory and in-situ tests, supplemented with a high degree of experience and sound engineering judgment, contributes to safe and economic design of structures in and on rock.

4-32. Properties of In-Situ Rocks

A perfectly elastic solid is truly an ideal material. However, in contradistinction, rocks in situ contain various structural imperfections (rock defects and various planes of weakness). Obviously, rock defects prevent rocks from behaving in a perfectly elastic manner. For these reasons, it is said that rocks are inelastic materials behaving inelastically under loading and unloading conditions.

Because rock in situ is fissured through and through, it possesses entirely different strength properties than those obtained from intact rock core testing in the laboratory. Laboratory test samples are almost invariably taken from rock between major discontinuities and are, therefore, not representative of the rock mass.

Elastic properties determined from testing rock core specimens are reliable only for low stress levels. For high stresses, the jointing and seam structure and the anisotropy of the rock must be considered. Also, the plastic creep from rock crack and crystal contact adjustment with time caused by various kinds of stresses in rock should be taken into account.

Nearly every problem in rock engineering involves either the compressibility of the rock mass in situ, or the in-situ strength of the rock. Because of the presence of fissures, cracks, joints, gaps, stratification, discontinuities, planes of weakness, and layer anisotropy, compressibility of in-situ rock is considerably greater than that of a laboratory intact rock test specimen.

Therefore, for purposes of design in rock, it is necessary to include in equations of engineering rock mechanics the appropriate numerical values representing the corresponding in-situ rock properties.

Whereas laboratory tests render the physical and mechanical strength properties of the intact rock matrix for a relatively small specimen free from obvious structural discontinuities, and under certain casual conditions also give an idea about the behavior of rock joint, the in-situ rock tests are appropriate for evaluating the primary stress field and the mechanical behavior of the rock mass under load (35, 36, 37, 44, 49, 55, 74, 84, 88—90, 91—93, 97, 98, 100, 101, 105, 106, 108, 113, 114).

The strength of in-situ rock depends primarily upon the extent (size, volume) of the stressed domain; e.g., the completed engineering structure may involve many and different kinds of rock weaknesses (84a). However, the in-situ test of a rock would ordinarily measure the gross effect of only a few geological defects. Thus, the strength properties of a rock as obtained from laboratory tests differ from those as they prevail in situ by the following conditions:

1. In situ, the rock is confined to some extent. This means that at the commencement of work in rock, its initial stress condition differs from that prevailing in the rock specimen during its testing in the laboratory.
2. In situ, the rock mass contains various kinds of rock defects, viz., macrogeological weaknesses.
3. In situ, the rock mass is exposed to various environmental factors (climate, temperature, water) not necessarily taken into account during laboratory testing.

The in-situ data elucidate also about the mechanisms of failure of the jointed rock mass, and they render actual numerical strength values for use in solving real problems in the design and construction in rock. One should be aware, however, that because rock is a very complex material to work with, the testing of the strength of a rock mass in situ is much more difficult to perform than the testing of other homogeneous construction materials.

The basic properties of an in-situ rock mass of interest to the engineer are the mechanical properties, such as, for example:

1. Deformability (rock deformation under load, and condition of failure),
2. Compressive strength,
3. Tensile strength,
4. Shear strength,
5. Bearing capacity,
6. Permeability, and
7. Internal stresses in rock.

Internal stresses in a rock mass result mainly because of:
1. The weight of the overburden rock material,
2. Pore-water pressure,
3. Residual tectonic stresses in rocks,
4. Thermal stresses in rocks, and
5. Stresses brought about by chemical processes in the rock, mostly by recrystallization of minerals.

4-33. Deformability

Some of the factors affecting deformability are:

1. *Geological details* such as stratification, various kinds of rock defects, hydrological conditions (groundwater, pore-water pressure), and
2. *Petrographical factors* such as kind of rock mineral, degree of rock alteration, fissures brought about by excavation in rock, and internal stresses.

Introduction of an opening in the rock mass means introducing a discontinuity in the body of the rock mass, thus changing the initial or primary stress into a secondary state of stress. Hence, a change in internal stress condition is brought about.

Generally, upon loading of rock, the deformations are greatly influenced by the closing of haircracks, fissures, voids, and joints; mechanical slippage along planes of discontinuity; and inelastic deformation. Other factors influencing rock deformability are changes in pore-water pressure, temperature, drying, and vibration effects from blasting.

Up to the recent past, rock deformations were calculated by means of modulus of elasticity values as obtained from laboratory tests on rock core specimens. However, during the course of time, observations made with executed structures in rock already in service revealed that actually in-situ rock deformations were much larger than those determined from laboratory test data. The variance in these matters was explained by the existence of rock defects and discontinuities in rock. This is so because a jointed rock deforms more than a sound rock core specimen. Hence, laboratory rock specimen test results do not represent the in-situ properties of the overall rock mass.

Thus, there arose the need for testing rocks in situ. This, in its turn, creates a major problem in research, namely, how to correlate in-situ test results with those obtained in the laboratory on rock core specimens. Depending upon the extent, amount, and distribution of joints and other defects in a rock mass, the modulus of deformation of rock may be quite different from its modulus of elasticity. Recall that the modulus of deformation is based on the total measured deformation (elastic plus inelastic).

There are two methods for determining the deformability of in-situ rock, namely: the *static method* and the *dynamic method*.

In the static test, relatively large static loads are applied on the rock surface. In the dynamic test, the velocity of propagation of elastic waves is measured.

In the static test, the in-situ deformation modulus is determined by means of the so-called jack-and-plate loading test or by the pressure chamber (tunnel) test.

Some of the in-situ static tests of rocks are:

1. Plate loading (bearing),
2. Jacking,
3. Pressure chamber (tunnel),
4. Compression,
5. Bore-hole deformation,
6. Tension,
7. Shear strength,
8. Torsion, and
9. Stressing of the rock in the interior of a bore hole by the so-called method of hydrofracturing of the rock.

4-34. In-Situ Testing of Rocks

The main *objectives* of in-situ testing of rocks are:

1. To assess the general suitability of a site for contemplating engineering work,
2. To facilitate adequate, safe, and economical design of geotechnical structures for that site,
3. To foresee and to provide against geotechnical problems during and after construction, and
4. To investigate any subsequent changes in site conditions, or any eventual failures which may occur during construction.

Plate Loading Test

The purpose of the plate loading test is to determine the extent to which the rock mass will deform under various externally applied loads. Sometimes it is possible to increase the loading until failure of a weak rock material comprised of blocks and slabs of rock.

In this method of test, a normal load is applied to an exposed, flat surface of the rock by means of a hydraulic jack, and the resulting displacement of the rock surface is measured.

The theoretical basis for this test is the well-known, modified BOUSSINESQ elasticity solution (19) for the distribution of stress and strain in a semi-infinite elastic medium subject to a surface load N:

$$w_o = \frac{m \cdot (1 - \mu^2) \cdot N}{E \cdot \sqrt{A}} \qquad (4\text{-}106)$$

where w_o = surface displacement of rock, viz., average deflection of a rectangular loading plate brought about by a total distributed load N on the plate;
m = displacement coefficient;
μ = Poisson's ratio;
E = Young's modulus of elasticity;
A = area of the loading plate; and
N = total normal surface load.

According to Reference 110, the displacement coefficient values of m for circular, square, and rectangular loading plates are those given in Table 4-16.

TABLE 4-16

Values of m

Form of loading plate	Side ratio a/b	m values	Reference
(1)	(2)	(3)	(4)
Circle	—	0.96	110
Square	1 : 1	0.95	
Rectangle	1 : 2	0.92	
	1 : 5	0.82	
	1 : 10	0.71	
	1 : 100	0.37	

Usually three to six plate loading tests are made at each location. Individual tests are usually separated by five plate diameters.

In each test, loads are usually cycled, and a secant modulus will usually be taken from the hysteresis loops at the stress level of design interest. Each test usually consists of four cycles of loading and unloading, with progressively higher loads and recording of the rock deflections. For example, in the first loading cycle, 25 % of the full load is applied; 50 % in the second cycle; and so on. A typical cyclic deformation graph from repeated loading is shown in Fig. 4-63.

In some rocks, when the load is cycled several times, the inelastic deformation increases.

Besides simply applying an increased load increment on the plate, it is also necessary to control the rate of loading because it, too, can affect the mechanical properties of a rock.

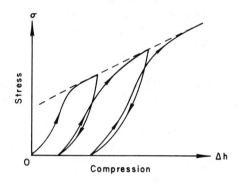

Fig. 4-63
Cyclic loading and unloading diagram

Small-scale in-situ tests confined to a relatively small volume of rock may give misleading results. Therefore, for large projects the tendency for in-situ testing is toward larger, comprehensive volumes of rock.

Depending upon the nature of the rock material tested, failure of rock can be recognized by a definite fracture, by a gradual continuous yielding, or by a sudden yielding of the rock.

Rock in situ may rupture as a result of:

1. Slip along planes of discontinuity,
2. Lack of cohesion, and
3. Plastic failure.

The physical constants used in estimating the load under which a rock mass in situ might fail are:

1. Insufficient angle of friction φ,
2. Insufficient cohesion c, and
3. The level of the stress applied to the rock, viz., exhaustion of its shear strength.

The plate loading test on rock provides a convenient means for determining the modulus of deformation as well as the in-situ strength of the rock.

Jacking Tests

Deformability of an in-situ mass of rock is usually determined by jacking (or jack-loading) tests (36, 49, 55a, 84, 89, 105) the principle of which is shown in Fig. 4-64. The test is usually carried out in excavated underground test galleries in rock (4, 21). Here the jacks are supported against the rock ceiling and floor of an underground opening. The load is applied by means of a jack placed on a flat surface of the rock, and the resulting deformation is measured. The data so obtained are utilized in computing the modulus of elasticity, viz., deformation modulus of the rock. The bearing plate in this test must be large enough to comprehend the structural features which contribute to the characteristics of the rock mass.

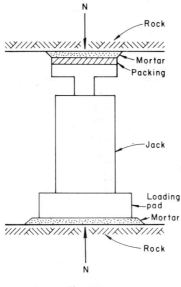

Fig. 4-64
Tunnel jacking test

Jack-loading tests have two serious disadvantages. First, no matter how carefully the excavation of the underground opening (tunnel, gallery, adit) or trench for the tests are made, excavation inevitably disturbs the rock mass surrounding the excavation. The second disadvantage is the necessity to limit the loaded area. This, in its turn, decreases the representativeness of the results and decreases the number of tested spots in the rock mass.

Cable Jacking Test

For performing plate bearing tests on flat, open surfaces of rock, STAGG (1968, Ref. 105) proposed that the jack reaction be taken up by cables anchored at suitable depth into predrilled bore holes in rock below the load-bearing plate (Fig. 4-65). It seems that this type of test would be more economical to perform than the simple plate loading test. Another advantage of the cable jacking test is that the rock can be tested at the exact location of the foundation, as well as in the directions in which the structural loads will actually be acting.

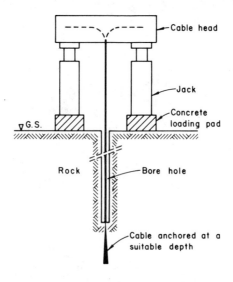

Fig. 4-65
Principle of cable jacking test

Pressure Chamber Test

The pressure chamber test (49, 55, 89, 90, 100, 105), known also by the term pressure tunnel test, consists essentially of sealing off a part of a tunnel by means of bulkheads. Prior to that, to avoid loss of water from tunnel to rock, a concrete lining is placed over the test tunnel surface of the rock, and the inside surface of the tunnel lining is coated with a waterproofing material for water tightness. Then the instruments are affixed for strain measurements. The chamber so prepared is then hydraulically pressurized, water usually being the pressurizing medium. Then the rock deflections brought about by the internal water pressure in the tunnel are remotely recorded at stations outside the sealed-off part of the tunnel. Thus in this kind of test the rock wall of the circular tunnel is subjected to a uniformly distributed radial loading, bringing about diametral deformation Δd. This diametral deformation is used to calculate the YOUNG's modulus of elasticity E:

$$\Delta d = 2 \cdot \frac{(1 + \mu)}{E} \cdot r_i \cdot p \quad (4\text{-}107)$$

and

$$E = \frac{(1 + \mu)}{\Delta d} \cdot d \cdot p \quad (4\text{-}108)$$

Herein, Δd = deformation of the rock measured behind the skin of the chamber;
μ = POISSON's ratio, usually obtained from laboratory tests on rock core specimens (it can also be obtained by geophysical test techniques by measuring the velocity of the shear wave within a rock mass);
r_i = $d/2$ = initial radius of tunnel;
d = diameter of tunnel; and
p = intensity of uniformly distributed radial, internal, hydrostatic pressure.

Because of the anisotropic nature of the rock, the E-values are to be determined for more than one direction within the rock mass.

From these measurements, pressure versus deformation diagrams are plotted, and the YOUNG's modulus of elasticity may be obtained.

One important feature of the pressure tunnel test method is that it brings about tensile hoop stresses in the rock. If this stress exceeds any residual compression of the rock, it may open radial cracks in the rock.

Some of the advantages of the pressure tunnel test are:
1. Rock deformations (deflections) can be measured in any radial direction.
2. Hydrostatic pressure on rock in the tunnel can be sustained for a long period of time.
3. The effect of the concrete lining on deformation modulus can be determined by testing lined and unlined rock in the same chamber.

The main disadvantage of this kind of in-situ test is that it is relatively expensive.

Large-Scale Compression Strength Test

The rock mass compressive strength is determined from tests on rock pillars. Their strength is a function of the rock material, the geometry of the pillar, its size, and the compressive strength of the rock strata between which the pillar is compressed.

Borehole Deformation Test

Deformability of rock in situ is also possible to test in boreholes by means of a device called the borehole deformation meter (39, 44, 45). In principle, the device consists of a shell split into two halves. It is inserted and lowered into the borehole. Then the shell is filled with oil and pressure is applied on the oil. Thereupon the split shell bears laterally against the rock, thus imparting pressure on the rock wall of the borehole. The principle of the working of a borehole deformation meter is sketched in Fig. 4-66. A sensitive transducer is used to measure the deformation of the rock wall

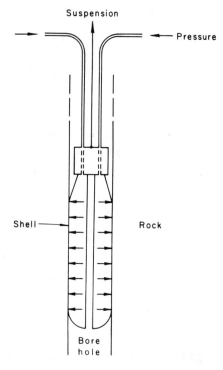

Fig. 4-66
Principle of the working of a borehole deformation meter

of the borehole. Various loading and unloading cycles can be performed by means of this meter, and the in-situ modulus of deformation can be determined.

In-Situ Tension Test

Any incompetent rock mass in its *natural state* cannot sustain tensile stresses because of the presence of joints, fissures, and other kinds of rock defects. Therefore, from a *structural* point of view, the tensile strength test in situ is not of too great importance, except in a layered roof of an underground opening where tensile stresses of considerable magnitude may have to be coped with. Tensile stresses in rock may also be of importance in the stability of rock slopes over pre-existing planar discontinuities, where the dipping rock layers may be subjected to tensile stresses and thus slope failures would be more probable.

4-35. In-Situ Shear Strength Test

The main purpose for performing in-situ rock shear strength tests is to obtain shear strength values that comprehend the effect of the various rock defects (4, 5).

The test area should be chosen to represent the area on which the structure will be built.

To test the shear strength of the in-situ rock mass, the rock specimen to use in such a test should be as large as possible. The size of the specimen should not be less than 40 cm square by 20 cm high. If the specimen is larger than 40 cm x 40 cm x 20 cm, the ratio of length, width, and height is usually taken as 2:2:1. Sometimes the size of the base area of the rock block is taken as 0.70 m x 0.70 m, or even as 1.0 m x 1.0 m. The shear test should be made on rock that can be trimmed without damage.

The in-situ shear strength tests are usually performed on rock blocks cut in the rock mass but kept attached to it in their bases. After the block has been properly prepared, the normal and tangential loads are applied to the block by means of hydraulic jacks (Fig. 4-67). In an underground opening, the jacks are supported against the ceiling and wall of rock.

The vertical jack supplies the normal load N on the forced shear plane s-s of the rock, whereas the second jack (horizontally installed, or installed at an inclination with the horizontal) supplies the shear load.

The shear thrust may be vectored at an angle of about $(\pi/4 - \varphi/2)$-degrees to the position of the shear plane. Thus an isolated block of rock is sheared off from the parent rock. During loading and unloading, measurements of deformations are recorded.

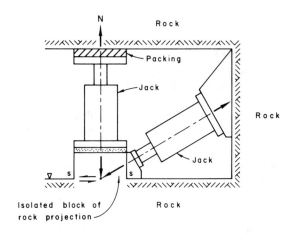

Fig. 4-67
In-situ shear strength test in an underground opening

This kind of in-situ shear strength test gives an idea of the magnitude of the angle φ of the shear resistance of the rock. It is assumed here that COULOMB's shear strength law applies:

$$s = \sigma_n \cdot \tan \varphi + c \qquad (4\text{-}109)$$

wherein s = shear strength,
σ_n = normal stress on the shear plane,
φ = angle of shear resistance of rock, and
c = cohesion of rock.

Because the shear strength of rock is given by two shearing strength parameters, φ and c, two equations are needed for their determination. Therefore tests are made with bearing plates or blocks of various sizes.

Depending upon the need in design, shear strength tests in situ may also be performed on oriented planes of separation or geological discontinuities such as planes of bedding, shistosity, jointing, and faults. In these planes of weakness, the least shear strength of a rock mass can thus be anticipated.

The in-situ shear strength of rock in an underground opening may also be determined as sketched in Fig. 4-68. This kind of a test arrangement has the advantage that both normal and shear stresses are applied on the ultimate plane, of shear failure, $s\text{-}s$, using only one jack.

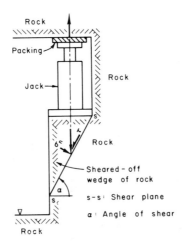

Fig. 4-68
Shear strength testing of rock in situ with one jack

The principle of in-situ shear and in-situ sliding tests is illustrated in Fig. 4-69. Generally, under small normal loads, the rock blocks shear off over a plane contact surface (Figs. 4-69, 2 to 7).

However, in evaluating heavily loaded (N), large-scale, direct, in-situ shear test results on rocks, one should be aware that no shear takes place along the contact plane between the block and rock, but rather over a gently curved sliding surface in the rock on the leeward side of the loaded block. This pertains also to testing of concrete blocks on rock (Fig. 4-69, 11). The same phenomenon has also been observed in the author's experiments with obliquely loaded, rigid foundation models on dry sand at relatively light, normal contact pressures (Fig. 4-70), Refs. (57-59).

For repeated tests, several shear blocks of rock are necessary.

Because of the irregular distribution of the various kinds of rock defects, the determination of the in-situ shear strength of a rock mass is a difficult task indeed.

Frequently, the sliding surfaces in a rock mass are not smooth, but are irregular, rough, and wavy. Often the sliding surfaces are filled with gouge (for example, geological debris such as silt, clay, sand, ground rock flour, calcite, and other kinds of decomposed rock material). Depending upon the degree of waviness, joints during a shearing, viz., sliding process, may open and/or close (Fig. 4-71). A relatively thin course of clay in the plane

Symbol of shearing system	Kind of in situ shear test	Test results
①	Compression confined $(\sigma_3 = \sigma_3)$ unconfined $(\sigma_3 = 0)$	Compressive strength. Deformation characteristics. Stress-strain diagram. Elastic properties. Shear strength parameters ϕ and c.
②	Shear homogenious material	Shear displacement. Shear strength. Parameters ϕ and c.
③	Shear parallel to bedding planes or laminations with σ_n; without σ_n	As under ②, but along bedding planes.

METHODS OF ROCK EXPLORATION

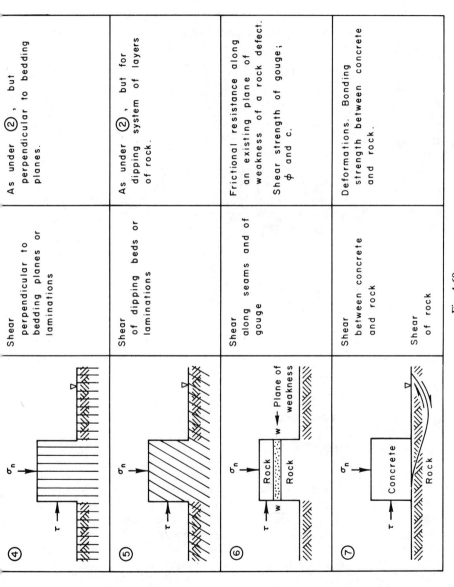

Fig. 4-69
Some principles of large-scale shear strength tests

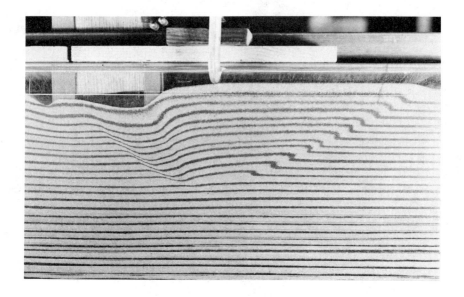

Fig. 4-70
Rupture surface pattern in dry sand brought about by an obliquely,
centrally loaded foundation model at a contact pressure of $\sigma_o = 1.25$ kg_f/cm^2
(Author's study, Refs. 57-59)

of weakness or discontinuity in rock, and fillings in the joints between beds and rock blocks, may considerably alter natural stress conditions in the rock mass, as well as bring about failures of stability such as those which were experienced at Malpasset dam (2, 3, 24, 52, 94), Vaiont reservoir (53, 70, 78, 79, 82, 83); the Madison Canyon (17), and at other sites.

Also, pore-water pressure affects sliding of rock masses on and along faults, joints, and bedding planes, thus impairing the stability of a layered rock structure. Thus, joints, shear zones, and other kinds of rock defects in a rock mass decrease the magnitude of the shear strength value much below the shear strength of the intact rock specimen of the same rock. Therefore, the in-situ shear strength of a rock is to be qualified as very anisotropic. Relative to design and construction in rock, this means that the most critical situation is one when the shear strength of the rock mass is governed by the shear strength, or resistance along the rock contact surfaces of a discontinuity. Therefore, rock defects of a rock mass generally cause scatter in the results of the shear strength test values. The difference in test results decreases with increase in size of the test area. Also, the in-situ strength increases with decrease in the amount of discontinuities (jointing).

Values of the shear stress parameter φ_r of residual shear strengths obtained from most types of rock, exclusive of clay shales, usually lie between $\varphi_r = 25°$ and $35°$, while the φ_r-values for most clay shales lie between $5°$ and $25°$ (91).

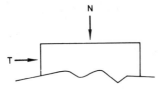

a) Shearing along planes of separation

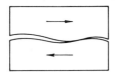

b) Opening of a joint upon sliding

Fig. 4-71
Rough and wavy sliding surfaces

Torsion Shear Test

The method of torsional shear test involves the application of a torque to a cylindrical rock core which is isolated without damage from the surrounding rock mass by means of a diamond rock core drill. The rock core so prepared is kept attached at its base to the rock mass (Fig. 4-72). The core is then gripped by the testing device and twisted. The angle of twist and the failure are measured and recorded. The test renders information about the test parameters c (cohesion), φ (angle of friction), and residual shear strength of the rock.

The torsion shear test is the only method of bringing about shear in a rock specimen without compression.

In order to set up a perfect torsion shear test, the influence of clamping or gripping effects and of all eventual rock core bending effects on test results should be eliminated.

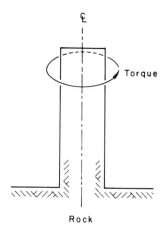

Fig. 4-72
Principle of in-situ torsion shear test of rock

4-36. Internal Stresses in a Rock Mass

The prerequisite for safe and effective design and work in rock is a comprehensive knowledge about the physical and mechanical properties of in-situ rock. These properties, and rock anisotropy, are known to vary with depth of the rock formation. The anisotropy of an in-situ rock mass may be evaluated by knowing its natural stress condition.

The main uses of rock stresses are in:

1. Designing of tunnel linings;
2. Designing of walls, pillars, roof arches, and domes for underground openings in rock;
3. Evaluation of rock stability of various kinds of structures in rock, including stability of rock slopes made in open cuts; and
4. Evaluation of stability of rock-abutment systems of high dams.

Internal, natural stresses in a rock mass in situ result mainly because of:

1. The weight of the overburden rock material;
2. Hydrostatic pressure;
3. Pore-water pressure;
4. Thermal stresses in the rock;
5. Residual tectonic stresses in the rock; and

METHODS OF ROCK EXPLORATION

6. Possible stresses brought about by chemical processes in the rock, mostly by recrystallization of rock minerals.

Factors affecting natural stresses in rock are:

1. Local stresses brought about by excavation in rock,
2. Effects of blasting,
3. Seismic forces,
4. Changes in hydrostatic pressure,
5. Changes in pore-water pressure,
6. Changes in rock thermal regimen, and
7. Variation and alteration in physical and chemical factors of the rock mass and its minerals.

In-Situ Test of Stresses in a Rock Mass

Any underground opening in rock is deformed by stresses in it (36). The opening deforms upon change in stress in the rock. However, it must be recognized that the concept of stress is a theoretical one, and that it is difficult to measure stress directly. In the same vein, therefore, it must be said that presently there is no way of measuring natural stresses in the interior of a rock mass without disturbing the rock at the location of measurement. However, in order to get a value for the stress at a point, other properties of the rock relating to stress are usually measured and calibrated. Thus, one resorts to measuring the stress by indirect methods. In these methods, usually elasticity of rock is assumed, strains in rock are measured, and strains and stresses are then related. The underground measurements further the understanding of the performance of rock around an underground opening, and provide for basic design data. Thus, the information about the stress field in the rock is derived from displacement, viz., strain, measurement, which can be used to measure changes in stress in rock around the opening. Hence, the analysis of strain is fundamental in the studies of deformation of the rock (95).

Some of the methods of measuring strains that lead to state of stress determination in rock are:

1. Direct strain measurement,
2. Dilatometer test,
3. Overcoring technique,
4. Flat-jack testing,
5. Electrical resistivity,
6. Seismic wave velocity, and
7. Photoelasticity analysis of stress.

Direct Strain Measurement

In this method, strain gauges are cemented on an exposed rock surface, or in a borehole, and the rock strains are measured.

Dilatometer Test

Deformation of in-situ rock mass can also be determined along the inside of boreholes by means of instruments called borehole deformation meters and dilatometers.

The dilatometer, developed in Portugal (ROCHA, 1966, Ref. 97), imparts a uniform pressure along a certain length (depth) of the borehole by means of a liquid contained between a metallic cylinder and a very deformable rubber or steel jacket. The jacket is applied against the rock wall of the borehole.

The diametrical deformation Δd of the diameter $d = 2r$ of the borehole calculates as:

$$\Delta d = 2 \frac{1+\mu}{E} \cdot r_i \cdot p \qquad (4\text{-}110)$$

From here, the modulus of elasticity E of the rock is:

$$E = 2 \frac{1+\mu}{\Delta d} \cdot r_i \cdot p \qquad (4\text{-}111)$$

where μ = POISSON's ratio,
   r_i = inside radius of the borehole, and
   p = intensity of dilatometer pressure.

The pressure p is measured by means of a pressure gauge, and the diametrical deformation of the borehole, viz., strains, is measured along the external face of the dilatometer by means of transducers.

Compared with the load test, measurements made by means of the dilatometer inside boreholes have the following advantages:
1. If diamond drilling is used, the rock mass remains almost undisturbed.
2. Because of the reduced cost and time required in performing dilatometer tests, many tests can be performed. Thus it is possible to obtain pertinent statistical data about deformability of an anisotropic rock mass.
3. Dilatometer tests can be performed at great depth.

4. Such tests can be performed under submerged condition.
5. Much higher pressures can be applied on the rock than in the load test.

A disadvantage of the dilatometer test is the small volume of rock involved in each such test.

Whereas a domain subjected to a jack-loading test is free from or decompressed of internal or initial stresses in the direction of the applied load, the volumes of rock involved in the dilatometer tests are in a state of radial compression (zero at the surface of contour of the hole) when the rock mass exhibits radial stresses. Therefore it should be expected that the deformability determined by means of dilatometers lies below the values obtained in jack-loading tests, because usually deformability decreases with increase in compressive stresses.

There have been several types of dilatometers developed in various countries. Those instruments in which a radial, uniform pressure is applied by means of a liquid on the rock wall of the borehole appear to be the best ones for measuring deformability in a rock mass. The other instruments induce more involved states of stress in the rock mass, thus making the interpretation of the test results somewhat difficult.

Flat-Jack Testing Method

The principle of in-situ state of stress measurement in a rock mass is: the relief of stress, and its recovery by means of hydraulic pressure. Such a test is accomplished by means of a flat jack. Essentially, the flat-jack testing technique may be described as set forth (98).

A narrow slot is cut into the rock wall, such as the wall of a tunnel, for example (Fig. 4-73). Prior to cutting of the slot, two displacement or deformation measuring pins are fixed into holes drilled into the rock, or cemented onto the surface of the rock wall. Upon cutting the slot, the stresses in the rock originally across the surfaces of the slot are relieved or relaxed on either side of the surfaces of the slot. Upon this stress relief, expansion of the rock into the slot takes place.

The rock expansion or displacement can be precision-measured either by the convergence of the rock across the slot, or by the convergence between the measuring pins on opposite sides of the slot. After the slot is cut, a thin flat jack is placed into the slot cut into the rock wall. The flat jack is filled with a fluid, usually oil, and pressure in the jack is applied. The pressure is increased until the displacements or deformations brought about by cutting the slot are cancelled (reversed, or until the measuring points return to their original position). This means that in such a case the flat-jack pressure p attained the value of the original normal stress σ_z of the rock initially existing in it before cutting the slot; i.e., $p = \sigma_z$. This pressure p, thus, represents the stress acting in the direction normal to the flat jack. Thus, in a way, the flat-jack testing method may be regarded as a null technique by means of which rock stresses can be measured directly.

Some of the advantages of the flat-jack testing method are:

1. The method is very simple and relatively effective in sound, hard rock.
2. Disturbance in the rock mass is reduced to a minimum.
3. It renders vertical and horizontal absolute stresses around an exposed face of rock.
4. Slots in rock can be cut, and flat-jack measurements can be made at any angle with the horizontal.

Some of the disadvantages of the flat-jack testing method for determining the state of stress within a rock mass at its surface in an underground opening are:

1. The initial state of stress in rock is changed already by excavation.
2. The method is limited to measuring of stresses near the exposed surface of the rock wall.
3. It is complicated by the effect of creep of the rock during the period between cutting of the slot and stress cancellation.

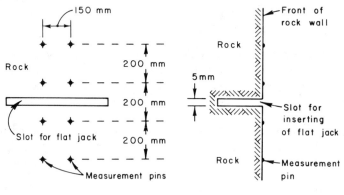

a) Front view of slot in rock

b) Section through slot in rock wall

Fig. 4-73
Slot in a wall of rock for performing a flat-jack test
(After ROCHA et al., Ref. 98)

Overcoring Technique

The overcoring technique is also known by the term borehole strain measuring technique. This technique is used to determine the "absolute stress" in the rock, according to LEEMAN and HAYES (1966, Ref. 74).

In essence, the technique involves measuring strains at the three points around the periphery of a borehole made in the "roof", or in the "sidewall", or at a point midway between the "roof" and the "sidewall". By using strain gauge rosettes of suitable configuration, sufficient strain measurements in three different planes are, in effect, then obtained.

This method of absolute stress measurement in situ may be described briefly as follows. An NX borehole is drilled into rock to a depth at which the stress is to be determined. Then an EX borehole is concentrically drilled for, say, 45 cm into the end of the NX borehole. Then three rosette strain gauges are glued around the periphery of the test borehole by means of a special tool in the EX portion of the borehole made in the rock, and spaced in positions at $\theta = 0°$ (at the roof of the borehole), at $\theta = \pi/2$ to the first, and at $\theta = 5\pi/4$ to the first rosette. Then strain readings are taken. After that, the EX portion of the borehole is overcored with an annulus (Fig. 4-74), using an NX casing crown. Then the hollow cylindrical core containing the

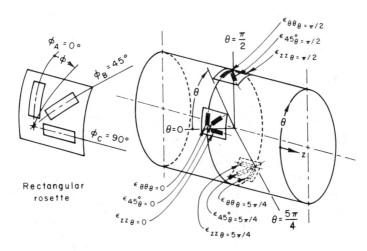

A. Positioning of strain gauge rosettes
ε = strain

Fig. 4-74
Overcoring technique
(After LEEMAN and HAYES, Ref. 74)

rosettes is removed and the stress-relieved strain readings are taken from each gauge. Each strain gauge rosette has three gauges to measure strains in the circumferential, axial, and 45-degree directions.

The standard sizes for casings, rods, core barrels and holes are compiled in Table 4-1.

It is advisable to use as large a diameter of core as economy will permit. The larger the diameter, the more accurate are the geological observations on jointing and fracturing. Very small diameters such as EX are to be avoided when possible: many widely spaced fractures may be missed entirely.

The difference in the strain readings before and after overcoring enables one to determine the complete stress in the rock, provided that the properties of the rock are known.

The removed core can be inspected in the laboratory for bonding condition of the strain gauge rosettes.

In this method, it is easy to measure the changes in strain. The method is a rapid one of obtaining absolute strain. Its accuracy, however, is open to question, particularly over a period of time.

According to LEEMAN and HAYES (1966), the term "complete state of stress" at a point in rock means the magnitudes and directions of the three principal stresses acting at the point (74).

Hydraulic Fracturing Test of Rock

Hydraulic fracturing of a borehole is used extensively in the oil and natural gas industry to stimulate yield from a well. The method of hydraulic fracturing of rock is also used in rock engineering.

In principle, the method of hydraulic fracturing of deep borehole rock walls should render the horizontal, tectonic, tangential, principal stress σ_t. In this test, a certain section of the borehole is sealed off with packers such that a fluid pressure can be applied to the bare wall of rock of the borehole between the packers only to a restricted range of depth of the hole, viz., thickness of a layer of rock. Then water (or any other liquid) is pumped into the borehole for fixed periods of time to a sufficient and successively higher pressure. This pressure p is increased until fracture of rock sets in the wall of the rock. This means that the tangential stress σ_t in the surface of the contour or periphery of the borehole becomes tensile. This tensile stress brings about fracture in the rock if this σ_t stress exceeds the tensile strength

METHODS OF ROCK EXPLORATION

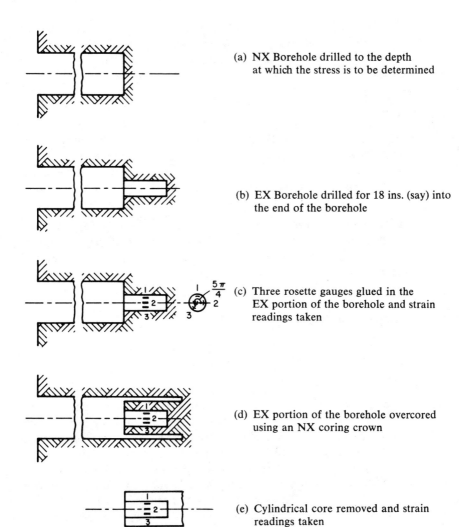

(a) NX Borehole drilled to the depth at which the stress is to be determined

(b) EX Borehole drilled for 18 ins. (say) into the end of the borehole

(c) Three rosette gauges glued in the EX portion of the borehole and strain readings taken

(d) EX portion of the borehole overcored using an NX coring crown

(e) Cylindrical core removed and strain readings taken

B. Sequence of overcoring

Fig. 4-74 (Continued)
Overcoring technique
(After Leeman and Hayes, Ref. 74)

σ_{t0} of the rock. In other words, a crack intersecting the borehole — which crack is kept "closed" by the initial compressive stress field in the rock — may become "opened" by the newly induced tensile stress after a hydraulic pressure p is applied.

As long as the flow of the fluid being pumped is laminar, the diagram of the amount of water acceptance Q by the rock plotted against the applied pressure p will be a straight line. Any sudden increase in the slope of this $(Q-p)$-diagram may indicate that rock fracturing takes place.

The cracks in the rock extend or propagate as the pressurized liquid flows into the cracks.

In rock mechanics, it is believed that the cracks propagate normal to the direction of the minor principal stress. Also, the pressure p of the liquid bringing about propagation of cracks may be considered as the magnitude of the minor principal stress in the rock. The pressure p for cracking the rock is somewhat larger than the magnitude of the minor principal stress.

Referring to the pertinent literature (FAIRHURST, 1964, Ref. 35) it must be said, though, that the method of hydraulic fracturing of rock as a means of its use for determining horizontal tectonic principal stresses in rock has aroused a great deal of discussion.

4-37. Permeability

Permeability is defined as that property of a porous material that permits the passage or seepage of fluids — such as water, or oil, or gas (air) — through its interconnecting voids.

The factor permeability of rock to water in rock masses is also of paramount importance, not to be forgotten in designing safe structures in rock. The most frequently observed flow in rock is the flow of water through joints and other rock defects. It should be remembered that the presence and flow of interstitial water (in pores, along faults, in fissures, in crushed zones, and in gouges) frequently bring about hydrostatic uplift and pore-water pressures. These pressures affect seriously the stability of foundations and rock abutments of dams and the roofs of underground openings, and may alter the internal stress condition in the rock. Knowledge about permeability of rock is also needed for executing injection and/or grouting projects in rock. More about permeability of rock to water may be found in Sections 4-4, 4-8, and 10-3, and in Refs. 26, 75, and 76.

The flow of water in rock in situ is determined by means of permeability tests.

Permeability tests of a rock mass in situ are usually made in a system of several boreholes by measuring loss of water to joints. When joints are closed and filled, and the rock is permeable, permeability tests render diagrams similar to those obtained for soils.

There are two kinds of borehole permeability tests:

1. Bottom-hole method, and
2. Packer method.

The bottom-hole permeability test consists of simply blocking one end of the hole and filling it with water. The time required for the water level to fall a given distance is then recorded.

In the packer test, several test holes, ~5 cm in diameter and at various depths, are drilled into the rock. These boreholes are sealed with rubber packers; then the holes so sealed are filled with water. The flow of water, the maximum pressure, and pressure changes in each hole are recorded.

Full-scale water permeability tests can also be made in aqueous tunnels. Such tests are supposed to indicate loss of water from a rock opening in the case of an aqueous tunnel, and the amount of ingress of water from the rock into the underground opening in the case of a vehicular tunnel, underground power plant, and other kinds of openings.

Permeability test results are also used in evaluating the feasibility of constructing a lined rather than an unlined tunnel.

There are three methods of water absorption tests, namely:

1. the single-packered,
2. the double-packered, and
3. the multi-packered tests.

A packer is a sealing device lowered into the borehole to form a watertight contact against the sides of the borehole, thus supposedly completely excluding water from higher horizons.

The use of a packer affords scanning exploration of the rock relative to its permeability at every desired elevation in the borehole.

The principle of the working of a single packer is shown in Fig. 4-75. A borehole is drilled into the rock, and the profile of the traversed rock material is logged. Then a watertight seal or packer, made of strong, elastic, impermeable material is lowered into the borehole to a desired depth. There the packer is activated from the surface to expand and to press watertight against the rock inside the borehole, thus excluding water from above. Water at a pressure, say 10 at. is forced through the lower end of the packer for a duration of five minutes into the free, unsealed test section of borehole. After that, the packer is raised 1 meter higher, and water under pressure is introduced. The permeability of the next section (1 m) is the difference between previously performed and the consecutive (adjacent) tests, and is recorded in liters per meter and per minute (liters/m · min). The test can also be performed by moving the packer downward, and setting it at the desired depth.

The double packer, Fig. 4-76, affords a further isolation of a test section in a borehole. Between the two packers, there is a perforated pipe from where water enters the test section of rock. Thus, with the double packer it is possible to test an isolated, closed, self-contained section of the borehole, so that the measured permeability data can be used directly.

In very fissured rock, it may happen that water would pass or escape through the system of fissures and cracks around the packer above and below the test section. This can be somewhat avoided if the packers are made of an appropriate length.

A device based on four packers has been suggested by LOUIS (75) to give

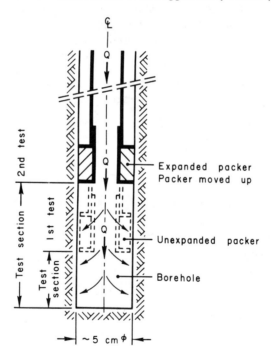

Fig. 4-75
Principle of a single packer

a better control of flow conditions during the permeability test. The central section between the second and third packers is the measuring test section, while the two outer sections act as flow barriers (see Fig. 4-77).

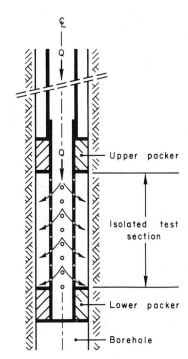

Fig. 4-76
Expanded double packer

1. Probe
2. Packer, 0.8 m long
3. Central measuring section, 2 to 5 m long
4. Outerflow barrier section, 2 m long
5. Radial flow lines from central section
6. Piezometers in lined holes

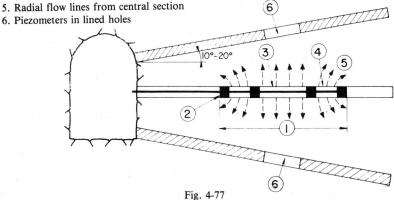

Fig. 4-77
A four-packer permeability test device after Louis (Ref. 75)

REFERENCES

1. ANDREAE, C., 1953, "Die Temperaturprognose im Tunnelbau," *Schweizerische Bauzeitung,* pp. 41—42.
2. Anonymous, 1959, "French Dam collapse", *"Engineering News-Record,* December 10, 1959, pp. 24—25.
3. Anonymous, 1959, "The Malpasset Dam Design." *Engineering,* December 18, 1959, pp. 642—643.
4. ASCE — *Symposium on Underground Rock Chambers,* an ASCE National Meeting held January 13—14, 1971, at Phoenix, Arizona. New York, N.Y.: American Society of Civil Engineers.
5. ASME — *Rock Mechanics Symposium,* A. L. SIKARSKIE (editor), 1973, The Winter Annual Meeting of the ASME, held November 11—15, 1973, at Detroit, Mich., The American Society of Mechanical Engineers.
6. ASTM — Special Technical Publication 420, 1966, *Testing Techniques for Rock Mechanics.* Philadelphia, Pa.: American Society for Testing and Materials.
7. ASTM — Special Technical Publication 429, 1967, *Determination of Stress in Rock,* A State-of-the-Art Report. Philadelphia, Pa.: American Society for Testing and Materials.
8. ASTM — Special Technical Publication 477, 1970, *Determination of the In-Situ Modulus of Deformation of Rock.* Philadelphia, Pa.: American Society for Testing and Materials, 1981.
9. ASTM — Special Technical Publication 479, 1970 (5th ed.), *Special Procedures for Testing Soil and Rock for Engineering Purposes.* Philadelphia, Pa.: American Society for Testing and Materials, 1981.
10. ASTM — Special Technical Publication 483, 1971, *Sampling of Soil and Rock.* Philadelphia, Pa.: American Society for Testing and Materials.
11. ASTM — Special Technical Publication 554, 1974, *Field Testing and Instrumentation of Rock.* Philadelphia, Pa.: American Society for Testing and Materials, 1981.
12. ASTM — Standard Definitions of Terms and Symbols Relating to Soil and Rock Mechanics, D 653-80, *1981 Annual Book of ASTM Standards,* Part 19. Philadelphia, Pa.: American Society for Testing and Materials, 1981, pp. 168-193.
12a. ASTM — Standard Test Method for Modulus of Rupture of Natural Building Stone, Designation C 99-52 (Reapproved 1976), *1981 Annual Book of ASTM Standards,* Part 19. Philadelphia, Pa.: American Society for Testing and Materials, 1981, pp. 4—6.
12b. ASTM — Standard Test Method for Compressive Strength of Natural Building Stone, Designation C 170-50 (Reapproved 1976). *1981 Annual Book of ASTM Standards,* Part 19. Philadelphia, Pa.: American Society for Testing and Materials, 1981, pp. 16—18.
13. ASTM — Standard Test Method for Unconfined Compressive Strength of Intact Rock Core Specimens, American National Standard ANS/ASTM D 2938—79, *1981 Annual Book of ASTM Standards,* Part 19. Philadelphia, Pa.: American Society for Testing and Materials, 1981, pp. 450—453.
14. ASTM — Standard Test Method for Direct Tensile Strength of Intact Rock Core Specimens, ANS/ASTM D 2936—78, *1981 Annual Book of ASTM Standards,* Part 19. Philadelphia, Pa.: American Society for Testing and Materials, 1981, pp. 440—444.

15. ASTM — Standard Test Method for Triaxial Compressive Strength of Undrained Rock Core Specimens without Pore Pressure Measurements, ASTM Designation D 2664—80 *1981 Annual Book of ASTM Standards*, Part 19. Philadelphia, Pa.: American Society for Testing and Materials, 1981, pp. 399—404.
16. BAIDYUK, B. V., 1967, *Mechanical Properties of Rocks at High Temperatures and Pressures*. New York, N.Y.: Consultants Bureau.
17. BARNEY, K. R., 1960, "Madison Canyon Slide," *Civil Engineering*, August 1960, pp. 72—75.
18. BIRCH, F., 1972, " Compressibility, Elastic Constants," in *American Institute of Physics Handbook*, 3rd ed. New York, N.Y.: McGraw-Hill Book Company, Table 7—15 on Rocks: Elastic Constants, pp. 100—170.
19. BOUSSINESQ, J. V., 1885, *Application des potentiels à l'étude de l'equilibre et du mouvement des solides élastiques*. Paris: Gautier-Villars.
20. BRACE, W. F., 1971, "Microcracks in Rock Systems," *Proceedings Southampton 1969 Civil Engineering Materials Conference*. London: John Wiley and Sons, Inc., Part I, pp. 187—204.
21. BREKKE, T. L., and F. A. JÖRSTAD (editors), 1970, *International Symposium on Large Underground Openings, Oslo, 1969*. Oslo: Universitetsforlaget.
22. BURWELL, E. B., and R. H. NESBIT, 1954, "The NX Borehole Camera." *Transactions of the American Institute of Mining Engineers*, 1954, Vol. 194, pp. 805—808.
23. *California Geology*, January, 1977.
24. CAMBEFORT, H., 1963, "The Malpasset Report," *Water Power*, April, 1963, Vol. 15, No. 4, pp. 137—138.
25. CASAGRANDE, A., and R. E. FADUM, 1939, *Notes on Soil Testing for Engineering Purposes*. Cambridge, Mass.: Harvard University Publication No. 268, 1939/40, Fig. II, p. 23.
26. CASAGRANDE, A., 1961, "Control of Seepage Through Foundations and Abutments of Dams." *Géotechnique* (London), September 1961, Vol. 11, No. 3, pp. 159—182.
27. CLARK, S. P. Jr., 1966, *Handbook of Physical Constants*. The Geological Society of America, Memoir 97.
28. COULOMB, C. A., 1776, "Essai sur une application des régles de maximis et minimis à quelques problémes de statique rélatifs à l'architecture." Paris: De l'Imprimerie Royale: *Mémoires de la mathématique et de physique*, présentés à l'Académie Royale des Sciences, par divers Savants, et lûs dans ses Assemblées, Vol. 7, Année 1773, pp. 346—384.
29. DARCY, H. P. G., 1856, *Les fontaines publiques de la ville de Dijon*. Paris: Victor Dalmont, Éditeur, 1856, pp. 570, 590, 594.
30. DUCKWORTH, W. H., 1951, "Precise Tensile Properties of Ceramic Bodies," *Journal of the American Ceramics Society*, Vol. 34, No. 1, pp. 1—9.
31. ESHBACH, O. W., (editor), 1952, *Handbook of Engineering Fundamentals*. New York, N.Y.: John Wiley and Sons, Inc., pp. 13—23 to 13—24.
32. EVERELL, M. D., HERGET, G., SAGE, R., and D. F. COATES, 1974, "Mechanical Properties of Rocks and Rock Masses," 3rd International Congress on Rock Mechanics at Denver, Colo., U.S.A.
33. FAIRHURST, C., 1961, "Laboratory Measurement of some Physical Properties of Rock," *Proceedings 4th Symposium on Rock Mechanics*, held 1961 at University Park, Pa., pp. 105—118.

34. FAIRHURST, C., 1964, "On the Validity of the 'Brazilian' Test for Brittle Materials," *International Journal for Rock Mechanics and Mining Science,* Vol. 1, No. 4, pp. 535—546.
35. FAIRHURST, C., 1964, "Measurement of In-Situ Rock Stresses, with Particular Reference to Hydraulic Fracturing," *Rock Mechanics and Engineering Geology,* Vol. 2, No. 3/4, pp. 129—147.
36. FAIRHURST, C., 1976, "In Situ Stresses Existence, Determination and Data Interpretation," *ASCE Rock Excavation Seminar,* held October 1976 in New York, N.Y., ASCE Metropolitan Section, Foundations and Soil Mechanics Group.
37. FARMER, I. W., 1968, *Engineering Properties of Rocks.* London: E. & F. N. Spon Ltd.
38. *Geothermal Hot Line,* 1978, Sacramento, California: Division of Oil and Gas. Periodical No. TRO2, July 1978, Vol. 8, No. 2, p. 8.
39. GOODMAN, R. E., VAN, K., and F. E. HEUZE, 1972, "Measurement of Rock Deformability in Boreholes," in *Basic and Applied Rock Mechanics,* Proceedings of the 10th Symposium on Rock Mechanics, New York: Intersociety Committee for Rock Mechanics, The Society of Mining Engineers of the American Institute of Mining Engineers in New York, pp. 523—555.
40. GRIFFITH, A. A., 1921, "The Phenomena of Rupture and Flow in Solids," *Philosophical Transactions,* Vol. 221, pp. 163—198.
41. GRIFFITH, A. A., 1924, "Theory of Rupture," Proceedings First International Congress on Applied Mechanics. Delft: J. Waltham, Jr. Press, pp. 55—64.
42. GRIFFITH, J. H., 1936, *Thermal Expansion of Typical American Rocks.* Ames, Iowa: Iowa State College of Agriculture and Mechanic Arts, Iowa Engineering Experiment Station, Vol. 35, No. 19, October 7, 1936, pp. 19—20.
43. GRIFFITH, J. H., 1937, *Physical Properties of Typical American Rocks.* Ames, Iowa: Iowa State College of Agriculture and Mechanic Arts, Iowa Engineering Experiment Station, Bulletin 131, Vol. 35, No. 42.
44. HAIMSON, B. C., 1974, "A Simple Method for Estimating In Situ Stresses at Great Depth," in *Field Testing and Instrumentation of Rock.* Philadelphia, Pa.: ASTM Special Technical Publication 554, pp. 156—182.
45. HARD, H. C. et al., 1972, *Flow and Fracture of Rocks: The Griggs Volume.* Washington, D. C.: American Geophysical Union, Geophysical Monograph 16.
46. HIX, C. F. Jr., and R. P. ALLEY, 1958, *Physical Laws and Effects.* New York, N.Y.: John Wiley and Sons, Inc., p. 64.
47. HOBBS, D. W., 1964, "The Tensile Strength of Rocks," *International Journal for Rock Mechanics and Mining Science,* Vol. 1, pp. 385—396.
48. HOLMAN, W. W., MCCORMACK, R. K., MINARD, J. P., and A. R. Jumikis, 1957, *Practical Applications of Engineering Soil Maps.* New Brunswick, New Jersey: Rutgers University — The State University of New Jersey. College of Engineering, Bureau of Engineering Research. Joint Project with the New Jersey State Highway Department in cooperation with Department of Commerce, U.S. Bureau of Public Roads. Engineering Research Bulletin No. 36, Report No. 22, pp. 74—75.
49. HORVATH, J., 1965, "A New Approach to the Determination of Stresses in the Earth's Crust and Strata Pressure on Tunnel Linings," *International Journal for Rock Mechanics and Mining Science,* Vol. 2, pp. 327—340.
50. HVORSLEV, M. J., 1949, *Subsurface Exploration and Sampling for Civil Engineering Purposes.* Vicksburg, Mississippi: Waterways Experiment Station. Report on a Research Project of the Committee on Sampling and Testing of the Soil Mechanics and Foundations Division of the American Society of Civil Engineers. November, 1949. Prepared for the American Society of Civil Engineers. Distributed by the Engineering Foundation, New York, N.Y.

51. International Society for Rock Mechanics, Commission on Standardization of Laboratory and Field Tests, 1972, *Suggested Methods for Determining the Uniaxial Compressive Strength of Rock Materials and the Point Load Strength Index.* Lisbon: Committee on Laboratory Tests, Documents No. 1 and No. 2, October and November, 1972.
52. JAEGER, C., 1963, "The Malpasset Report," *Water Power,* February 1963, pp. 55—61.
53. JAEGER, CH., 1965, a) "The Vaiont Rock Slide — I". *Water Power,* March 1965, Vol. 17, No. 3, pp. 110—111. b) "The Vaiont Rock Slide, II". *Water Power,* April 1965, Vol. 17, No. 4, pp. 142—144.
54. JAEGER, J. C., and E. R. HOSKINS, 1966, "Rock Failure under the Confined Brazilian Test", *Journal of Geophysical Research,* Vol. 71, No. 10, pp. 2651—2659.
55. JAEGER, J. C., and N. G. COOK, 1969, *Fundamentals of Rock Mechanics.* London: Chapman and Hall, p. 82.
55a. JAEGER, J. C., 1976, *Fundamentals of Rock Mechanics* (2nd ed.). London: Chapman and Hall; New York, N.Y.: John Wiley and Sons, Inc.
56. JOHNSON, A. M., 1970, *Physical Processes in Geology.* San Francisco, Calif.: Freeman, Cooper and Company, p. 202.
57. JUMIKIS, A. R., 1956, "Rupture Surface in Sand under Oblique Loads." *Proceedings ASCE,* Soil Mechanics and Foundations Division, Paper 861, Vol. 82, pp. SM-1 to SM-6.
58. JUMIKIS, A. R., 1961. "The Shape of Rupture Surface in Dry Sand," *Proceedings, 5th International Conference on Soil Mechanics and Foundation Engineering,* held July 17—22, 1961 in Paris, France. Paper 3A, pp. 693—698.
59. JUMIKIS, A. R., 1965, *Stability Analyses of Soil-Foundation Systems.* New Brunswick, N. J.: Rutgers University, College of Engineering, Bureau of Engineering Research, Engineering Research Publication 44 (55 pages).
60. JUMIKIS, A. R., 1967, *Introduction to Soil Mechanics.* New York, N.Y.: D. Van Nostrand Company, pp. 341-344.
60a. JUMIKIS, A. R., 1964, *Mechanics of Soils* (Fundamentals for Advanced Study). Princeton, New Jersey: D. Van Nostrand Company, Inc.
61. JUMIKIS, A. R., 1969, *Theoretical Soil Mechanics.* New York, N.Y.: Van Nostrand Reinhold Company, pp. 271—285.
62. JUMIKIS, A. R., 1969, *Experimental Studies on Moisture Transfer in a Silty Soil upon Freezing as a Function of Porosity.* New Brunswick, N. J.: Rutgers University, College of Engineering, Engineering Research Bulletin 49, pp. 30—32, 40.
63. JUMIKIS, A. R., 1973, *Settlement Tables for Centrically Loaded Rigid Circular Footings on Multilayered Soil Systems.* New Brunswick, N. J.: Rutgers University, College of Engineering, Engineering Research Bulletin 54, p. 21.
64. JUMIKIS, A. R., 1977, *Thermal Geotechnics.* New Brunswick, New Jersey: Rutgers University Press, pp. 320—328.
65. JUMIKIS, A. R., 1978, "Engineering Soil Maps." *Proceedings of the 3rd International Congress on Engineering Geology — I.A.E.G.,* held September 4—8, 1978 in Madrid, Spain. Section 1, Vol. 2, pp. 228—234.
66. JUMIKIS, A. R., 1982, *Soil Mechanics.* Melbourne, Florida: R. E. Krieger Publishing Co., Inc. (in press).
67. JUMIKIS, A. R., and W. A. SLUSARCHUK, 1973, "Electrical Parameters of Some Frost-Prone Soils," in *Advanced Concepts and Techniques in the Study of Snow and Ice Resources,* a United States contribution to the International Hydrological Decade. Washington, D.C.: National Academy of Sciences, 1974, pp. 765—781.

68. JUMIKIS, A. R. and A. A. JUMIKIS, 1975, *Red Brunswick Shale and its Engineering Aspects.* New Brunswick, N.J.: Rutgers University, College of Engineering, Engineering Research Bulletin 55, p. 51.
69. KÁRMAN, VON, T., 1911, "Festigkeitsversuche unter allseitigem Druck," *Zeitschrift des Vereins Deutscher Ingenieure,* Vol. 55, pp. 1749—1757.
70. KIERSCH, G. A., 1964, "Vaiont Reservoir Disaster," *Civil Engineering,* March, 1964, Vol. 34, pp. 32—39.
71. KOVÁRI, K., Editor, 1977, *Proceedings of the International Symposium* held April 4—6, 1977 at Zürich (two volumes). Rotterdam: A. A. Balkema, 1977.
72. KREBS, E., 1967, "Optical Surveying with a Borehole Periscope." *Mining Magazine,* 1967, Vol. 116, pp. 390—399.
73. KRYNINE, D. P., and W. R. JUDD, 1957, *Principles of Engineering Geology and Geotechnics.* New York, N.Y.: McGraw-Hill Book Company, Inc., pp. 23—24.
74. LEEMAN, E. R., and D. J. HAYES, 1966, "A Technique for Determining the Complete State of Stress in Rock Using a Single Borehole," *Proceedings First Congress of the International Society of Rock Mechanics,* held September 25 to October 1, 1966, at Lisbon, Vol. 2, Paper 4.3, pp. 17—24.
75. LONDE, P., 1973, *La mécanique des roches et les fondations des grand barrages. — Rock Mechanics and Dam Foundation Design.* Paris: International Commission on Large Dams (CIGB — ICOLD), pp. 32, 33. (Reference to C. Louis, 1970, Hydraulic Triple Probe to Determine the Directional Hydraulic Conductivity of Porous or Jointed Rock). London: Imperial College, Report D 12.
76. LUGEON, M., 1933, *Barrages et Géologie.* Methodes des recherches, terrassement et imperméabilization. Lausanne: Librarie de l'université F. Rouge et Cie. S. A.
77. MANGHNANI, M. H., SCHREIBER, E., and N. SOGA, 1968, "Use of Ultrasonic Interferometry Technique for Studying Elastic Properties of Rocks," *Journal of Geophysical Research,* 1968, Vol. 73, No. 2, pp. 824—826.
78. MARCELLO, C., 1964, "Quelques considérations a propos des accidents et ruptures de barrages." (La catastrophe du Vajont). *Transactions of the 8th International Congress on Large Dams,* held May 4—8, 1964, at Edinburgh, Vol. 5, pp. 557—572.
79. MENCL, V., 1966, "Mechanics of Landslides with Noncircular Slip Surfaces with Special Reference on the Vaiont Slide," *Géotechnique,* Vol. 16, No. 4, pp. 329—337.
80. MOHR, O., 1900, in *Zeitschrift des Vereins Deutscher Ingenieure,* Vol. 44.
81. MOOS, A., and F. DE QUERVAIN, 1948, *Technische Gesteinskunde.* Basel: Birkhäuser.
82. MÜLLER, L., 1964, "The Rock Slide in the Vaiont Valley," *Felsmechanik und Ingenieurgeologie,* Vol. 2, pp. 148—228.
83. MÜLLER, L., 1968, "New Considerations on the Vaiont Slide," *Rock Mechanics and Engineering Geology,* Vol. 6. Wien—New York, N.Y.: Springer-Verlag.
84. MÜLLER, L., (editor), 1974, *Rock Mechanics,* A course held at the Department of Mechanics of Solids of the International Center for Mechanical Sciences at Udine, Italy. Wien—New York, N.Y.: Springer.
84a. MÜLLER, L., 1965, "Safety in Rock Engineering" ("Die Sicherheit im Felsbau"), 15th Colloquium of the Austrian Regional Group of the International Society for Rock Mechanics, held September 24—25, 1964, at Salzburg. Wien: Springer.
85. MUMPTON, F. A., 1975, "Commercial Uses of Natural Zeolites." *Industrial Minerals and Rocks,* 4th ed., edited by S. J. Lefond. American Institute of Mining, Metallurgical and Petroleum Engineers, 1975, pp. 1262—1274.

86. MUMPTON, F. A., and W. C. ORMSBY, 1976, "Morphology of Zeolites in Sedimentary Rocks by Scanning Electron Micrography." *Clays and Clay Minerals*, 1976, Vol. 24, pp. 1—23.
87. NÁDAI, A., 1950, *Theory of Flow and Fracture of Solids,* Vol. 1 (2nd ed.). New York, N.Y.: McGraw-Hill Book Company, Inc., Chapter 24, p. 379.
88. OBERT, L., WINDES, S. L., and W. I. DUVALL, 1946, *Standardized Tests for Determining the Physical Properties of Mine Rock,* Bureau of Mines Report 3891.
89. OBERT, L., 1962, "In-Situ Determination of Stress in Rock," *Mining Engineering,* Vol. 14.
90. OBERT, L., and W. I. DUVALL, 1967, *Rock Mechanics and the Design of Structures in Rock.* New York, N.Y.: John Wiley and Sons, Inc.
91. PATTON, F. D., 1968, Lecture notes on "The Determination of Shear Strength of Rock Masses," Lecture given on May 7, 1968, at the Seminar on Rock Mechanics in Civil Engineering Practice to the Metropolitan Section of ASCE, Foundations and Soil Mechanics Group, in New York City, p. 12.
92. PRICE, N. J., 1966, *Fault and Joint Development in Brittle and Semibrittle Solids.* Oxford: Pergamon Press.
93. *Proceedings Geotechnical Conference Oslo, 1967* on Shear Strength Properties of Natural Soils and Rocks. Oslo: Norwegian Geotechnical Institute, Vol. 1, 1967.
94. PULS, L. G., 1963, "The Malpasset Report," *Water Power,* June 1963, Vol. 15, No. 6, pp. 228—230.
95. ROBERTS., A., 1968, "The Measurement of Strain and Stress in Rock Mass," in Chapter 6 of *Rock Mechanics in Engineering Practice,* Stagg, K. G. and O. C. Zienkiewicz, Editors. London—New York, N.Y.: John Wiley and Sons, pp. 190—191.
96. ROBERTSON, E. C., 1959, "Experimental Study of the Shear Strength of Rocks," *Bulletin of the Geological Society of America,* Vol. 66, October 1959.
97. ROCHA, M., 1966, "Determination of the Deformability of Rock Masses along Boreholes," *Proceedings First Congress of the International Society of Rock Mechanics,* held from September 25 to October 1, 1966, at Lisbon, Vol. 1, Paper 3.77, pp. 697—704.
98. ROCHA, M. et al., 1966, "A New Technique for Applying the Method of the Flat Jack in the Determination of Stresses inside Rock Masses," *Proceedings First Congress of the International Society of Rock Mechanics,* held September 25 to October 1, 1966 at Lisbon, Vol. 2, Paper 4.10, pp. 57—65.
99. ROWE, P. W., 1964, "Importance of Free Ends in Triaxial Testing," *Proceedings ASCE,* Soil Mechanics and Foundations Division, Vol. 90, pp. 1—77.
100. SERAFIM, J. L., 1962, *Internal Stresses in Galleries.* Lisbon: Laboratorio Nacional de Engenharia Civil, Technical Paper No. 204.
101. SHARMA, P. V., 1976, *Geophysical Methods in Geology.* Amsterdam-New York: Elsevier Scientific Publishing Company.
102. SINGER, F. L., 1951, *Strength of Materials.* New York, N.Y.: Harper and Brothers, Publishers, p. 69.
103. *Smithsonian Physical Tables,* 1920, 7th ed., Washington, D. C.: Vol. 71, No. 1, Publication 2359, p. 229.
104. SNOW, D. T., 1968, "Rock Fracture Spacings, Openings and Porosities." *Proceedings ASCE,* 1968, Vol. 94, No. M1.
105. STAGG, K. G., 1968, "In Situ Tests on the Rock Mass," in *Rock Mechanics in Engineering Practice,* K. G. STAGG and O. C. ZIENKIEWICZ (editors). London—New York, N.Y.: John Wiley and Sons, Inc., pp. 135—137.

106. Standford Research Institute, 1966, "On the Strength of Rock," *SRI Journal*, No. 9, March 1966, pp. 12—14.
107. SZÉCHY, K., 1966, *The Art of Tunnelling*. Budapest: Akadémiai Kiadó, p. 76.
108. TERZAGHI, K., 1962, "Measurement of Stresses in Rock," *Géotechnique* (London), Vol. 12, No. 2, pp. 105—124.
109. THOMAS, L. J., 1973, *An Introduction to Mining: Exploration, Feasibility, Extraction, Rock Mechanics*. Sydney: Hicks, Smith and Sons.
110. TIMOSHENKO, S., and J. N. GOODIER, 1951, *Theory of Elasticity*, 2nd. ed., New York, N.Y.: McGraw-Hill Book Company, Inc.
111. TRESCA, H., 1864, 1867, "Mémoire sur l'écoulement des corps solides soumis à des fortes pressions," Paris: *Comptes Rendus*, Académie des Sciences, July-December, Vol. 59, 1864, pp. 754—758; Vol. 64, 1867, p. 809.
112. U.S. Bureau of Reclamation, 1954, *Physical Properties of Typical Foundation Rocks*, Concrete Laboratory Report SP 39.
113. VUTUKURI, V. S., LAMA, R. D., and S. S. SALUJA, 1974, *Handbook on Mechanical Properties of Rocks*, Vol. 1, Clausthal, Germany: Trans Tech Publications.
114. VUTUKURI, V. S., and R. D. LAMA, 1978, *Handbook on Mechanical Properties of Rocks*, Vol. II, Vol. III and Vol. IV. Clausthal, Germany: Trans Tech Publications.
115. WINKLER, H. G. F., 1943, "Über die Thixotropie des Montmorillonits." *Kolloid-Zeitschrift*, 1943, Vol. 105, No. 1.
116. ZEMANEK, J., 1968, "The Borehole Televiewer — a New Logging Concept for Fracture Location and Other Types of Borehole Inspection." *Society of Petroleum Engineers*. Houston, Texas. September, 1968.

OTHER RELATED REFERENCES

ALEXANDER, L. G., 1960, "Field and Laboratory Tests in Rock Mechanics." *Proceedings of the 3rd Australia—New Zealand Conference on Soil Mechanics and Foundation Engineering*, 1960, pp. 161—168.

ASTM — Special Technical Publication 392, 1965, *Instruments and Apparatus for Soil and Rock Mechanics*. Philadelphia, Pa.: American Society for Testing and Materials.

ASTM — Special Technical Publication 479, 1970, *Special Procedures for Testing Soil and Rock for Engineering Purposes*. Philadelphia, Pa.: American Society for Testing and Materials.

ASTM — Special Technical Publication 601, 1976, *National Symposium on Fracture Mechanics*, 9th Symposium held at the University of Pittsburgh, Pa., in 1975. Philadelphia, Pa.: American Society for Testing and Materials.

ASTM — Special Technical Publication 654, 1978, *Dynamic Geotechnical Testing*. Philadelphia, Pa.: American Society for Testing and Materials.

BARKAN, D. D., 1962, *Dynamics of Bases and Foundations*. Translated from the Russian by L. DRASHEVSKA. New York, N.Y.: McGraw-Hill Book Company.

BÅTH, M., 1968, *Mathematical Aspects of Seismology*. Amsterdam—New York, N.Y.: Elsevier Publishing Company.

BORST, R. L., and W. D. KELLER, 1969, "Scanning Electron Micrographs of API Reference Clay Minerals and Other Selected Samples." Proceedings of the International Conference, Tokyo, 1969, Vol. 1.

BROCK, D., 1974, *Elementary Engineering Fracture Mechanics*. Leyden: Nordhoff International Publishing.

BULLEN, K. E., 1963, *An Introduction to the Theory of Seismology*. Cambridge University Press.

COATES, D. F., 1970, *Rock Mechanics Principles*. Ottawa, Canada: Information Canada, Mines Branch Monograph 874.

CORDING, E. J., 1968, "Stability of Large Underground Openings at the Nevada Test Site." Preprint of paper presented at the Second Space-Age Conference of the ASCE, Los Angeles, April 25, 1968.

DIMAS, J., SAVINI, T., and W. WEYERMANN, 1978, *Rock Treatment of the Canelles Dam Foundations*. Zürich: A Publication by RODIO in Collaboration with the Institute for Engineering Research Foundation Kollbrunner-Rodio, No. 42, May, 1978.

DONATH, F. A., 1964, *Effect of Loading Rate on the Deformation Behavior of Rocks Subjected to Triaxial Compression*, ARO-D Report No. 3055: 1, July 31, 1964, New York, N.Y.: Columbia University (81 pages).

FAIRHURST, C., (editor), 1967, *Failure and Breakage of Rock, Proceedings 8th Symposium on Rock Mechanics*, American Institute of Mining Engineers, pp. 237—302.

FUMAGALLY, E., 1973, *Statistical and Geomechanical Models*. New York, N.Y.: Springer-Verlag.

GREGORY, C. E., 1979, *Explosives for North American Engineers*. Second Edition. Clausthal, Germany: Trans Tech Publications.

GRETENER, P. E., 1965, "Can the State of Stress Be Determined from Hydraulic Fracturing Data," *Journal of Geophysical Research*, Vol. 70, pp. 6205—6215.

GRIFFITHS, D. H., and R. F. KING, 1965, *Applied Geophysics for Engineers and Geologists*. Oxford-New York, N.Y.: Pergamon Press.

GRIGGS, D., 1939, "Creep of Rocks," *Journal of Geology*, Vol. 47, No. 3, pp. 225—251.

GRIGGS, D., and J. HANDIN, 1960, "Observations on Fracture and a Hypothesis of Earthquakes," in *Rock Deformation*. New York, N.Y.: The Geological Society of America, Memoir 79.

GROBBELAAR, C., 1970, *A Theory of the Strength of Pillars*. Johannesburg: Voortrekkerpers.

HEISKANEN, W. A., and VENING, MEINERZ, F. A., 1958, *The Earth and Its Gravity Field*. New York, N.Y.: McGraw-Hill Book Company.

International Society for Rock Mechanics, 1972, *Proceedings, Percolation Through Fissured Rock*, a Symposium held in 1972 at Stuttgart. Essen: Deutsche Gesellschaft für Erd- und Grundbau.

KEHLE, R. O., 1964, "The Determination of Tectonic Stresses through Analysis of Hydraulic Well Fracturing," *Journal of Geophysical Research*, Vol. 69, pp. 259—273.

KOMARNITSKII, N. I., 1968, *Zones and Planes of Weakness in Rocks and Slope Stability*. New York, N.Y.: Consultants Bureau, p. 15.

KRSMANOVIĆ, D., 1967, "Initial and Residual Shear Strength of Hard Rocks," *Géotechnique* (London), Vol. 17, pp. 145—160.

LAMA, R. D., and V. S. VUTUKURI, 1978, *Handbook on Mechanical Properties of Rocks*, Vol. I, II, III and IV. Clausthal, Germany: Trans Tech Publications.

LEE, W. H, K.. (ed.), 1965, *Terrestrial Heat Flow*. Geophysical Monograph 8. Washington, D. C.: American Geophysical Union.

LEGGET, R. F., 1962, *Geology and Engineering* (2nd ed.). New York, N.Y.: McGraw Hill Book Company.

LEGGET, R. F., 1973, *Cities and Geology*. New York, N.Y.: McGraw-Hill Book Company.

LEGGET, R. F., 1979, "Geology and Geotechnical Engineering." *Proceedings ASCE, Journal of the Geotechnical Engineering Division*. Paper 14444, March 1979, Vol. 105, No. GT 3, pp. 342—391.

LOMNITZ, C., 1956, "Creep Measurement in Igneous Rocks," *Journal of Geology*, Vol. 64, No. 5, pp. 473—479.

LORENZ, H., 1960, *Grundbau Dynamik*. Berlin: Springer.

LOVE, A. E. H., 1911, "Some Problems of Geodynamics." Adams Prize Essay. Cambridge University Press.

LOVE, A. E. H., 1944, *A Treatise on the Mathematical Theory of Elasticity* (4th ed.). New York, N.Y.: Dover Publications.

MOHR, F., 1963, *Gebirgsmechanik*. Goslar: H. Hübner.

MÜLLER, L., 1965, "Safety in Rock Engineering" ("Die Sicherheit im Felsbau"), 15th Colloquium of the Austrian Regional Group of the International Society of Rock Mechanics, held September 24—25, 1964, at Salzburg. Wien: Springer.

MUSCAT, M., 1937, *Flow of Homogeneous Fluids Through Porous Media*. New York, N.Y.: McGraw-Hill Book Company, Inc.

NEWMARK, N. M., and E. ROSENBLUETH, 1971, *Fundamentals of Earthquake Engineering*. Englewood Cliffs, New Jersey: Prentice-Hall, Inc.

OBERT, L., 1964, "In Situ Stresses in Rock, Rainier Mesa, Nevada Test Site, Operations Nougat and Storax:" WT-1869, U.S. Bureau of Mines, College Park, Md.

POVLOVIĆ, M., 1970, "Determining of Quasihomogeneous Zones of Elasticity and Deformability Characteristics of Rock Mass in Tunnel on the Basis of In-Situ Investigations," *Proceedings Second Congress of the International Society for Rock Mechanics*, held in 1970 at Belgrade, Yugoslavia.

PERSEN, L. N., 1975, *Rock Dynamics and Geophysical Introduction to Stress Waves in Rocks*. Amsterdam-New York, N.Y.: Elsevier Scientific Publishing Company.

Proceedings of a Conference on Site Exploration in Rock for Underground Design and Construction, held March 29—31, 1978 at Alexandria, Virginia. Springfield, Virginia: The National Technical Information Service. Sponsored by the U.S. Department of Transportation and the ASCE, National Capital Section, Washington, D. C. Final Report, July 1979 (Report No. FHWA — TS — 79 — 221). (98 pages).

RAYLEIGH (J. W. STRUTT), 1855, "On Waves Propagated Along the Plane Surface of an Elastic Solid." Proceedings of the London Mathematical Society, 1885, Vol. 17.

RAYLEIGH, 1877 and 1945, *The Theory of Sound*. New York, N.Y.: Dover Publications, 1945.

RICHART, F. E., Jr., J. R. HALL, Jr., and R. D. WOODS, 1970, *Vibrations of Soils and Foundations*. Englewood Cliffs, New Jersey: Prentice-Hall, Inc.

ROBERTS, A. F., 1977, *Geotechnology*. Oxford: Pergamon Press.

ROCHA, M., 1965, *Some Problems on Failure of Rock Mass*. Lisbon: Laboratório Nacional de Engenharia Civil, Memória No. 258, p. 7.

RODRIGUES, F. P., 1970, "Anisotropy of Rocks: Most Probable Surfaces of the Ultimate Stresses and of the Moduli of Elasticity," *Proceedings Second Congress of the International Society for Rock Mechanics*, Belgrade, Yugoslavia.

ROLFE, S. T., and J. M. BARSON, 1977, *Fracture and Fatigue Control in Structures: Applications of Fracture Mechanics.* Englewood Cliffs, N.J.: Prentice-Hall, Inc.

SCHEIDEGGER, A. E., 1976, *Foundations of Geophysics.* Amsterdam—New York, N.Y.: Elsevier Scientific Publishing Company.

SCHULTZE, E., and H. MUHS, 1967, *Bodenuntersuchungen für Ingenieurbauten.* Berlin-Heidelberg-New York, N.Y.: Springer-Verlag.

SINGH, D. P., 1975, "A Study of Creep of Rock." *International Journal of Mechanics and Mining Science,* 1975, Vol. 12, No. 9, pp. 271-275.

SNYDER, J. L., 1977, *Vibratory Densification of Fly Ash.* Rutgers Ph. D. Thesis. New Brunswick, New Jersey: Department of Civil Engineering, Rutgers University, May 1977 (217 pages).

STACEY, F. D., 1969, *Physics of the Earth.* New York, N.Y.: John Wiley and Sons.

TER-STEPANIAN, G., 1966, "Types of Depth Creep of Slopes in Rock Masses." *Proceedings First Congress of the International Society for Rock Mechanics,* Lisbon, Vol. 2, pp. 157—160.

TERZAGHI, K., 1962, "Does Foundation Technology Really Lag?" A letter to the Editor of the *Engineering News-Record*, February 15, 1962, pp. 58—59.

U.S. Corps of Engineers, Department of the Army, 1948, *Subsurface Investigation Geophysical Explorations, Engineering Manual.* Civil Works Construction, Part 118, Chapter 2, September 1948.

WENNER, F., 1916, *A Method of Measuring Earth Resistivity.* Washington, D.C.: U.S. Bureau of Standards Bulletin No. 12, 1916; No. 3. Scientific Papers of the Bureau of Standards Vol. 258.

WÖHLBIER, H., and D. HENNING, 1969, " Effect of Preliminary Heat Treatment on the Shear Strength of Kaolinite Clay." Highway Research Board's Special Report 103, on *Effects of Temperature and Heat on Engineering Behavior of Soils.* Proceedings of an International Conference held at Washington, D.C., January 16, 1969, with the Support of the National Science Foundation; pp. 287—300.

CHAPTER 5

ROCK MASS PROPERTIES

5-1. Mechanical Defects of Rocks

All rock mass properties may be grouped into two large groups, namely:

1. Macroscale, and
2. Microscale.

Macroscale rock mass properties are essentially properties of the component.

Microscale rock properties are properties of the substance. In evaluating static and dynamic conditions within the body of the rock *en masse*, engineers consider natural inferiorities or weaknesses (mechanical defects) of rocks. The inferior properties of a rock mass affect underground construction operations in rock, and performance of structures built on or in rock.

Without exception, large masses of all kinds of rocks, regardless of the type, have more or less conspicuous mechanical or structural defects or natural inferiorities that influence adversely the strength of rock *en masse* and engineering operations in rock. These defects consist of more or less closely spaced fractures and other *planes of weakness* in rock. The existing weaknesses may be naturally produced systems of more or less closely spaced fractures in rock *in existence,* such as

1. Fractures, cracks and haircracks;
2. Fissures;
3. Bedding planes, laminations, schistosity, partings;
4. Stratification;
5. Joints;

6. Fault planes and zones, crushed zones;
7. Folds;
8. Voids;
9. Cavities;
10. Seams and interbeds of weak and plastically unstable rocks, aquifers, clays, and shales; and
11. Ancient slip planes and other possible weaknesses.

Planes and zones of weakness in rocks are classified as existing and/or potential ones.

Bedding Planes

Bedding planes are the division planes in bedded deposits such as sedimentary and stratified rocks, which planes separate the individual layers, beds, or strata.

Cleavage Planes

Cleavage is the tendency to split or cleave along definite, smooth, parallel, closely spaced planes. As applied to rocks, cleavage is the property of splitting into thin, parallel sheets. The cleavage plane is the plane in rocks or minerals along which cleavage takes place.

A *parting* is a thin layer of deposited and altered material (carbonaceous or other organic material) separating beds in sedimentary or metamorphic rocks.

Separation is a relatively fresh break along a bedding plane or between beds in sedimentary or metamorphic rocks. Separations are usually man-made: they develop as a consequence of mining.

Schistose rock: Any crystalline rock whose constituent minerals have a more or less parallel arrangement.

If the walls of the joint are separate, the term *fissure* is sometimes used.

If a rock has innate mechanical defects such as bedding or cleavage planes, the joints and faults constitute an additional source of weakness.

Potential planes and zones of weakness in rock are artificially created weaknesses in rock sequences. These may be brought about by man's activities during broad-scale, intensive blasting operations in rock; the possibility of influx of water in and seepage through fissures, cracks, joints, and fault zones in rock; and possibly by other factors.

Naturally and/or artificially produced planes and zones of weakness in rocks affect the stability of underground openings and the stability of excavation slopes made in rocks, and provide for passageways for the underground flow of water.

Investigations of planes and zones of weakness caused by man's activities have been pursued very meagerly; they have concentrated mainly on vibration cracks and on zones and fractures produced by frost wedging.

Stratification

Stratification pertains to layering or bedding of geological materials that readily separate along bedding planes because of different kinds, arrangement, sizes, and assortment of material, or because of some geological processes — for example, interruption of deposition, or other changes affecting deposition.

Within layers, minor units about 5 mm or less in thickness are called *laminae*. A laminae-containing deposit is considered to be laminated.

Lamination of rock strata also constitutes a discontinuity in the geological medium. Hence, it is considered to be a mechanical weakness of rock *en masse*.

Separation is a break between beds along bedding planes.

As to rock engineering, horizontal, relatively thick rock strata are sometimes suitable for excavating small underground openings, such as driving small adits and narrow underground passageways, depending upon the intensity of the jointing. Here, in bridging the narrow span, the thick rock roof stratum acts like a beam or slab (Fig. 5-1a). Thin, fissured rock strata are unsuitable for construction of wide underground openings in rock (Fig. 5-1b). In such a case, the underground opening of a pointed arch roof design may be satisfactory (Fig. 5-1c).

Seams (Interbeds)

Stratified, horizontal rock layers on a thin seam of plastic clay, when undercut by an underground opening, have been observed to slide out of the vertical alignment into the opening (Fig. 5-2). This is so because of the removal of the lateral restraint brought about by the excavation of the rock.

Seams of plastic clay and of calcite may cause a stratified, dipped rock mass above the seam to slide down into the excavation (Fig. 5-3).

5-2. Fractures

A fracture is a fresh break in the continuity of a body of rock, not attended

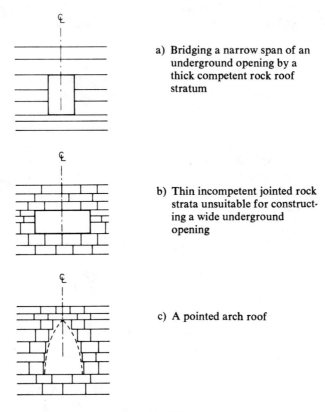

a) Bridging a narrow span of an underground opening by a thick competent rock roof stratum

b) Thin incompetent jointed rock strata unsuitable for constructing a wide underground opening

c) A pointed arch roof

Fig. 5-1
Underground openings in level rock strata

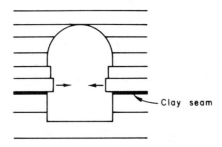

Fig. 5-2
Movement of rock strata over a clay seam

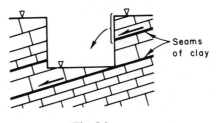

Fig. 5-3
Sliding over clay seams

by a displacement and not oriented in a regular system. Fractures may be open or closed but not bonded. Fractures are often man-made (for example, those caused by blasting).

Fractures may be kilometers in length, thousands of meters deep, and proportionally wide. Or else, fractures may be microscopic in all dimensions. Fractures in rock provide one of the easiest ways for water to enter.

The main types of fractures are *joints* and *faults*.

Generally, both joints and faults are surfaces of parting, or fractures—elements of rock weakness.

A *joint* is a crack or fracture in a rock along which no, or no noticeable, displacement has occurred. A *fault* is a fracture along which there has been some displacement.

Faults are named and classified according to inclination of the fault surface, and by the direction and relative movement of the rocks along the fault plane.

5-3. Joints

A joint is a simple fracture of geologic origin in the continuity of a rock mass. Joints in rock occur in a system of sets, but never singly.

If the walls of the joint are separated, the term fissure is sometimes used. Both joints and faults are geologic elements of practical interest to rock mechanics and rock engineering, i. e., foundation and hydraulic structures engineering.

Besides of ease or difficulty in working in rock; besides of *cost* of rockworks; and besides that groundwater flows along the rock and on the surfaces of joints and faults, rock alteration processes, too, take place. Much of the excessive leakage of water into an underground opening may take place along faults. This situation creates the problem of leaky and wet underground openings and tunnels.

Joints are formed by tensile stresses caused by contraction resulting from cooling of magma and of lava flows. The origin of planar joint sets is really not well known. However, almost every rock contains joints.

A joint can be *open* or *closed*. A *closed* joint is one whose walls are in contact. Closed joints may be almost invisible. Yet they constitue surfaces along which there is no resistance against separation. There are also tiny haircracks present in rocks.

Joints may be continuous or discontinuous.

Joints and joint systems have attracted the attention of builders ever since rock was used in engineering structures and as a structure-supporting material. In underground openings, joints affect the extent to which their sides, walls and the roof must be supported. Joints respond to explosions applied static and dynamic loads, and seepage of water. Because joints are one of the major causes of excessive *overbreak* of rock excavations (Fig. 5-4) and because joints constitute potential rock-slide hazards in unlined, undercut rock formations, and create water-trouble in rock excavations (Fig. 5-5), jointing always deserves careful attention and careful exploration.

Also, sand and other kinds of gouges tend to flow into an underground opening as the excavation is worked through a fault zone. A gouge is finely abraded material between the walls of a fault, the result of a grinding movement.

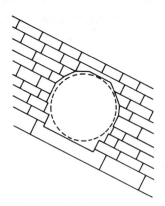

Fig. 5-4
Illustrating overbreak around a tunnel driven through a highly jointed stratified rock

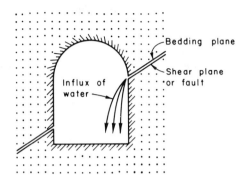

Fig. 5-5
Water influx from a bedding plane (or a joint, or fault zone)
into an underground opening

Generally, joints subdivide rock masses into individual blocks to form a three-dimensional network that presents the rock as an uncemented, cohesionless aggregate of cuboid blocks somewhat comparable to closely fitted blocks in a dry masonry wall (imbrication).

The term *imbrication* means the lying of blocks lapped over each other in rock strata in a regular order so as to form an imbricate pattern surface. Some simple imbrication joint patterns are sketched in Fig. 5-6.

The influence of joints and faults on the failure of a rock mass is now widely recognized. Because the geometry of the joint pattern has a decisive effect on the strength of the rock mass to resist structural loads, the jointing patterns require some kind of characterization.

The need for rating and characterization of jointing and fissuring of rocks has been voiced at the First International Congress of Rock Mechanics held in Lisbon in 1966.

Assuming that in some instances the rock mass may be considered as being fractured in a regular network of rock blocks (which blocks are sometimes imbricated), ROCHA (27) applied the concept of *imbrication i* for characterizing jointing in rocks (Fig. 5-6).

In particular, the more imbricated the blocks, the higher their shear strength, because rupture surfaces have not only to follow the joints, but also to cut the material of the blocks. A jointing system determines the ease of rock excavation by means of a power shovel or by means of explosives. The spacing of joints in quarries determines the largest size of blocks of sound rock that can be quarried for cut block stones for construction purposes.

Because joints are of geological origin, there are usually some weathering and decompositon products (gouge) on the joint surfaces which in some instances may bond the joint.

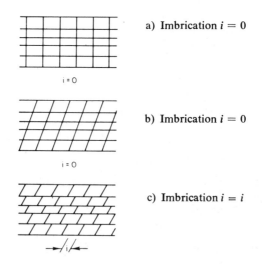

Fig. 5-6
Some simple imbrication joint patterns in rock masses

Popov (after KOMARNITSKII, 1968) suggested that the effect of jointing in rocks on their strength property may be characterized by a *coefficient of weakness*. It is the ratio of the cohesive force in the weak rock to the cohesive force in the same rock with no visible traces of such weakness. On the basis of experimental data, Popov devised a table (Table 5-1) showing the coefficient of weakness of rocks as a function of rock jointing characteristic.

5-4. Faults

If the rock masses on opposite sides of a rupture are relatively displaced or shifted, the break is called a fault. Thus, a fault is a fracture in a rock formation along which there has been movement parallel to the surface of the fracture. Faults occur in regions of tectonic disturbance. A fault, as a result of tectonic disturbance of a geological material, is a break or fracture

TABLE 5-1

Coefficient of Weakness of Rocks as a Function of Rock-Jointing Characteristics

(after S. I. Popov)*

Jointing characteristics in rock	Coefficient of Weakness	
	Limiting values	Average values
1	2	3
Dense network of fractures, in all directions of layered rock to individual uncemented blocks	0.0 – 0.01	0.0005
Dense network of open fractures in all directions	0.001 – 0.02	0.005
Dense jointing	0.01 – 0.04	0.02
Jointing above average	0.04 – 0.08	0.06
Average jointing (open and closed fractures every 20–30 cm)	0.08 – 0.12	0.1
Rocks with below-average jointing	0.12 – 0.9	0.2
Network of deep joints every 30–50 cm; insignificant number of open fractures	0.3 – 0.4	0.35
Little-jointed rocks; closed fractures	0.4 – 0.6	0.5
Microfractures almost absent	0.6 – 0.8	0.7
Monolithic rock with no sign of jointing	0.8 – 1.0	0.9

* This table is reprinted with the permission of the PLENUM PUBLISHING CORPORATION from N. I. KOMARNITSKII, *Zones and Planes of Weakness in Rocks and Slope Stability*, p. 15, Table 1. Translated from the Russian by Consultants Bureau, New York, in 1968, and copyrighted by the Plenum Publishing Corporation.

(or zone of fractures or zone of weakness) in the continuity of a body or a system of rock (of a geological formation of rocks).

Mechanically, faults are shear failures of rock which result from tensile, compressive, or torsional stresses acting on a rock mass. Along the fracture or plane of failure, or fault plane, viz., shear plane, there takes place a relative shear displacement of the adjoining rock masses parallel to one another.

The shear displacement may be a few centimeters long, or run as long as many kilometers.

If the break is complex and the fractures occupy a wide strip, the fault plane becomes a fault zone.

In some instances the rock adjacent to faults is completely crushed. Such rock constitutes a *crushed zone*.

A fault sets in where the continuity of rock (or soil) layers is interrupted by failure in shear of the rock due to compressive forces, i. e., when the rock is strained past the breaking point and yields along a crack or series of cracks, so that corresponding points on the two sides become distinctly offset (Fig. 5-7a). One side of the so-sheared rock mass may rise or sink, or move laterally with respect to the other, depending on the nature of disturbance.

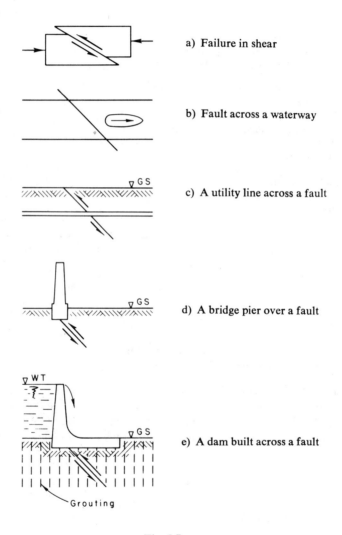

a) Failure in shear

b) Fault across a waterway

c) A utility line across a fault

d) A bridge pier over a fault

e) A dam built across a fault

Fig. 5-7
Structures across faults

Faults may be vertical, horizontal, or a combination thereof. Some fault structures are illustrated in Fig. 5-8.

In rock engineering, one distinguishes between *active or live faults* and *passive or dead faults*. An active fault is one in which movements or shear displacements have occured along it during the time of recorded history of man. In contradistinction, a passive fault is one in which no recorded slipping in the recorded history of man has taken place, but remains in static

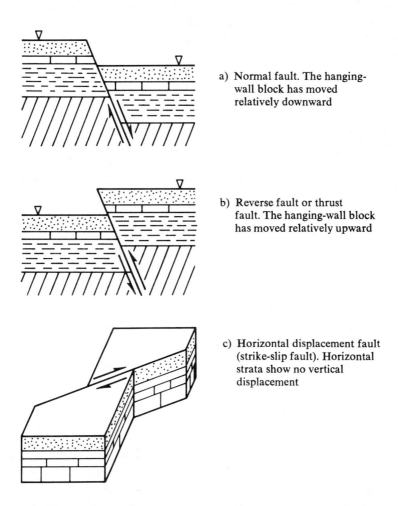

a) Normal fault. The hanging-wall block has moved relatively downward

b) Reverse fault or thrust fault. The hanging-wall block has moved relatively upward

c) Horizontal displacement fault (strike-slip fault). Horizontal strata show no vertical displacement

Fig. 5-8
Principal kinds of faults

condition. The latter, however, is not a certainty because earthquake vibrations may cause the fault to open or to become active. Under structural load, static or dynamic, the rock mass can slide easily along the fault planes or surfaces.

In seismic regions, faults give rise to an additional problem; namely, when built across faults, structures such as tunnels, dams, bridges, and pipelines may experience, and may have experienced, distress and/or total destruction if subsequent movements of faults would occur, or have occured (Figs. 5-7b to 5-7e). Also, earthquake tremors and man-induced vibrations may cause a fault to open or to become active. Hence, in geotechnical engineering, faults constitute an undesirable hazard in working in rock.

5-5. Folds

A layered rock system, originally in a horizontal position, may be folded into folds of successive degrees of intensity of the partaking forces. Folds result from forces acting tangentially to the earth's surface (Fig. 5-9). The nature of the rock and its strength determine whether the strata of a rock formation will rupture or fold. The folds may assume a wide variety of shapes. An upwarped or upfolded segment of the earth's crust is an anticline. A downwarped or downfolded segment is a syncline.

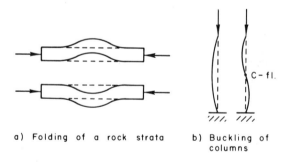

a) Folding of a rock strata   b) Buckling of columns

Fig. 5-9
Formation of folds

In geotechnical engineering, the synclines may cause problems of water accumulation. Also, rock folding is usually accompanied by fissures in anticlines and synclines. Along the crest of anticlines tension cracks are usually formed in the upper fibers where the outer rock strata are under tension (Fig. 5-10). In synclines, the lower fibers of the rock strata are under tension, hence the cracks at the bottom (Fig. 5-11).

ROCK MASS PROPERTIES

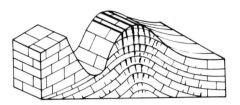

Fig. 5-10
Cracks in an anticline

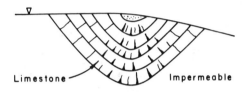

Fig. 5-11
Cracks in a syncline

Sliding of rock masses, too, is associated with folds.

Besides mechanical defects in folded rocks, one should also consider the effect of surcharge pressure imparted on a structure by folded rocks. This may be vizualised from the following examples.

If the alignment of a tunnel is traced normal to the strike under an anticline or syncline of a folded rock system, there is a variation in rock surcharge pressure p on the tunnel from rock masses from above. These variations are qualitatively sketched in Figs. 5-12 and 5-13. Figure 5-12 shows variation in surcharge pressure on a tunnel driven through an anticline. Notice the arching effect of the convexly curved rock stratification on the magnitude of pressure at the ends (basis) of the arch at both ends and at the middle of the tunnel.

If a tunnel is driven through a syncline (Fig. 5-13), large rock pressures are imparted in the middle part of the tunnel. Here the large stresses in the rock strata transmitted to the tunnel may complicate the driving and construction of the tunnel technically as well as economically. Also, large quantities of water may accumulate in the syncline. Hence, saturated layers can be anticipated here, requiring elaborate drainage systems.

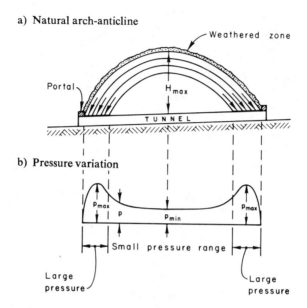

Fig. 5-12
Variation in surcharge pressure on a tunnel driven through an anticline

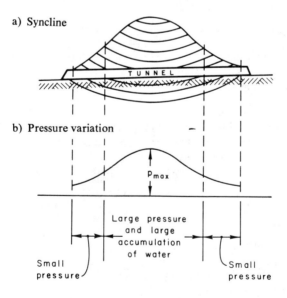

Fig. 5-13
Variation in surcharge pressure on a tunnel driven through a syncline

ROCK MASS PROPERTIES

Figure 5-14 shows possible positions of a tunnel in horizontal rock formations.

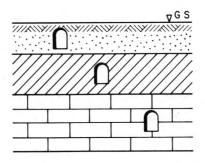

Fig. 5-14
Tunnel positions in horizontal strata

Figure 5-15 shows positions of a tunnel parallel with the strike in steeply dipping strata.

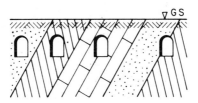

Fig. 5-15
Tunnel positions in steeply dipping strata

Figure 5-16 illustrates three possible variants of tunnel location:

1. In the syncline; large pressures and large accumulation of water; saturated rock strata may be anticipated;
2. In the anticline; natural arch; least pressure on tunnel; and
3. Increased stresses in layers, which stresses complicate the driving and construction of the tunnel.

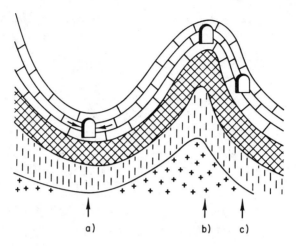

Fig. 5-16
Locations of tunnels in syncline and anticline

The conditions of folded rock systems and their relation to stability of the water-impounding massive dams are sketched in Fig. 5-17. In the case of a dam founded on an anticline (Fig. 5-17a) one expects a relatively small uplift pressure. However, the dam appears less safe against lateral sliding. The case where a dam is founded on a syncline (Fig. 5-17b) presents an unfavorable uplift condition for the dam; it is, though, relatively safe against sliding.

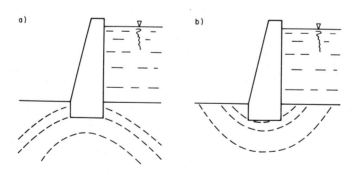

Fig. 5-17
Effect of rock folding on performance of a dam

Cavities and Voids

Sinkholes and other sorts of cavities are most commonly formed in limestone, gypsum, and salt. The collapsing of roofs of the cavities by structural loads means distress to the structure if founded on such a cavity-containing rock. Thus, cavities consitute mechanical defects of the rocks concerned.

Also, large voids and great porosity of rocks mean weaknesses requiring grouting or chemical injection, especially for hydraulic structures such as hydraulic power plants, aqueous pressure tunnels, and the like.

Porosity

Pores or voids in rock, too, are very important forms of non-uniformity of a rock structure. The largest porosities have been observed in sedimentary rocks. Porosity decreases the strength of the rock because, at the places of pores or voids there is a lack of bond (cementation) in the rock.

The foregoing review of mechanical weaknesses of rocks reveals that rock properties, like soil properties, vary from place to place and from stratum to stratum. Also, they vary from sample to sample even if the sample is obtained from the same rock formation. In this respect, and considering the many kinds of weaknesses a rock material may contain, one discovers that fissured, jointed, faulted, and folded rocks as a construction material for transmitting and distributing of structural loads to the rock ground, especially where the transmission of the horizontal components of inclined loads are concerned, are a nonhomogeneous, anisotropic material of bewildering complexity.

Hence, the nature of the rock mass and its complex structure call increasingly for specialized approaches in rock mechanics and rock engineering for design of engineering structures on and in rock.

5-6. Summary on Rock Mass Weaknesses

The significance of rock in dam and foundation engineering comes to the fore from the fact that more than 80 % of failures of massive dams are attributable to insufficient or incompetent strength of the rock, and/or deficient laying of foundations of such structures.

The rock should be accepted by the engineer as given by Mother Nature in its natural state with all of its own weaknesses, as determined by stratification, cracks, fissures and other mechanical rock defects, weathering, and the like. Of course, if economically and technically feasible, the strength properties of the rock can be improved, thus increasing the stability of the dam or foundation or underground opening, whichever the case may be.

Foremostly, it is the macroscopic structural anisotropy of the rocks, as governed by stratification, fissuring, and jointing, which frequently causes considerable deformations of the underground rock, but for which the analytical comprehension is not satisfactorily possible as yet.

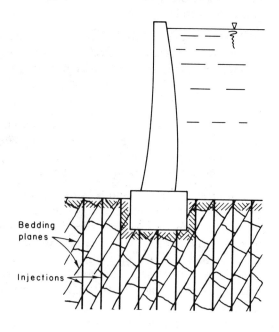

Fig. 5-18
Rock-dam system

Based on what was said above, let us fix a mental picture of a water-impounding concrete dam founded on or in a rock. A dam (barrage) and its supporting rock (Fig. 5-18) should form a static, integral unit; i. e., a barrage should be united with its supporting rock by a "solid marriage." Unfortunately, however, it usually turns out to be an unequal marriage, because:

She, whose name is Madame Gracia Archy Barrage, is young,
    formfit, graceful, and beautiful in shape.

He, whose name is Señor Rock, is usually age-old and weathered.

She: Smooth, slender, well-built, and of good-looking stature.

He: His interior is disturbed by tectonic processes in the past; his face is creased and deeply furrowed by chasms, caverns, cracks, and fissure wrinkles, or swollen up with folds.

She: Is getting quickly tall and strong (grown-up) and lives frequently on a large pedestal; and according to the opinions of geotechnical engineers and concrete technologists, the finest for her is just good enough.

He: Upon him rests all the applied loads. He bears them patiently, and fortunately for him, he is smarter than all the calculations and analyses, and he knows how to help himself (visualize stress distribution in rock). However, if the rock is weak, it may be given antiweakness medicine-injections. If in spite of this "medical help" the Rock constitution loses his internal balance (equilibrium), the rock simply breaks, ruptures — he is finished. Subsequently, and tragically, she also, Madame Barrage, loses her existence.

Nowadays, earth and concrete dams become higher, longer, bigger, thus increasing in their weight. And so increase, also, the forces which are transmitted into the underground. Over and above this, engineers are faced with the necessity to select and adopt more unfavorable construction sites because most of the best sites have been built up already. Therefore, an analytical treatment of the stresses in the underground rock medium, a technological stress measurement in situ, and testing of rock strength become now, more and more an urgent necessity than before.

Relative to the dam, underground opening, and structural foundation-supporting rock, one reckons with the following forces:

1. Self-weight of structure,
2. Weight of rock,
3. Geological forces brought about by tectonic processes and seismic action,
4. Hydrostatic uplift,
5. Hydrostatic shear, induced by the load of the impounded water,
6. Seismic forces, and
7. Reactions brought about by the structure.

The higher the dam, the greater are these forces. The highest concrete dam in the world today is the Grande Dixence dam, 281 m high (in Switzerland).

Influence of Rock on Stability of Dams and Foundations

The basic concept in rock mechanics studies and design pertaining to subsurface engineering in rock (domes, tunnels, shafts, foundations) is the consideration of the rock and the structure as a functionally and statically coherent, integral entity. The prerequisite for this, besides knowledge of geological structure, stratigraphy, and rock weaknesses, is knowledge of the physical and mechanical properties of the rock under consideration.

The mutual interaction of rock and structural foundation systems is of utmost importance and should not be underestimated. The mutual interaction forms the basis for giving the form to the foundation and to its proportioning, viz., sizing the structure itself. This basis, in its turn, depends upon the strength properties of the foundation material itself, and the foundation-supporting rock. Also, in tunnel and shaft engineering, the rock pressure on such underground-opening structures, and its stress and strain influence on the rock in the vicinity of the tunnel and/or shaft opening, come to the fore. Thus, rock mechanics is a discipline subject to all concepts of rock testing, technical evaluation, and engineering application. The concepts are based on exact bases of theoretical and applied mechanics, soil mechanics principles, and foundation engineering.

5-7. Grouting

Grouting is a process of injecting under pressure a slurry of fluid grout, or other suitable materials into the mass of a defective rock formation through a borehole to mend fissures and cracks in the hope that all fissures, joints and cavities will be sealed off against water in rock.

Grout is a pumpable slurry of neat cement; or a mixture of neat cement and fine sand; or clay-cement; or bentonite; or pozzolan; or polyurethane foam; or a resin-asphalt emulsion; or certain chemicals. When appropriate, sand, clay, rock flour, fine gravel and other inert material can be used as fillers.

The grout is forced into the borehole to seal crevices, fissures, cracks, and joints in rock to prevent groundwater from seeping or flowing into an excavation, or from seeping through rock defects underneath a dam. As the grout hardens, it closes most of the rock fissures.

To accelerate setting time of the grout, additives such as calcium chloride or sodium silicate are added to the mix of the grout. The setting time of cement grout can be retarded by a gypsum additive.

Fine-particled bentonite clay increases plasticity of the grout. Most cement grouts now also contain flyash, which is inexpensive. Because of its spherical particles, flyash facilitates the ease of pumping of the grout, and reacts pozzolanically with lime.

The best grout is the stiffest mix that can be grouted effectively. This must be ascertained experimentally.

A historical survey on grouting in engineering may be found in

References 11 and 12. For a bibliography about grouting see References 2, 3, 5, 10, 13 and 18.

The purpose of grouting rock is to improve such rock properties as

1. increasing rock strength and bearing capacity
2. reducing of rock deformations
3. controlling of the hydraulic regimen in, and permeability to water of, the rock.

The injected grout solidifies the rock giving it a more or less monolithic structure. Grouting increases the modulus of elasticity of the rock. It cuts down the amount of discharge of the seepage water, and, in conjunction with a judiciously installed drainage system, grouting may also contribute to the reduction of uplift pressure on hydraulic structures. All these improvements in rock properties contribute to the stability of the structure-rock system.

Grouting is used for sealing of permeable igneous rocks, porous foundation-supporting rocks such as coral limestone, cavernous (karstic) limestone, siliceous sandstone, quartzite, and other kinds of permeable rocks to keep water out of open excavations as well as to provide for watertight diversion aprons underneath hydraulic structures such as hydraulic power plants, aqueous pressure tunnels, dams, vehicular tunnels, locks, docks, bridge foundations, or to reduce hydrostatic uplift on such structures.

Fig. 5—19 shows cement grouting on the main curtain at Cedar Springs dam near San Bernardino, California. The grouted rocks are here fractured granite, diorite, gneiss.

The pressure to use in grouting operations should be a safe one in order to avoid lifting up the rock above the jointing, or widening of fissures.

Grouting is also practiced for

a) transferring of stresses from pressure tunnels to the surrounding rock
b) control of pore-water pressures in rock
c) repairing of blasting damage in rock around underground openings
d) sealing off water and consolidating weak rock ahead when driving of tunnels, and
e) for stabilizing rock slopes.

The afore-mentioned rock improvements are achieved by

1. consolidation grouting
2. grouting of rock joints, faults, fissures and cavities

Fig. 5—19
Cement grouting on the main curtain at Cedar Springs dam near San Bernandino, California — May 1969. Rock: fractured granite, diorite, gneiss.
(Photo courtesy of W. W. Peak, California Department of Water Resources)

3. flushing out clay film and gouge from rock defects and subsequently concreting with grout under an appropriate pressure
4. establishing grout curtains or aprons (9, 28)
5. providing for drainage facilities in rock
6. surface stabilization of rock, and
7. rock bolting (see Chapter 11).

Consolidation Grouting

Fissured rock formations are sealed and strengthened by a method known as the consolidation or blanket grouting. Injection of cement grout through shallow boreholes into open fissures and cracks makes the rock mass stronger. The injection increases the modulus of elasticity of the in-situ rock, and decreases irreversible deformations.

Grout Curtain

A grout curtain is a zone of treated rock, normally a vertical wall about 3 to 10 m thick. The curtain is installed to a considerable depth and length, and should be of low permeability within the rock mass on the upstream side near the heal of the dam, and on the flanks of the dam. The overall permeability of the grouted zone is less than that of the untreated rock on either side of the dam.

Grout curtains lengthen the seepage path thus cutting down the quantity of seepage water round underneath a dam or any other kind of hydraulic structure to an acceptable amount (9), and decrease uplift pressure.

The main grout curtain is usually grouted prior to concreting the dam foundation. The grout curtain is formed in single or multiple rows of injection holes. Ideally, its purpose is meant to withstand a full hydrostatic pressure on the upstream side of the curtain. In reality, however, the state of the matter is different, namely: the grout cannot and does not penetrate into all of the fine fissures and haircracks, or to force out pervious sandy or clayey gouge from the larger cracks. For a large scale operation, chemical grouting, and flushing out the gouge from cracks, joints and fissures may turn out to be unreasonably expensive. Besides, impermeability of the grout curtain may be greatly impaired by local defects in the grout curtain itself, as has been pointed out by A. Casagrande (7). In the case of finely fissured rocks, this view is now generally shared by the engineering profession (22).

Where justified, for proper hydrostatic uplift control on hydraulic structures, grout curtains are used in conjunction with adequate drainage facilities of the rock-dam-water system. It is because drainage is often an important factor in the stability of the structure-supporting rock and thus the structure itself. Grouting and drainage are two mutually complementary techniques, although one or the other may be used alone.

The need for grouting is established by obtaining rock core samples from the site of the contemplated foundation work and by studying their strength and permeability properties (4).

The shortcoming of grouting is "working blind", because there is little control of where the grout is moving. Therefore complete filling of all rock voids is impossible to insure.

According to JÄHDE (14, 15), the idea of whether an injection is or is not successful may be obtained from the characteristic forms of the time-pressure diagrams plotted during the process of injection (Fig. 5-20a, b, and c). For example,

> Fig. 5-20a shows that pressure increases slowly and uniformly until the pump capacity, or the allowable injection pressure is attained. This may be interpreted that the injection is successful.
>
> Fig. 5-20b indicates that after the initial pressure did increase, the pressure drops. This may mean that the grout has "broken out". For example, a clay gouge, filling a crack that might have ended in the free atmosphere, has been expelled out of the crack. Accordingly, the injection is unsuccessful.
>
> Fig. 5-20c conveys the idea that after an initial increase in pressure, the pressure drops, and then increases slowly again. This may be interpreted that after the occurrence as in the case b), the crack, or seam, or joint did subsequently close. The injection is successful.

The effectiveness of the grouting operation is usually verified by making check borings in the grouted zone and examining rock cores extracted from these boreholes. The presence of sufficient grout in all seams, cavities and fissures filled may be construed as a successful grouting operation. Also, if the grouted boreholes for checking do not take water or grout under pressure, then this may be considered as an indication that the results of the grouting operation are satisfactory. It is also customary to evaluate the improvement in the rock mechanical properties by means of the modulus of elasticity of the rock. The modulus of elasticity and the deformation modulus can be determined in the laboratory on the grouted rock check cores, as well as in situ.

Bituminous Grouting

In jointed and fissured rocks through which water flows, sometimes the cement grout may be carried away by water before the grout can set. In such a case, heated bituminous materials are injected into seams, fissures, cracks, and voids of rock to cut off infiltrating water. After the flow has been stopped, cement grouting operations may follow.

The great relative ease of permeation of *cold,* liquid bituminous *emulsions,* as compared with hot bituminous grout, affords utilizing *emulsions* for stabilization of sandy soils.

To retard seepage and to cut down future maintenance work, a quick-setting chemical grout may be used in conjunction with cement and asphalt grouting (24).

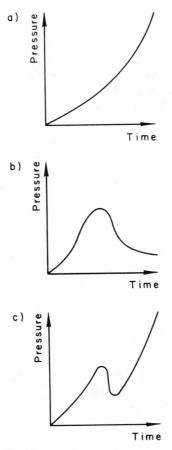

Fig. 5-20
Nature of time-pressure diagrams during the process of injection

Clay Grouting

Pressure grouting using materials other than cement has been practiced for a long time but generally has been regarded as a substitute for neat cement grouting. Clay grout and other kinds of colloidal suspensions are used to fill seams and cavities in rock, where low hydrostatic pressure heads prevail. Clay grout does not contribute much to the strength properties of the rock or soil. However, clay suspensions offer great resistance to seepage.

The use of soil as a grouting material in soil-water mixes, or with the addition of cement, is similar in many respects to the use of neat cement grout, because these are all basically suspension grouts in which solid material is suspended in water.

Chemical Solidification of Soil

Since 1925, one of the best known methods of chemical soil stabilization is that by the process patented by JOOSTEN (16, 17). In Joosten's method, two types of chemical solutions are consecutively injected into the soil under pressure: first, a solution of silicic acid (water glass, liquid glass), $Na_2 \cdot n \cdot SiO_2 + H_2O$, and second, a strong solution of calcium chloride (an electrolyte), $CaCl_2$ (Fig. 5-21). Silicic acid (sodium silicate) is a

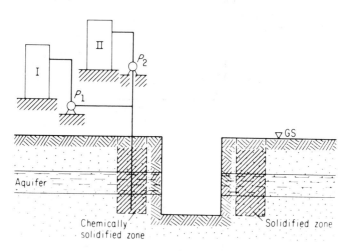

Fig. 5-21
Principle of chemical injection by JOOSTEN's method:
I. first chemical
II. second chemical

colloidal solution, coating the soil particles. Then, as the injection pipes are withdrawn, the second chemical is introduced. The latter — a salt solution — accelerates the transition from sol to gel. This silicating method of soil stabilization is applicable also in the presence of groundwater in the soil ("petrification of 'quick condition' of sand") where the coefficient of permeability k of the soil to groundwater is high (between $k = 1.5 \times 10^{-2}$ cm/s and $k = 1.0 \times 10^{-1}$ cm/s). This transforms a sand from its quick condition into an impervious material. There is no dilution of the chemicals by groundwater.

The strength attained by the Joosten's method of chemically solidified soil material depends upon the type and properties of the soil. Under normal conditions, strength of approximately 0.74 MN/m² (7.5 kg_f/cm²) to 2.94 kN/m² (30.0 kg_f/cm² or 140 psi to 425 psi) have been obtained in sand; approximately 3.92 to 4.9 MN/m² (40 to 50 kg_f/cm² or 570 to 710 psi) in gravel, with strength up to approximately 8.82 MN/m² (90 kg_f/cm² or ~ 1280 psi) under the most favorable conditions (16).

JOOSTEN's method has been also applied for sealing geologic rock formations in tunnels and mines against the influx of water (26). JOOSTEN's method has been developed further and improved by JÄHDE (14, 15).

Chemical grouts such as silica gels or synthetic resins can be used in finely fissured rock.

The American Cyanamid's AM-9 chemical grout is essentially a mixture of two organic monomers — acrylamide and methylenebisacrylamide — in proportions which produce very stiff gels from dilute, aqueous solutions when properly catalyzed (19).

The AM-9 gel is a rubbery, elastic material which by itself has negligible load-supporting capacity. However, it is substantially impermeable to water ($k \approx 10^{-10}$ cm/s). Hence, this type of grout is used primarily for seepage control of water and not as a load-bearing grout because of its low strength.

Among other chemical grouts, the chrome-lignin, winsol resins, and epoxy resins have been studied and used. Whereas successful results have been achieved with epoxy resin and polyester resin grouts, in the summary of his report ERICKSON wrote (10):

"The life of chemical grouts is not definitely known as they have been in use for only a relatively short time. It is possible that the chemical reaction may proceed until the grout becomes brittle, or the grout may eventually lose strength. A long-time study to determine the terminal condition is needed."

A bibliography on chemical grouting is given in Reference 3.

5-8. Lugeon Water Pressure Test

To measure the perviousness of rock, the Lugeon method of water pressure test in a borehole is used (23). This test does not give permeability coefficient values k. The test does, however, afford a quantitative comparison when evaluating the permeability of a rock in situ. Water pressure tests with single, double and multiple packers based on the Lugeon method are preferred in testing rock permeability in situ.

The Lugeon test is usually performed for establishing criteria for grouting of rocks.

In 1933, Professor MAURICE LUGEON (23) of the University of Lausanne, a prominent Engineer-Geologist, whose specialty was geological preparatory work for foundations of massive dams on hard rocks, put forward rules for grouting of concrete dam-supporting rocks with a cement grout. His criterion for groutability is still used today on many dam projects.

The criterion is based on the so-called *Lugeon unit*. The *Lugeon* unit is obtained from water injection and absorption test in situ. The Lugeon unit corresponds to 1 liter of water per minute absorbed by the rock from a one-meter test length of borehole under pressure of $10\,\mathrm{kg_f/cm^2}$ of 10 min. duration (23, page 87). According to Lugeon, a rock absorbing less than one Lugeon unit of water is considered to be reasonably water tight, and no grouting is needed.

One Lugeon unit is approximately equivalent to a coefficient of permeability k to water of $k = 10^{-5}\,\mathrm{cm/s}$.

The correlation between one Lugeon unit for a 5 m test length of the borehole is:

1 Lugeon unit corresponds to q = 5 liters/min and e = 0.1 mm
10 Lugeon units correspond to q = 50 liters/min and e = 0.2 mm
100 Lugeon units correspond to q = 500 liters/min and e = 0.5 mm.

Here q = discharge, and
e = width of fissure.

The allowable permeability values in rocks vary considerably from one country to another. The following empirical values are on record (20):

Terzaghi (1929): 0.05 liter/min · m per m of water column
Lugeon (1933): 1 liter/min · m at 10 $\mathrm{kg_f/cm^2}$ for heads $> 30\,\mathrm{m}$
 3 liter/min · m at 10 $\mathrm{kg_f/cm^2}$ for heads $< 30\,\mathrm{m}$
Jähde (1953): 0.1 liter/min · m at 3 at gauge pressure (15).

These figures have been obtained by experience. From these figures, one cannot derive mean permeability values k of rock masses.

5-9. Gunite

Gunite is a sprayed concrete, a mortar, projected from an air gun. Gunite is applied to rock surface in thin layers by an air jet.

The "gun," a mechanical feeder, mixer and compressor constitute the set of the basic equipment for spraying the mortar. Compressed air and the dry concrete mix are fed to the gun. It jets them out through a nozzle provided with a perforated manifold. Water flowing through the perforations is mixed with the dry mix before it is ejected. Because sprayed concrete (gunite) can be placed with a low water-cement ratio, it usually has high compressive strength. The method is especially useful for building up shapes without a form on one side. Gunite may be effective in sealing off and/or cutting down seepage from minute fissures of rock under low pressure head. Also, swelling ground containing anhydrite or bentonite can be treated with gunite.

5-10. Shotcrete

Shotcrete is a pneumatically applied cement mortar containing aggregates of up to 25 mm in size. Shotcrete is applied in about a 7 to 15 cm thick

Fig. 5-22
Spraying shotcrete to cover rock surface and encase steel ribs. Washington Metro tunnel. (Photo courtesy of ROBERT S. O'NEIL, P.E., Senior Vice President of De Leuw, Cather and Company, Washington, D.C.)

mass of concrete projected on the uneven rock surface. The shotcrete layer forms a coherent, solid bond with the rock. Fig. 5-22 shows the spraying of shotcrete in the Washington, D.C. Metro tunnel.

Shotcrete is used alone, or combined with a steel mesh reinforcement and/or rock bolts. Shotcrete is applied to the rock immediately after the opening up of a new face of rock, after excavation, and pinned back to the rock, provides an effective protection against weathering. Shotcrete fills all holes and depressions, penetrates into cracks, and turns a rock of mediocre strength into a stable one, resulting in a marked increase in the stiffness of the shotcrete "skin" of the rock. In tunnels, shotcrete lining is flexible enough to accomodate itself to deformations of the rock mass without failing.

Shotcreting, especially when combined with rock bolting, has proved to be excellent as a temporary support for all qualities of rock.

Shotcrete for tunnel lining is now used all over the world (8). It has found a considerable acceptance for the use in underground powerhouse

Fig. 5-23
With shotcrete stabilized steep excavation slopes in Triassic shale.
(Photo by ANDRIS JUMIKIS)

caverns, initial or temporary support and final linings for tunnels and shafts (1, 6, 25), and for protection of rock slopes (see Fig. 5-23).

For checking the strength of shotcrete linings, small cores are drilled out of the lining, and subjected to compression tests (25).

Fig. 5-24 shows a shotcreted rock slope along a highway.

Fig. 5-24
Shotcreted rock slope along a highway.
(Photo by ANDRIS A. JUMIKIS)

REFERENCES

1. American Concrete Institute and ASCE, 1972, "Use of Shotcrete for Underground Structural Support." *Proc. Engineering Foundation Conference,* ACI Publication SP-45, 1972 (467 pages).
2. ASCE Committee on Grouting, 1962, Progress Report of the Task Committee on Cement Grouting. *Proc. ASCE, Journal of the Soil Mechanics and Foundations Division.* April 1962, No. SM 2, pp. 49—98.
3. ASCE Committee on Grouting, 1966, "Bibliography on Chemical Grouting." Third Progress Report, Committee on Grouting, ASCE, (J. P. Elston, Chairman). *Proc. ASCE, Journal of the Soil Mechanics and Foundations Division,* November 1966, Vol. 62, No. SM 6, pp. 39—66.
4. BERNATZIK, W., 1942, "Die Zementeinpressung und ihre Überprüfung durch Zementbohrkerne," *Die Bautechnik,* No. 13/14, March 27, 1942, pp. 121—125. No. 15, April 3, 1942, pp. 135—138.
5. BOWEN, R., 1975, *Grouting in Engineering Practice.* New York, N.Y.: John Wiley and Sons.
6. BREKKE, T. L., 1972, "Shotcrete in Hard Rock Tunnelling." *Bull. Assoc. Eng. Geol.,* Vol. 9, No. 3, pp. 241—264.
7. CASAGRANDE, A., 1961, "Control of Seepage Through Foundations and Abutments of Dams." *Géotechnique* (London), Sept. 1961, Vol. 11, No. 3, pp. 159—182.
8. CECIL III, O. S., 1970, "Shotcrete Support in Rock Tunnels in Scandinavia." *Civil Engineering — ASCE,* January 1970, Vol. 11, pp. 74—79.
9. DIMAS, J., SAVINI, T., and W. WEYERMANN, 1978, *Rock Treatment of the Canelles Dam Foundations.* Zürich: A Publication by Rodio in Collaboration with the Institute for Engineering Research Foundation Kollbrunner-Rodio. May, 1978, No. 42 (37 pages).
10. ERICKSON, H. B., 1968, "Strengthening Rock by Injection of Chemical Grout." *Proc. ASCE, Journal of the Soil Mechanics and Foundations Division,* January 1968, Vol. 94, No. SM 1, pp. 159—175.
11. GLOSSOP, R., 1960, "The Invention and Development of Injection Processes. Part I: 1802—1850." *Géotechnique* (London), 1960, Vol. 10, No. 3, pp. 91—100.
12. GLOSSOP, R., 1961, "The Invention and Development of Injection Processes. Part II: 1850—1960." *Géotechnique* (London), 1961, Vol. 11, No. 4, pp. 255—279.
13. *Grouts and Drilling Muds in Engineering Practice.* A Symposium organized in May 1963 by the British National Society of the International Society of Soil Mechanics and Foundation Engineering at the Institution of Civil Engineers. London: Butterworths, 1963 (236 pages).
14. JÄHDE, H., 1937, "Die Abdichtung des Untergrundes beim Talsperrenbau." *Beton und Eisen,* 1937, No. 12, p. 193.
15. JÄHDE, H., 1953, *Injektionen zur Verbesserung von Baugrund und Bauwerk.* Berlin: VEB-Verlag, Technik, 1953, p. 14.
16. JOHNSON, A. W., 1948, translated from the German after Schütz on "New Method of Increasing the Stability and Tightness of Earth." *Highway Research Abstracts.* Washington, D.C.: Highway Research Board, March, 1948.
17. JOOSTEN, H. J., 1953, *Das Joosten-Verfahren zur chemischen Bodenverfestigung und Abdichtungen seiner Entwicklung von 1925 bis heute.* The Netherlands: privately published, 1953.
18. JUMIKIS, A. R., *Foundation Engineering.* Scranton, Pa.: INTEXT International Publishers, pp. 279—290.

19. KAROL, R. H., 1968, "Chemical Grouting Technology." *Proc. ASCE, Journal of the Soil Mechanics and Foundations Division.* January, 1968, Vol. 94, No. SM 1, pp. 175—204.
20. KOENIG, H. W., and K. H. HEITFELD, 1964, "Permeability and Grouting of Rock Foundations and Dams." *Transactions of the 8th International Congress on Large Dams.* Congress held May 4—8, 1964 at Edinburgh. Vol. 5, pp. 581—597.
21. KOMARNITSKII, N. I., 1968, *Zones and Planes of Weakness in Rocks and Stability.* Translated from the Russian by Consultants Bureau, New York, New York: Copyrighted by Plenum Publishing Corporation. Table 1, p. 15.
22. LONDE, P., 1973, *La Mécanique des Roches et les Fondations des Grand Barrages — Rock Mechanics and Dam Foundation Design.* Paris: International Commission on Large Dams (CIGB — ICOLD), 1973, p. 96.
23. LUGEON, M., 1933, *Barrages et Géologie.* Methodes des recherches, terrassement et imperméabilization. Lausanne: Librarie de l'université F. Rouge et Cie. S. A., p. 87.
24. MOORE, J. T., 1965, "Controlling Leakage from Cowans Ford Dam." *Civil Engineering,* June 1965, Vol. 35. No. 6, pp. 52—55.
25. PACKHAM, G. R., 1976, "Shotcrete Discussions in Maryland." *Tunnels and Tunneling.* November—December 1976, Vol. 8, No. 7, pp. 23—24.
26. PYNNONEN, R. O., and A. D. LOOK, 1958, "Chemical Solidification of Soil in Tunneling at a Minnesota Iron-Ore Mine." *Bureau of Mines Circular,* No. 7846. Washington, D. C.: U. S. Department of the Interior, 1958.
27. ROCHA, M., 1965, Some Problems on Failure of Rock Masses, Memória No. 258. Lisbon: Laboratório Nacional de Engenharia Civil.
28. SIMMONDS, A. W., 1953, "Final Foundation Treatment at Hoover Dam." *Transactions ASCE,* 1953, Vol. 118, Paper 2537, pp. 78—79.

PART 3
STRESSES IN ROCK ABOUT CIRCULAR UNDERGROUND OPENINGS

CHAPTER 6

STRESS FIELDS

6-1. Force Field

Any design and engineering analysis in rock is a two-phase process, namely:
1. Definition of the force field acting upon the construction material (such as soil or rock, for example), and
2. Determination of the reaction of the material to that force field.

Thus, in the design of an engineering structure, one deals with the analysis of stresses acting within the structural elements, as well as with the various properties of the structural materials.

A *force field* is a region of space or a medium where at every point at any time there acts a definite, coordinated force — the so-called field force. Thereby, the coordination of the field force to the points in space may be real or potential.

Examples of field force are:
1. Real field force: specific gravity; and
2. Potential field force: electrical field.

In statics, one limits oneself to stationary force fields:

$$X = x(x, y, z),$$
$$Y = y(x, y, z),$$
$$Z = z(x, y, z).$$

If the line of action of the field force for all points passes through a fixed center, one speaks, then, of a central force, viz., the central field.

A plane force field is expressed as

$$\left.\begin{array}{l} X = x(x,y) \\ Y = y(x,y) \\ Z = 0. \end{array}\right\} \quad \text{force plane}$$

Force (stress) fields may result from a variety of causes, one of which is the direct application of load.

6-2. Primary Stresses in Sound Rock

The cause of stresses in sound (virgin) undisturbed rock is the weight of the superimposed, or overburden rock, sometimes spoken of as the primary, or initial, or natural, or roof pressure of rock *en masse*. It varies in magnitude with depth from the ground surface. Near the surface, natural rock stresses are influenced by the weight of the rock, tectonic forces (caused by folding of the earth's crust), jointing, fractures, restraint against lateral expansion, and hard spots.

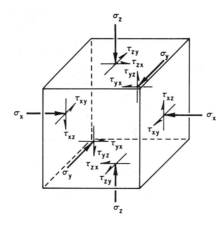

Fig. 6-1
Stress components in sound rock

Figure 6-1 illustrates general stress components in sound, homogeneous, isotropic rock. If the normal stresses are principal stresses, then the shear stresses vanish.

The magnitude and direction of the *initial* or primary stress in the rock — viz., the state of stress in the rock before excavation — depends upon the stress field. The state of stress in rock is disturbed or altered by excavations for underground openings and engineering structures. Excavations in rock bring about a new distribution of secondary or *induced* stresses within the rock around the excavations.

The amount and depth of this alteration are influenced by blasting and the local stress produced by the shape of the opening. According to test results as reported by TALOBRE (1957, Ref. 7 Chapter 6), the depth over which the stresses are altered extends about one-half the width of the excavation. This assumes, of course, that the length of the excavation is several times greater than the width.

Assuming the rock is a homogeneous material, and knowing its exact elastic properties, the calculations of stresses in rock massifs, weakened by openings, can be made somewhat reasonably by means of the theory of elasticity.

It should be said, however, that in practice these assumptions are very seldom met.

Imagine an elementary cube. On this cube, there act the principal vertical stresses σ_z and the principal horizontal stresses. All these stresses act mutually perpendicularly.

Assuming that the rock mass is elastic and homogeneous (having in all directions the same physical-mechanical properties), both horizontal stresses σ_x and σ_y must be equal.

According to Professor HEIM's (1) hypothesis, rock stress at great depths is the same in all directions. Professor HEIM's theory, according to which the pressure at any point in the interior of a rock massif is the same in all directions (hydrostatic pressure distribution hypothesis), and the magnitude of the pressure is calculated as the pressure of the overburden rock ($\downarrow \sigma_v = \gamma \cdot z =$ geostatic pressure), may be regarded as a close approximation to real conditions in the rock massif at great depth z.

"Of all the assumptions as to the horizontal pressure, that of equal pressure in any direction is the most probable."(1)

It must be realized, though, that:

1. The usual depths in underground rock engineering are too small for HEIM's pressure distribution at "great depth".
2. Thermal and other factors may transform the rock into a plastic state (gneiss, slate).
3. At usual depths, there is a difference between the magnitude of vertical and horizontal stresses, resulting in shear stresses.

4. Usually the largest stresses are not the vertical ones (by observation).
5. Stratification of rocks render inclined stresses.
6. Stresses in rock depend upon tectonic activity of the earth's crust (mountain-building forces).

Because the rock lateral confinement prevents the cube from expanding laterally under the action of the vertical stress σ_z, the sum of the strains in the elastic hemispace must be equal to zero:

$$\frac{1}{m} \cdot \frac{\sigma_z}{E} + \frac{1}{m} \cdot \frac{\sigma_y}{E} - \frac{\sigma_x}{E} = 0 \qquad (6\text{-}1)$$

or

$$\sigma_z + \sigma_y - m \cdot \sigma_x = 0 \qquad (6\text{-}2)$$

It follows that, if $\sigma_x = \sigma_y$, then

$$\sigma_x = \sigma_y = \frac{\sigma_z}{m-1} = \frac{\mu}{1-\mu} \cdot \sigma_z \qquad (6\text{-}3)$$

where $m = 1/\mu$ = Poisson's number, an elastic property of a material. It is valid for the elastic interval of deformations. From seismic measurements the average value of m of the earth's crust is $m = 3.7$;

$\mu = 1/m$ = Poisson's ratio;

E = modulus of elasticity of rock as tested.

The modulus of elasticity E_c of the confined rock in situ may be ascertained as follows.

The vertical strain (compressive) is:

$$\varepsilon_v = \frac{p_v}{E_c} = \frac{p_v}{E} - \frac{2p_v}{Em(m-1)} \qquad (6\text{-}4)$$

Here E is the modulus of elasticity as obtained from an unconfined compression test, for example. Therefore:

$$E_c = E \frac{m(m-1)}{(m+1)(m-2)} \qquad (6\text{-}5)$$

For most rocks, Poisson's ratio μ is between 0.20 and $\mu = 0.33$, and $\mu/(1-\mu) = \lambda_o$ should lie between 0.25 and 0.50.

For a compilation of Poisson's ratio for various kinds of soils and rocks, refer to Jumikis (3—5) and Jumikis and Jumikis (5).

TABLE 6-1

Lateral Pressure Coefficients λ_o for Various Types of Rocks

Rock	Poisson's ratio μ	Poisson's number m	$\lambda_o = \dfrac{\mu}{1-\mu} = \dfrac{1}{m-1}$
1	2	3	4
Diabase	0.33	3.00	0.500
Dolomite	0.083 – 0.20	12.00 – 5.00	0.091 – 0.250
Gabbro	0.125 – 0.20	8.00 – 5.00	0.143 – 0.250
Gneiss	0.11	3.30 – 6.60	0.303 – 0.152
Granite	0.15 – 0.24	0.66 – 4.16	0.177 – 0.312
Limestone	0.16 – 0.23	6.25 – 4.35	0.190 – 0.299
Limestone	0.14 – 0.17	7.00 – 6.00	0.170 – 0.200
Marble	0.25 – 0.38	4.00 – 2.63	0.333 – 0.613
Sandstone	0.17	5.88 – 2.80	0.205 – 0.556
Schist	0.08 – 0.20	12.50 – 5.00	0.087 – 0.250
Shale	0.11 – 0.54	9.10 – 1.85	0.123 – 1.176
Tuff	0.11	9.10	0.123

Table 6-1 shows computed λ_o-values of some common rock types. However, most of the in-situ measured values of $\sigma_h/\sigma_v = \lambda_o = \mu/(1-\mu)$ lie between 0.5 to 0.8 for hard rock, and between 0.8 and 1.0 for soft or inelastic rocks such as shale or salt (OBERT and DUVALL, 1967, Ref. 6).

If for limestone $m \approx 6$-7, then the horizontal stress σ_x is:

$$\sigma_x = (0.20)\,\sigma_z \text{ to } (0.17)\,\sigma_z$$

i.e., the horizontal stress σ is 17% to 20% of the vertical stress.

Assuming a gravitational stress field, the weight of the superimposed overburden in undisturbed condition exerts a vertical pressure σ_z at any point below the ground surface:

$$\sigma_z = \gamma \cdot h$$

where γ = unit weight of rock, and
h = thickness of rock layer.

From Eq. (6-3) one sees that the smaller m is, the larger the horizontal stress becomes.

It should be said here again that in order to specify completely a gravitational stress field, it must be assumed that:

1. The rock is linear-elastic, isotropic and homogeneous.
2. The lateral rock constraint or confinement is complete.
3. There are no stresses of tectonic origin such as those accompanying folding, shrinkage, or other distortions of the earth's crust.

In some instances, these tectonic stresses, together with the effects caused by inelasticity, inhomogeneity, and anisotropy, are known to affect the gravitational force field considerably. The rule that the largest principal stress in rock is generally acting vertically may be upset by tectonic stresses. For example, it has been reported that in the power plant cavern of the Snowy Mountain power scheme, Australia, the horizontal stresses were about 20% higher than the vertical stresses because of tectonic stresses.

6-3. In-Situ State of Stress

The in-situ state of vertical stress at a depth h, which depth is small as compared with the radius of the earth, is expressed as:

$$\sigma_z = \sigma_v = \gamma h \tag{6-6}$$

and the coordinated lateral stress is:

$$\sigma_x = \sigma_y = \sigma_h = [\mu/(1-\mu)] \cdot \sigma_v = \lambda_o \cdot \sigma_v \tag{6-7}$$

where $\sigma_v = \sigma_z$ = vertical, normal stress component;
$\sigma_h = \sigma_x = \sigma_y$ = horizontal, normal stress component;
h = vertical depth below ground surface;

$$\lambda_o = \mu/(1-\mu) = \frac{1}{m-1} \tag{6-8}$$

σ_h/σ_v = ratio of horizontal to vertical stress. It is a constant characterizing the stress field;
$\lambda_o \cdot \sigma_v$ = pressure at rest.

Here σ_v and σ_h are applied stresses.

In these discussions, compressive stresses are designated as positive and tensile stresses as negative.

Physico-geometrically, the stress components σ_x, σ_y and σ_z represent a stress ellipsoid.

When the lateral rock constraint is not completely rigid, λ_o will attain higher values than an ideally complete rigid constraint. However, the real value of λ_o for rock is difficult to determine (5).

6-4. Stress Fields

The stress field, sometimes referred to as the pre-existing stress, or initial or primary stress, or as the stress of the medium, is the state of stress that exists in the rock prior to mining or prior to any other kind of rock excavation or boring.

According to OBERT and DUVALL (6), three kinds of stress fields in rock may exist.

The three basic kinds of assumed stress fields are:

A *uniaxial* (unidirectional) *stress field*, a *biaxial* (two directional) *stress field*, and a *hydrostatic stress field*.

A uniaxial, unconstrained stress field is one where the medium (rock material) is subjected to compressive or tensile stresses in one direction only. Here $\lambda_0 = 0$. This state of stress would be encountered in rock at shallow depth below ground surface, and also near free vertical surfaces.

A biaxial stress field is one where the rock medium is subjected to compressive or tensile stresses in two mutually perpendicular directions. Here $\lambda_0 = 1/3$. This state of stress condition would be encountered at greater depth over a wide range of depth, depending upon type of rock, and at a depth of approximately 1000 m and more.

If in Eq. (6-8) one introduces $\mu = \dfrac{1}{m} = 1/4$, then $\lambda_0 = 1/3$. Such a state of stress pertains to the condition of a complete lateral constraint (no lateral deformation) in rock whose $\mu = 1/4$.

A hydrostatic stress field is one where the rock medium is subjected to equal stresses in three mutually perpendicular directions. Here $\lambda_0 = 1.0$. This means that σ_z, σ_x, and σ_y are all of equal magnitude of pressure. Such a condition when $\lambda_0 = 1.0$ may be encountered at great depth, and in semi-viscous or plastic rocks.

The stress fields are expressed analytically as:

$$\sigma_v = \gamma \cdot h$$
$$\sigma_h = \lambda_0 \cdot \sigma_v \qquad (6\text{-}9)$$

These three kinds of stress fields are sketched in Fig. 6-2.

Knowledge about the state of stress in rock is necessary in designing structures in rock and in evaluating the stability of underground structures.

HORVATH (1965, Ref. 2) put to the fore the idea that for a rock with certain characteristics of yield stress, unit weight, and POISSON's ratio, there

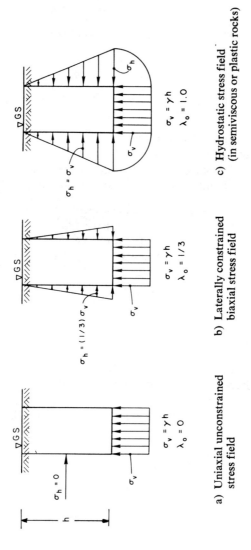

Fig. 6-2
Three assumed kinds of stress fields

is a limiting depth above which the lateral stresses may be computed on the basis of the theory of elasticity, but below which the horizontal principal stresses σ_2 and σ_3 may be derived from a plastic yield criterion, as:

$$\sigma_2 = \sigma_3 = \sigma_1 - \sigma_y = \gamma \cdot h - \sigma_y \qquad (6\text{-}10)$$

where $\sigma_1 = \gamma h$, and

σ_y = yield stress.

Most underground structures are built at depths where the hydrostatic stress condition is not attained. Hence, for the purpose of stress analysis in rock around an opening (such as a tunnel, for example), it is necessary to know the lateral pressure coefficient at rest, λ_0. Otherwise, as is frequently done, λ_0 is calculated by Eq. (6-8) by means of a known, or assumed, POISSON's number m for a particular rock.

Because of the existence of faults, joints, folds, stratification, anisotropy, and mechanical heterogeneity of rocks, the stresses in rocks can vary quite a lot from one point to another.

In rock mechanics and rock engineering, the engineer cannot choose a rock material with known mechanical properties and known initial state of stress because they are unknown.

Also, in rock engineering, usually the engineer has no wide latitude for selecting an ideal construction site.

There is no real justification for assuming that $\sigma_h = \lambda_0 \cdot \sigma_v = [\mu/(1-\mu)](\gamma h)$. Any one geological event could have altered the horizontal stress considerably from the above value. During the geological past, the stress that exists today may have been acted upon by various geological processes. For example, in a tectonically disturbed area of active regional subsidence, the center of the area would undergo a compressive strain, while the periphery and stresses would be under tensile strain. Also, faulting, folding, and other geological processes would bring about certain states of stress, which stresses may differ considerably from those as evaluated by means of the theory of elasticity. Rheological processes in, and weathering of, rocks would alter the initial stress field also.

It must be realized that the state of stress in rock is the sum of forces and processes of various origins and natures.

Practically, a reliable stress field can be determined only by in-situ measurement. Hence, there is a need for determining quality indices of rocks by means of which the engineer could quickly and economically test, evaluate, and characterize the stress field in rock and the mechanical properties of a rock mass. These indices are needed for assessment of depths of foundations and underground openings, as well as for stability of rock slopes. In other words, they are important in every geotechnical undertaking.

REFERENCES

1. HEIM, A., 1912, "Zur Frage der Gebirgs- und Gesteinsfestigkeit," *Schweizerische Bauzeitung,* February, 1912, Vol. 24.
2. HORVATH, J., 1965, "A New Approach to the Determination of Stresses in the Earth's Crust and Strata Pressure on Tunnel Linings." *International Journal for Rock Mechanics and Mining Science,* Vol. 2, pp. 327—340.
3. JUMIKIS, A. R., 1973, *Settlement Tables for Centrally Loaded Rigid Circular Footings on Multilayered Soil Systems*. New Brunswick, New Jersey: Rutgers University, College of Engineering, Bureau of Engineering Research Bulletin 54.
4. JUMIKIS, A. R., 1973, "Settlement Influence-Value Chart for Rigid Circular Foundations." Washington, D.C.: Highway Research Board Record, No. 457, National Research Council — National Academy of Sciences — National Academy of Engineering, pp. 27—38.
5. JUMIKIS, A. R., and A. A. JUMIKIS, 1975, *Red Brunswick Shale and its Engineering Aspects*. New Brunswick, New Jersey: Rutgers University, College of Engineering. Engineering Research Bulletin 55.
6. OBERT, L., and W. I. Duvall, 1967, *Rock Mechanics and the Design of Structures in Rock*. New York, N.Y.: John Wiley and Sons, Inc., p. 495.
7. TALOBRE, J. A., 1957, *La Mécanique des Roches*. Paris: Dunod.

CHAPTER 7

ELASTIC STRESS ANALYSIS IN ROCK AROUND UNDERGROUND OPENINGS

7-1. Underground Openings

One distinguishes between openings of horizontal underground structures with circular or elliptical cross-sections (tunnels, adits), vertical openings with circular cross-section (shafts, borings), and inclined tunnels (aqueous pressure penstocks in rock, for example).

Underground power plants and railroad yards (Oslo) are unaffected by topographic and climatic conditions such as freezing and avalanches. The surface landscape need not be destroyed. Such underground structures are also relatively more bombproof than surface generating stations.

The pressure tunnels may be unlined or lined with a shell. In unlined tunnels, the inherent strength of the rock is utilized.

Incompetent rock may be reinforced for the purpose of transmitting the lining pressure to the rock.

7-2. Secondary Stress Conditions in Rock

Generally, by an underground structure one refers to any excavated or natural subsurface opening, or system of openings, that is virtually self-supporting.

A self-supporting system is one in which the structural stresses are carried on the walls, pillars, and other unexcavated parts of the opening.

By the term "subsurface" or "underground" openings in rock is understood not merely adits, galleries and tunnels for vehicular traffic, aqueous pressure tunnels, and tunnels for river diversion, but also excavations in rock for shafts, subsurface factories, underground power plants, transformer chambers, garages, storage places, warehouses, shelters, arsenals, fortifications, and other possible uses. The purpose of tunnels is to provide for a) passageways for vehicular traffic, such as roads and railroads through mountains and under bodies of water; b) open and pressure conduits for water, and access to various underground spaces and structures.

7-3. Stresses in a Thick-Walled Cylinder

As an outset for stress analysis in rock, let us first avail ourselves of stresses in a thick-walled cylinder (3, 4, 5, 6, 7, 8).

Consider a circular, annular disk sliced out from a long, horizontal tunnel. The inside radius of the disk is r_i, its outside radius is r_o, and r is a variable radius (Fig. 7-1). The thick-walled cylinder is subjected to a uniform inside or internal pressure p_i, and to a uniform outside or external pressure p_o. These pressures induce in the wall, at any point, normal radial and normal tangential stresses, σ_r and σ_t, respectively.

The problem of determining the radial and tangential stresses at any point on a thick-walled cylinder, as a function of the applied pressures and radii involved, was solved by the French elastician GABRIEL LAMÉ in 1833. The elements of a thick-walled cylinder are shown in Fig. 7-1. The thick wall may be thought of as composed of thin, annular shells.

The radial stresses in the shell are designed as σ_r. The tangential stresses in the shell are designated as σ_t. Compressive stresses are indicated as positive. Tensile stresses are designated as negative.

A relationship between the radial and tangential stresses can be obtained from the equilibrium condition by summation of projections of all forces along a certain axis. The summation of forces must be equal to zero.

Referring to Fig. 7-1b, set $\Sigma F_r = 0$ along radius r to the volume element dV:

$$\Sigma F_r = \sigma_r \cdot r dw - \left(\sigma_r + \frac{d\sigma_r}{dr} \cdot dr\right)\left(r + dr\right) dw + 2 \cdot \sigma_t \cdot dr \cdot \frac{dw}{2} \quad (7\text{-}1)$$
$$-\sigma_r \cdot dr - d\sigma_r \cdot r - \sigma_t \cdot dr = 0$$

$$\sigma_r + r \cdot \frac{d\sigma_r}{dr} = \sigma_t$$

ELASTIC STRESS ANALYSIS

a) Cross-section of a thick-walled cylinder

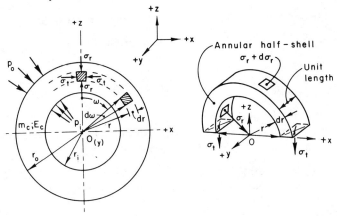

b) Forces on an elementary volume dV

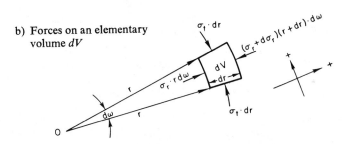

c) Displacements

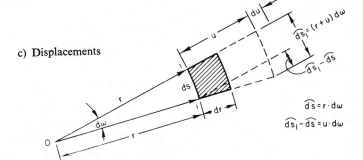

Fig. 7-1
Elements of a thick-walled cylinder

or
$$\sigma_t - \sigma_r - r \cdot \frac{d\sigma_r}{dr} = 0. \tag{7-2}$$

This is the equilibrium equation for the volume element dV.

Equation (7-2) can also be written in a differential form:

$$\sigma_t = \frac{d}{dr}(r \cdot \sigma_r) \tag{7-3}$$

Another relationship between radial and tangential stresses is obtained from the assumption that plane cross-sections remain plane, and hence the longitudinal strain ε_y is constant for all fibers.

Applying the extended HOOKE's law for triaxial stress (2), based on the plane stress condition, the longitudinal strain ϵ_y is written as:

$$\varepsilon_y = \frac{1}{E}\left[\sigma_y - \frac{1}{m}(\sigma_r + \sigma_t)\right] \tag{7-4}$$

where $m = 1/\mu =$ POISSON's number and $\mu = 1/m =$ POISSON's ratio. Because m, E, ε_y, and σ_y are all constant quantities:

$$\sigma_r + \sigma_t = \text{const} \tag{7-5}$$

throughout the cross-section of the thick-walled cylinder.

If $\varepsilon_y = 0$, then:

$$\sigma_y = \frac{1}{m}(\sigma_r + \sigma_t) \tag{7-6}$$

The elasticity equations for the plane stress condition (remember the long cylinder) expressing the other strains are:

$$\varepsilon_r = \frac{1}{E}\left(\sigma_r - \frac{1}{m} \cdot \sigma_t\right) \tag{7-7}$$

$$\varepsilon_t = \frac{1}{E}\left(\sigma_t - \frac{1}{m} \cdot \sigma_r\right) \tag{7-8}$$

From the geometry (Fig. 7-1c), the interior side i-i of the elementary volume is displaced by an amount u, and the outer one by $u + du$. Thus, the original thickness dr of the annular, elementary shell changes by an amount du. The elastic, radial strain component is:

$$\varepsilon_r = \frac{du}{dr} \tag{7-9}$$

ELASTIC STRESS ANALYSIS

The elastic tangential strain component ε_t in a symmetrical system in the tangential direction is computed as follows:

$$\varepsilon_t = \frac{ds_1 - ds}{ds} = \frac{(r+u)\,dw - r\,dw}{r\,dw} = \frac{u}{r} \tag{7-10}$$

$$\therefore \quad u = r \cdot \varepsilon_t \tag{7-11}$$

$$\therefore \quad \varepsilon_r = \frac{du}{dr} = \frac{\partial(r \cdot \varepsilon_t)}{\partial r} = \varepsilon_t + r \cdot \frac{\partial \varepsilon_t}{\partial r} \tag{7-12}$$

$$\therefore \quad \varepsilon_r - \varepsilon_t - r \frac{\partial \varepsilon_t}{\partial r} = 0. \tag{7-13}$$

From Equations (7-7), (7-8), (7-9) and (7-10), obtain the radial and tangential stress components as

$$\sigma_r = E \cdot \frac{m}{m^2 - 1}\left(m \cdot \frac{du}{dr} + \frac{u}{r}\right) \tag{7-14}$$

$$\sigma_t = E \cdot \frac{m}{m^2 - 1}\left(\frac{du}{dr} + m \cdot \frac{u}{r}\right) \tag{7-15}$$

Substitution of Eqs. (7-14) and (15) into Eq. (7-2) renders the EULER's differential equation for the radial displacement u:

$$\frac{du}{dr} - \frac{u}{r} - r \cdot \frac{d\left(\frac{u}{r}\right)}{dr} = 0$$

or

$$r^2 \cdot \frac{d^2u}{dr^2} + r \cdot \frac{du}{dr} - u = 0 \tag{7-16}$$

With $u = f(r)$, the elastic strain components ε_r and ε_t, as well as the corresponding elastic normal stress components σ_r and σ_t, can be calculated. Therefore, our immediate task is to determine the expression of $u = f(r)$.

Let the outset for the solution of EULER's homogeneous, linear differential equation of the second order be:

$$u = A \cdot r + \frac{B}{r} \tag{7-17}$$

where A and B are constants to be determined.

Divide by r:

$$\frac{u}{r} = A + \frac{B}{r^2} \qquad (7\text{-}18)$$

Differentiate u (Eq. 7-17) with respect to r:

$$\frac{du}{dr} = A - \frac{B}{r^2} \qquad (7\text{-}19)$$

The boundary conditions are:

1) when $r = r_i$, $\sigma_r = p_i$
2) when $r = r_o$, $\sigma_r = p_o$

To determine the stress components as a function of r, substitute u- and (u/r)-values from Eqs. (7-17) and (7-18) into Eqs. (7-14) and (7-15) to obtain

$$\sigma_r = E \cdot \frac{m}{m^2 - 1} \left[A(m+1) - \frac{B}{r^2}(m-1) \right] \qquad (7\text{-}20)$$

$$\sigma_t = E \cdot \frac{m}{m^2 - 1} \left[A(m+1) + \frac{B}{r^2}(m-1) \right] \qquad (7\text{-}21)$$

The constants A and B will now be determined from Eqs. (7-20) and (7-21) by means of the known boundary conditions:

$$A = \frac{m-1}{m \cdot E} \cdot \frac{1}{a^2 - 1}(p_o a^2 - p_i) \qquad (7\text{-}22)$$

$$B = \frac{m+1}{m \cdot E} \cdot \frac{r_o^2}{a^2 - 1}(p_o - p_i) \qquad (7\text{-}23)$$

where

$$a = \frac{r_o}{r_i} \qquad (7\text{-}24)$$

With the known constants A and B, the radial displacement u of any point on the thick-walled cylinder, by Eq. (7-17), is:

$$u = \frac{1}{E} \cdot \frac{1}{a^2 - 1} \left[\frac{m-1}{m}(p_o \cdot a^2 - p_i) r + \frac{1}{r} \cdot \frac{m+1}{m} \cdot \frac{r_o^2}{a^2 - 1}(p_o - p_i) \right] \qquad (7\text{-}25)$$

By setting in Eq. (7-25), $r = r_i$ and $r = r_o$, the displacements of the inner and outer contour of the disk, u_i and u_o, respectively, can be calculated.

ELASTIC STRESS ANALYSIS

With the known values of A and B, LAMÉ elastic stresses are:

$$\sigma_r = p_o \frac{a^2 - \alpha^2}{a^2 - 1} + p_i \frac{\alpha^2 - 1}{a^2 - 1} \quad \text{(compression)} \tag{7-26}$$

$$\sigma_t = p_o \frac{a^2 + \alpha^2}{a^2 - 1} - p_i \frac{\alpha^2 + 1}{a^2 - 1} \tag{7-27}$$

Here $a = r_o/r_i$, and

$$\alpha = r_o/r \tag{7-28}$$

Notice, that for the inner contour, where $r = r_i$:

$$\alpha = r_o/r_i = a \tag{7-29}$$

and

$$\sigma_{ri} = p_i \tag{7-30}$$

$$\sigma_{ti} = p_o \frac{2a^2}{a^2 - 1} - p_i \frac{a^2 + 1}{a^2 - 1} \tag{7-31}$$

For the outer contour of the thick-walled cylinder:

$$\alpha = r_o/r_o = 1 \tag{7-32}$$

$$\sigma_{ro} = p_o \tag{7-33}$$

$$\sigma_{to} = p_o \frac{a^2 + 1}{a^2 - 1} - p_i \frac{2}{a^2 - 1} \tag{7-34}$$

With $r_o \to \infty$, and for inside pressure p_i only (i. e., when $p_o = 0$), by Eqs. (7-26) and (7-27):

$$\sigma_r = p_i \frac{r_i^2}{r^2} \quad \text{(C)} \tag{7-35}$$

$$\sigma_t = -p_i \frac{r_i^2}{r^2} \quad \text{(T)} \tag{7-36}$$

On the tunnel contour, where $r = r_i$:

$$\sigma_{r_{max}} = p_i \tag{7-37}$$

$$\sigma_{t_{min}} = -p_i \tag{7-38}$$

When $r = r_i$, the strain ε_t in tangential direction is:

$$\varepsilon_t = \frac{2\pi(r + \Delta r) - 2\pi r}{2\pi r} = \frac{\Delta r}{r} \qquad (7\text{-}39)$$

$$\therefore \Delta r = \varepsilon_t \cdot r_i = -\frac{1}{E_{rock}} \cdot \left(\frac{m+1}{m}\right) \cdot p_i \cdot r_i \qquad (7\text{-}40)$$

where m is the POISSON's number for rock.

7-4. Theoretical Basis for Stress Analysis in Rock

In a plane stress condition, all stresses induced in rock around a circular opening (such as a tunnel) act in the cross-sectional planes of the long structure. Perpendicular to the cross-sectional slice areas, no stresses are transferred.

At this point, it should be remembered from Section 6-3, that the primary stresses, such as $\sigma_z = \gamma h$ and $\sigma_h = \sigma_x = \sigma_y = [1/(m-1)]\, \sigma_r = \lambda_o \cdot \sigma_r$, are the primary stresses in an ideal, undisturbed medium of rock.

However, the stress field in a rock changes by the presence of an underground opening within the rock. Also, a stress concentration occurs in the rock around the opening.

After excavating a tunnel or making any other cavity in the rock, vertical and horizontal forces around the opening are set free.

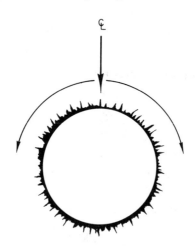

Fig. 7-2
By-passing pressure

These forces, prior to excavation, were carried by the rock. The pressure of the overburden rock must now, after excavation, bypass the opening in the rock, and must be transferred through the adjacent rock (Fig. 7-2). The rock around the opening (say tunnel) loses its support, and can now imperceptibly begin to deform.

An underground opening brings about a lateral squeeze of vertical force lines (Fig. 7-3). This results in a large stress concentration at the abutments *a* and *b* (springings of a vault). Also, one should be aware that between the tunnel vault and the natural arching of rock tension zones form at the crown and invert (at the top and bottom of the tunnel) of the vault.

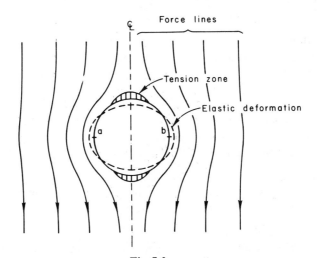

Fig. 7-3
Concentration of force lines, elastic deformation, and tension zones around a circular underground opening

Thus, the stress condition around the opening changes considerably, resulting here in the so-called induced, secondary stress condition.

Upon transition from the primary to the secondary stress condition, elastic and plastic deformations of rock occur.

The reasons for inducing secondary stress conditions in rock are:

1. Loosening of the rock mass around cavities;
2. Weight of the overburden rock mass;
3. Tectonic forces, and

4. Volumetric expansion of the rock mass by thermal effects, or by swelling brought about by the action of physical or physico-chemical processes.

Depending upon the kind and properties of the rock, secondary stress conditions require the use of different construction methods. All of the above-mentioned four secondary stress (pressure) conditions may also occur simultaneously.

Relative to probable secondary stress conditions in rock, they may set in:

1. In sound rock;
2. In pseudo-solid, soft, and weathered rock; and
3. In loose rock, viz., soil.

Sound rock transfers externally applied load by beam action limited by beam deflection. In loose rocks and in soil, load transfer takes place, most likely, by friction that develops during mass displacement. In order to maintain the underground opening, the intrusion of the rock into the opening must be prevented by rock-supporting structures. The load acting upon the supports is referred to as rock pressure. Remember: *"Ut tensio, sic vis."* This old principle has also been recognized at the First International Congress on Rock Mechanics in 1966 held in Lisbon, Portugal.

Stresses in rocks that are weakened by subsurface openings, which stresses remain below the elastic limit, in special cases may be calculated theoretically by means of the theory of elasticity.

The theoretical analysis of elastic, plane stress conditions in a circular plate of infinite extent, in which plate there prevailed an assumed homogeneous stress condition prior to introducing a circular opening, is a relatively simple one (1, 8). This assumption, in reality, applies to the rock, the better, the deeper is the opening below the ground surface. This may be so because at great depth, the consideration of stress increase vertically by the force of the rock mass may be disregarded in the region under consideration.

The opinion that inhomogeneity and anisotropy are insignificant at large pressures of the rock mass does not apply, because upon excavating an underground opening there sets in reduction or *relaxation* of stresses in the rock mass.

Assuming that:

1. The rock mass is an ideally elastic, homogeneous, and isotropic medium;
2. Hooke's law between stress and strain is valid;

3. The thickness of rock overburden is large as compared with the diameter of the tunnel, so that σ_v and σ_h can be considered to be constant within the influence zone of the tunnel, i. e., $\sigma_v = \gamma_h$, and $\sigma_h = \lambda_o \cdot \gamma \cdot h$;

the secondary stress conditions in rock can be computed by means of theory of elasticity.

In a very hard rock, with no planes of separation, the stress condition, as calculated, is approximately true, and the stress distribution equals the theoretical stress distribution within the elastic, homogeneous, and isotropic medium. These assumptions, however, do not apply in layered and schistose rocks.

In studying stresses in polar systems such as horizontal tunnels with circular cross-section, it is convenient to perform calculations in the polar, or cylindrical, coordinate system.

As a start of analysis, assume now a thick-walled, horizontal tunnel from which a vertical slice perpendicular to the tunnel axis is sliced off at a point that is far from both ends of the tunnel.

Now let the outer radius r_o of the cylinder approach infinity. Thus, the rock boundaries around the opening are infinite, representing the case of stress condition in rock about a freshly excavated, unlined tunnel or underground opening (Fig. 7-4).

The stresses acting perpendicularly to the cross-sectional plane are usually not of much interest in the secondary stress analysis of circular tunnels in rock. Therefore, for purpose of simplicity, tunnel problems in rock are analyzed under the assumption of the state of plane stress.

Obviously, after the fresh excavation of the opening, the primary stresses undergo alteration in the vicinity of the tunnel circumference, because the rock tends to fill the opening as soon as the driving and excavation of the tunnel commence.

Because the deformations of the slice in the direction of the longitudinal axis of the horizontal tunnel are counteracted in the same measure presumably by the deformations of the adjacent rock slices in contact with the slice under consideration, the stress analysis can be based on the mechanics of a circular disk by means of elasticity equations (LAMÉ equations for a thick-walled cylinder).

Calculations of stresses in a rock massif weakened by openings and cavities require the use of the theory of elasticity and the theory of plasticity.

322　　　　　　　　　ROCK MECHANICS

a) Nomenclature

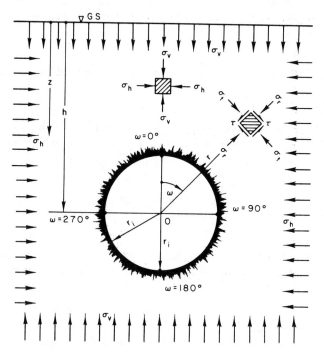

b) Elementary volume

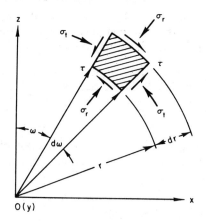

Fig. 7-4
Tunnel-rock system in a biaxial stress field;
σ_v and σ_h are primary stresses

The radial, tangential, and shear stresses are given by the theory of elasticity (5, 8) as:

$$\sigma_r = \frac{\sigma_v}{2}\left[(1+\lambda_0)(1-a^2) + (1-\lambda_0)(1+3a^4-4a^2)\cdot\cos 2\omega\right] \quad (7\text{-}41)$$

$$\sigma_t = \frac{\sigma_v}{2}\left[(1+\lambda_0)(1+a^2) - (1-\lambda_0)(1+3a^4)\cdot\cos 2\omega\right] \quad (7\text{-}42)$$

$$\tau_{rt} = -\frac{\sigma_v}{2}(1-\lambda_0)(1-3a^4+2a^2)\cdot\sin 2\omega \quad (7\text{-}43)$$

where $\sigma_v = \gamma \cdot h$ = vertical overburden stress,
γ = unit weight of rock,
h = thickness of overburden,
r_i = inside radius of opening,
r = variable radius,
ω = amplitude = angle between vertical (z) axis and radius r,
$\lambda_0 = \dfrac{1}{m-1} = \dfrac{\mu}{1-\mu} = \sigma_h/\sigma_v$ = rock lateral pressure coefficient at rest,
$a = r_i/r$ for brevity. $\quad (7\text{-}44)$

Equations (7-41) to (7-43) show that the extreme values of these stresses around the contour of the tunnel opening are determined by the $\cos 2\omega$ and $\sin 2\omega$ functions. Hence, the stress diagrams about the horizontal and vertical axes must be symmetrical. Especially informative for the stability of the tunnel perimeter are the extreme tangential stresses.

The extreme values of the tangential stress along the perimeter ($r = r_i$) of the circular tunnel opening can be determined by setting $d\sigma_t/d\omega = 0$

1. When $\omega = 0$ and $\omega = \pi$ (crest and invert):

$$\sigma_{t\,\min} = -\sigma_v(1-3\lambda_0) \quad (7\text{-}45)$$

i. e., tension when $\lambda_0 < \dfrac{1}{3}$

2. When $\omega = \pm \pi/2$:

$$\sigma_{t\,\max} = \sigma_v(3-\lambda_0) \quad (7\text{-}46)$$

Depending upon the magnitude of λ_0, these calculations render along the vertical and horizontal direction (along the x- and z-axes) the course of stresses as shown in Fig. 7-5.

a) Tangential principal stress

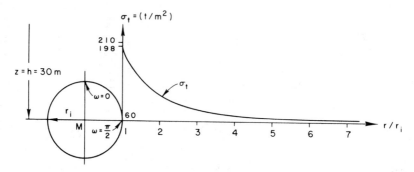

b) Radial principal stress

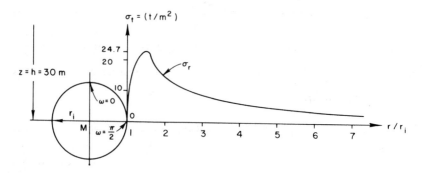

Fig. 7-5
Stress distribution diagrams around a circular opening in rock
for $\gamma = 2.2$ t/m³, $h = 30$ m, and $\omega = 90°$

Theoretically, the factor λ_o varies from $\lambda_o = 0$ to $\lambda_o = 1.0$ (see Chapter 6 on stress fields).

Typical stress distribution around circular openings in rock are shown in Figs. 7-5a and 7-5b. In each case, especially high stresses prevail along the periphery or contour of the tunnel. σ_t stresses at the springings ($\omega = 90°$) attain two to three times the amount of the average vertical stresses σ_v. For $\omega = 90°$, $\tau_{rt} = 0$ (by Eq. 7-43).

If the horizontal stress is of the same magnitude as the vertical stress, then all stresses at any point around the opening are symmetrical, i. e., a compressive stress of magnitude of $2\sigma_v$. If there is no horizontal stress, then $\sigma_h = 0$ and $\lambda_o = 0$. In such a case, there may occur considerable tensile stresses in a zone between the tunnel vault and the natural arching of rock;

ELASTIC STRESS ANALYSIS

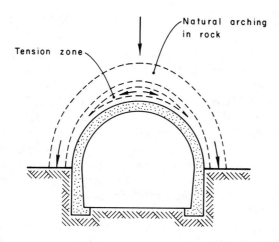

Fig. 7-6
Tension zone above an excavated underground opening below
the natural arching

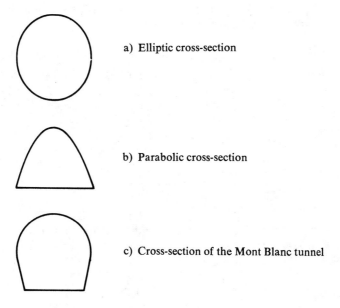

Fig. 7-7
Some cross-sections of underground openings

at the roof point the magnitude of these tensile stresses may attain the value of $-\sigma_v$.

Between these two extreme values, there is an intermediate case where the magnitude of the stresses at the roof point ($\omega = 0$) has a value of zero. This occurs when $\lambda_0 = 1/3$, i. e., when the horizontal stress has a value of one-third of that of the vertical stress, corresponding to a POISSON's ratio of $\mu = 0.25$, or $m = 4.0$. A lateral deformation which produces such a high stress may be brought about by a massive granite, basalt, diorite, augite, or limestone. For all other rocks, m is less than 4.0; hence, there exists frequently above the vertix (crown) a zone of tension. This tension zone above a vault of the circularly arched tunnel frequently brings about cracks in the rock, so that the blocks separated from one another above the crown lose their lateral support and are free-hanging (Fig. 7-6). Such free-hanging rock fragments are very dangerous, and should be avoided by choosing a cross-section other than circular (Fig. 7-7).

7-5. Specialization of Stress Equations

Of particular interest in rock mechanics is the dependence of the secondary stress condition as a function of λ_o.

a) $\lambda_0 = 0$. This case represents the case of a uniaxial, primary stress condition. When $\lambda_0 = 0$, Eqs. (7-41), (7-42), and (7-43) take the form:

$$\sigma_r = \frac{\sigma_v}{2}\left[(1 - a^2) + (1 - 4a^2 + 3a^4)\cos 2\omega\right] \quad (7\text{-}47)$$

$$\sigma_t = \frac{\sigma_v}{2}\left[(1 + a^2) - (1 + 3a^4)\cos 2\omega\right] \quad (7\text{-}48)$$

$$\tau = \tau_{rt} = +\frac{\sigma_v}{2}(3a^4 - 2a^2 - 1)\sin 2\omega \quad (7\text{-}49)$$

When $\omega = 90°$ and $\omega = 270°$, then at the rim of the opening where $a = r_i/r = 1.0$, there is an increase in the tangential stress up to the value of $\sigma_t = 3\sigma_v$. Thus, for $a = 1.0$:

$$\left.\begin{array}{l}\sigma_r = 0 \\ \sigma_t = 3\sigma_v \\ \tau = 0\end{array}\right\} \quad (7\text{-}50)$$

When $\omega = 0$ (at the crown), the prevailing stresses in rock are:

$$\begin{aligned} \sigma_r &= 0 \\ \sigma_t &= -\sigma_v \\ \tau &= 0 \end{aligned} \right\} \qquad (7\text{-}51)$$

b) $\lambda_0 = 1/3$. If there should be no tensile stresses at the crown, i. e., $\sigma_t = 0$, then for $\omega = 0$, $r = r_i$, and $a = 1$, the following stress equation results:

$$\sigma_t = \frac{\sigma_v}{2}\left[2\,(1 + \lambda_0) - 4\,(1 - \lambda_0)\right] = 0 \qquad (7\text{-}52)$$

From this equation for the assumption as before, $\lambda_0 = 1/3$.

When $\lambda_0 = 1/3$, then $m = 4$; i. e., when Poisson's number $m \leq 4$, there are no tensile stresses at the crown. For these conditions, when $\lambda_0 = 1/3$ and $\sigma_t = 0$, the set of stress equations is:

$$\sigma_r = \frac{\sigma_v}{3}\left[2\,(1 - a^2) + (1 - 4a^2 + 3a^4)\cos 2\omega\right] \qquad (7\text{-}53)$$

$$\sigma_t = \frac{\sigma_v}{3}\left[2\,(1 + a^2) - (1 + 3a^4)\cos 2\omega\right] \qquad (7\text{-}54)$$

$$\tau = \frac{\sigma_v}{3}[3a^4 - 2a^2 - 1]\sin 2\omega \qquad (7\text{-}55)$$

c) $\underline{\lambda_0 = 1.0}$. In this case, $\sigma = \sigma_v = \sigma_h$, and the stresses are:

$$\sigma_r = \sigma\,(1 - a^2) \qquad (7\text{-}56)$$

$$\sigma_t = \sigma\,(1 + a^2) \qquad (7\text{-}57)$$

$$\tau = 0 \qquad (7\text{-}58)$$

At the rim, or contour of the opening, $a = 1$, and:

$$\sigma_r = 0 \qquad \sigma_t = 2\sigma_v \qquad \tau = 0 \qquad (7\text{-}59)$$

The largest and least values of the σ_r stresses are at the position where $\omega = 0°$ and $\omega = 90°$.

Hence, this specialization (c) of Eqs. (7-41) and (7-42) for $\lambda_0 = 1.0$ and $a = r_i/r = 1.0$ at the contour shows that the value of the radial stress σ_r is equal to zero, and that the value of the tangential stress σ_t attains a value twice the original value of the primary stress, i. e.:

$$\sigma_t = \sigma_v\left[1 + \left(\frac{r_i}{r}\right)^2\right] = 2\cdot\sigma_v = 2\cdot(\gamma h) \qquad (7\text{-}60)$$

Figure 7-8 shows the theoretical distribution of the σ_r- and σ_t-stresses after the excavation of a circular opening in elastic rock. The asymptote for both stresses (σ_r and σ_t) is the same—namely, γh.

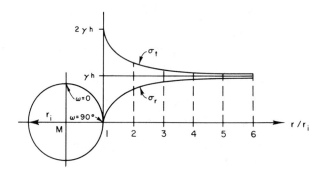

Fig. 7-8
Stress diagrams around an unlined tunnel for $\lambda_0 = 1.0$ at $\omega = 90°$

From this figure, it can be observed that beyond the abscissa whose value is $r/r_i > 4$, the stress increment induced is less than approximately 6%.

Hence, practically, the extent of the disturbed zone may be taken as proximately twice the width of the opening.

The softer and the more fissured or plastically yielding the rock material is, the farther the maximum stress moves from the contour of the opening into the rock.

The biaxial stress component equations show that the stresses around the opening are independent of the elastic constants of the rock material, and of the radius of the opening. The radius enters into these equations only as a dimensionless ratio r_i/r, which specifies the distance from the boundary or contour of the opening.

7-6. Discussion of Equations

1. The magnitude of the stresses depends upon the depth of the opening, viz., upon σ_v, as well as upon Poisson's number m

$$[\lambda_0 = 1/(m - 1)] \qquad (7\text{-}61)$$

2. By equations (7-41) and (7-42), and Fig. 7-4, the largest and the smallest values of σ_r and σ_t occur in the vertical and horizontal sections,

respectively, where $\omega = 0°$ and $\omega = 90°$:

$$\sigma_{r_{\omega=0°}} = \frac{\sigma_v}{2}\left[(1+\lambda_0)(1-a^2)+(1-\lambda_0)(1+3a^4-4a^2)\right] \quad (7\text{-}41\,\text{a})$$

$$\sigma_{t_{\omega=0°}} = \frac{\sigma_v}{2}\left[(1+\lambda_0)(1+a^2)-(1-\lambda_0)(1+3a^4)\right] \quad (7\text{-}42\,\text{a})$$

and

$$\sigma_{r_{\omega=90°}} = \frac{\sigma_v}{2}\left[(1+\lambda_0)(1-a^2)-(1-\lambda_0)(1+3a^4-4a^2)\right] \quad (7\text{-}41\,\text{b})$$

$$\sigma_{t_{\omega=90°}} = \frac{\sigma_v}{2}\left[(1+\lambda_0)(1+a^2)+(1-\lambda_0)(1+3a^4)\right] \quad (7\text{-}42\,\text{b})$$

3. The shear stresses are at a maximum when $\omega = 45°$; by Eq. (7-43):

$$\tau_{rt_{max}} = -\frac{\sigma_v}{2}(1-\lambda_0)(1-3a^4+2a^2) \quad (7\text{-}43\,\text{a})$$

4. At the contour of the circular opening where $r = r_i$, the radial stresses σ_r have a value equal to zero.

5. At the contour of the circular opening, where $r = r_i$, the tangential normal stresses σ_t assume the following values:

a) at the crown ($\omega = 0°$):

$$\sigma_{t_{\omega=0°}} = -\sigma_v(1-3\lambda_0) \quad [\text{T}] \quad (7\text{-}42\,\text{c})$$

b) at the abutment ($\omega = 90°$):

$$\sigma_{t_{\omega=90°}} = \sigma_v(3-\lambda_0) \quad [\text{C}] \quad (7\text{-}42\,\text{d})$$

i. e., the tangential stress at the abutment is always a compressive stress.

6. At the contour of the circular opening, the tangential stress (σ_t) is independent of the diameter of the opening. This means that by decreasing the diameter of the opening, the stresses at the abutment cannot be reduced.

7. The tangential stress σ_t along the contour of the circular opening is not constant. When POISSON's number $m = 4$, the stress at the crown is equal to $\sigma_t = $ zero, and at the abutment it is

$$\sigma_t = (2.66)\,\sigma_v. \quad (7\text{-}62)$$

8. If $m < 4$, then at the crown there prevails a compressive tangential stress. If $m > 4$, then there prevails a tensile stress.

9. The tangential stress σ_t at the abutment is always a compressive stress.

10. When $m = 3$, $m = 4$, $m = 5$, and $m = 7$, the tangential stresses σ_t at the circular contour as shown in Table 7-1 result.

11. Between the crown and the abutment there is a place where the tangential stress $\sigma_t = 0$. Its position depends upon POISSON's number m. The angle $\omega_{\sigma_t} = 0$ is calculated by Eq. (7-42), introducing $r = r_i$ and setting the equation equal to zero to obtain:

$$\cos 2\omega = \frac{1 + \lambda_o}{2(1 - \lambda_o)} = \frac{m}{2(m-2)} \tag{7-63}$$

12. Commonly, the tensile strength of a rock is considerably less than its compressive strength. Frequently it is:

$$\sigma_{\text{tensile}} \approx (0.1)\, \sigma_{\text{compressive}} \tag{7-64}$$

13. Also, it has been generally observed in practice that the larger the POISSON's number, the less is the tensile strength. This observation may give one an idea for choosing the necessary method of construction in rock relative to the kind of support (propping, anchor bolting, and the like).

14. The ratio of any one of the stress components at a point to one of the applied stresses is termed the stress concentration. The maximum positive stress concentration and the minimum negative stress concentration are called *critical stress concentrations* as defined by OBERT and DUVALL (5).

TABLE 7-1

Tangential Stresses σ_t at Contour of a Circular, Horizontal Tunnel in an Elastic Rock Mass

POISSON's Number m	Tangential stresses, σ_t	
	At the crown, $\omega = 0°$	At the abutment, $\omega = 90°$
1	2	3
3	$+0.50 \cdot \sigma_v$ [C]	$2.50 \cdot \sigma_v$
4	0	$2.66 \cdot \sigma_v$
5	$-0.25 \cdot \sigma_v$ [T]	$2.75 \cdot \sigma_v$
7	$-0.50 \cdot \sigma_v$ [T]	$2.83 \cdot \sigma_v$

ELASTIC STRESS ANALYSIS

7-7. Elastic Deformations in Sound Rock Upon Excavation of a Cavity

Stress is always associated with an accompanying relative deformation of the material called the strain.

Upon excavation of a large cavity in rock, there takes place a transition from a primary stress condition in rock to a secondary stress condition, accompanied by relatively large elastic deformations of the rock material. This is so because upon making large openings in the rock, the process of a complete stress relaxation requires a long time.

In the deformation analysis that follows, assume again:

σ_v = primary stress
$\sigma_h = \lambda_o \cdot \sigma_v$ = horizontal stress

Further, assume that the thickness h of the overburden is large as compared with the radius of the cavity $(h \gg r)$.

In the plane stress problem when $\omega = 0$, the secondary stresses by Eqs. (7-41), (7-42), and (7-43) are:

$$\sigma_r = \frac{\sigma_v}{2}\left[(1+\lambda_o)(1-a^2) + (1-\lambda_o)(1+3a^4-4a^2)\right] \quad (7\text{-}65)$$

$$\sigma_t = \frac{\sigma_v}{2}\left[(1+\lambda_o)(1+a^2) - (1-\lambda_o)(1+3a^4)\right] \quad (7\text{-}66)$$

$$\tau = 0 \quad (7\text{-}67)$$

Upon loading, the hatched rock element A at $\omega = 0$ in Fig. 7-9 distance r from the center, whose side is dr, undergoes a radial deformation Δdr after excavation. Its radial strain is, by HOOKE's law:

$$\frac{\Delta dr}{dr} = \varepsilon_r = \frac{dw}{dr} = \frac{1}{E_{rock}}\left(\bar{\sigma}_{r0} - \frac{1}{m_r}\cdot\bar{\sigma}_{t0}\right) \quad (7\text{-}68)$$

where w = displacement, and
m_r = POISSON's number of the rock.

One should be aware that in deformation, the stress difference between primary and secondary stresses are of significance:

$$\bar{\sigma}_{r0} = \sigma_v - \sigma_{r0} \quad (7\text{-}69)$$

$$\bar{\sigma}_{t0} = \sigma_h - \sigma_{t0} \quad (7\text{-}70)$$

332   ROCK MECHANICS

The zeros as indexes to the stresses indicate that these stresses pertain to $\omega = 0$, i. e., they pertain to the crown-point.

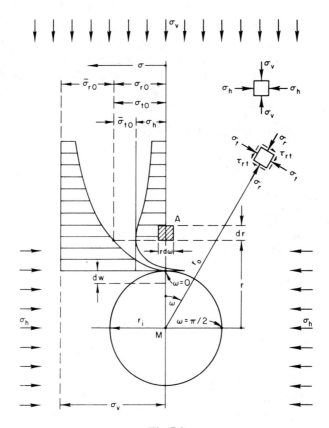

Fig. 7-9
Primary and secondary elastic principal stresses for determining elastic displacements w of the crown

Elastic Displacement w at Crest ($\omega = 0$).

By means of the stresses as given by Eqs. (7-69) and (7-70), the displacement w of the crown-point at the crest calculates as:

$$w = \int_{r_i}^{r=\infty} \Delta dr = \frac{1}{E_{\text{rock}}} \cdot \int_{r_i}^{\infty} \left[(\sigma_v - \sigma_{r0}) - \frac{1}{m_r}(\sigma_v - \sigma_{t0}) \right] dr \qquad (7\text{-}71)$$

ELASTIC STRESS ANALYSIS

With equations (7-65) and (7-66), and with $\lambda_o = \sigma_h/\sigma_v = 1/(m_r - 1)$, integration yields the elastic displacement w as:

$$w = \frac{\sigma_v \cdot r_i}{E_{rock}} \cdot \frac{2 \cdot m_r^2 - 3 \cdot m_r + 1}{m_r(m_r - 1)} \tag{7-72}$$

$$= \beta \cdot \frac{\sigma_v \cdot r_i}{E_r} \tag{7-72a}$$

where

$$\beta = \frac{1 - 3m_r + 2m_r^2}{m_r(m_r - 1)} \tag{7-73}$$

Elastic Displacement u at Abutment ($\omega = 90°$).

Analogically, as before, from Fig. 7-10

$$u = \int_{r_i}^{\infty} \Delta dr = \frac{1}{E_r} \int_{r_i}^{\infty} \left[(\sigma_h - \sigma_{r\,90°}) - \frac{1}{m_r}(\sigma_v - \sigma_{t\,90°}) \right] dr, \tag{7-74}$$

and

$$u = -\frac{\sigma_v \cdot r_i}{E_r} \cdot \frac{1 + m_r^2 - 4m_r}{m_r(m_r - 1)} \tag{7-75}$$

or

$$u = -\xi \cdot \frac{\sigma_v \cdot r_i}{E_r} \tag{7-75a}$$

where

$$\xi = \frac{1 + m_r^2 - 4m_r}{m_r(m_r - 1)} \tag{7-76}$$

The w- and u-deformations indicate that a circular opening in an elastic rock deforms elliptically: the vertical diameter decreases in length; the horizontal diameter increases in length (see Fig. 7-3).

7-8. Zone of Elastic Tangential Tensile Stresses

To delineate the tensile stress zone around the circular opening made in elastic rock, set the tangential stress equation, Eq. (7-42), equal to zero:

$\sigma_t = 0$. The evaluation of the $\sigma_t = 0$ equation renders the sought delineation zone or boundary line:

$$(1 + \lambda_o)(1 + a^2) - (1 - \lambda_o)(1 + 3a^4) \cos 2\omega = 0 \qquad (7\text{-}77)$$

One sees that this tensile stress is a function of Poisson's number $m = f(\lambda_o)$ only. Here $a = r_i/r$.

At the crown where $\omega = 0$,

$$a^4 - a^2 \cdot \frac{1 + \lambda_o}{3(1 - \lambda_o)} - \frac{2 \cdot \lambda_o}{3(1 - \lambda_o)} = 0 \qquad (7\text{-}78)$$

Solving for $a = r_i/r$, compute r for a given radius of circle. The difference, $r - r_i = d_{\omega=0}$, is the thickness of the tensile zone above the crest.

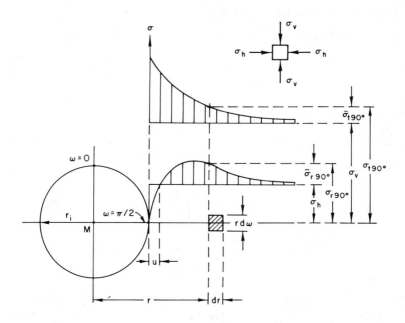

Fig. 7-10
Primary and secondary principal stresses for determining elastic displacement u of the abutment

REFERENCES

1. Jumikis, A. R., 1956, "Rupture Surface in Sand Under Oblique Loads." *Proc. ASCE, Journal of the Soil Mechanics and Foundations Division,* 1956, Vol. 82, Paper 861, pp. SM-1 to SM-6.
2. Jumikis, A. R., 1969, *Theoretical Soil Mechanics.* New York, N.Y.: Van Nostrand Reinhold Company, pp. 41—55.
3. Kastner, H., 1971, *Statik des Tunnel- und Stollenbaues auf der Grundlage geomechanischer Erkenntnisse.* Berlin: Springer-Verlag.
4. Mindlin, R. D., 1939, "Stress Distribution Around a Tunnel." *Proc. ASCE, Journal of the Soil Mechanics and Foundations Division.* April, 1939, pp. 619—642.
5. Obert, L., and W. I. Duvall, 1967, *Rock Mechanics and the Design of Structures in Rock.* New York, N.Y.: John Wiley and Sons, Inc., p. 495.
6. Széchy, K., 1966, *The Art of Tunnelling.* Budapest: Académiai Kiadó, pp. 83—84.
7. Terzaghi, K., and F. E. Richart, Jr., 1952, "Stresses in Rock about Cavities." Géotechnique (London), Vol. 3, No. 2, p. 57.
8. Timoshenko, S., and J. N. Goodier, 1951, *Theory of Elasticity.* New York, N.Y.: McGraw-Hill Book Company, Inc., pp. 78—81.

CHAPTER 8

PLASTIC ZONES IN ROCK AROUND UNDERGROUND OPENINGS

8-1. Concept of Plastic Zone

In the discussion that follows, it is assumed that:

1. The rock material in the immediate vicinity of the tunnel is an ideally plastic material;
2. The rock lateral pressure coefficient $\lambda_o = 1.0$ and $\sigma_h = \sigma_v$, i.e., geostatic pressure distribution is assumed; and
3. Initially the rock mass was at elastic equilibrium. When the compressive strength of the rock is exceeded, around the underground opening there forms a plastic zone (Fig. 8-1).

The plastic zone is characterized by the exhaustion of the internal shear resistance of the rock up to its limit. Whether the process of yielding or plastic deformation of the rock still remains within the stage of plastic deformation movement, or whether this plastic deformation attains the point of rupture, depends upon the nature of the rock mass.

Now, for the sake of discussion, assume as valid that the plastic deformation takes place just below the limit of ultimate rupture of the rock (point C on the stress-strain diagram in Fig. 8-2).

8-2. Derivation of Plasticity Condition in Rock

Assuming a direct transition from the ideally elastic state of the rock to the ideally plastic one, let us establish equilibrium conditions that have equal validity for the elastic as well as the plastic zone of the rock mass.

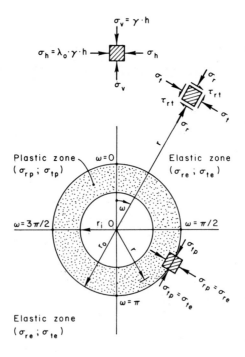

Fig. 8-1
Elastic and plastic zones

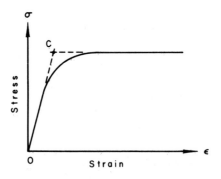

Fig. 8-2
Stress-strain diagram of an ideal plastic material

PLASTIC ZONES AROUND OPENINGS 339

For axial, central, symmetrical stress distribution around a circular opening whose radius is $r = $ const, there is no change in stress with ω. Therefore, in polar coordinates, the stress components, considering the effect of the weight of the rock, are independent of ω. From the theory of elasticity (3, 5), the stress components for equilibrium condition are:

$$\sigma_r = \frac{1}{r} \cdot \frac{d\Phi}{dr} \tag{8-1}$$

$$\sigma_t = \frac{d^2\Phi}{dr^2} \tag{8-2}$$

$$\tau = 0 \tag{8-3}$$

In these equations, the symbol Φ represents AIRY's stress function. However, in order to describe the stress condition in rock, equilibrium equations (8-1) and (8-2) alone are insufficient, although they are necessary. The use of the AIRY's stress function opens the way to determine the stresses if one more conditions is given. Recall that for the elastic zone, HOOKE's law of proportionality between stress and strain (*"ut tensio, sic vis"*) is valid. For

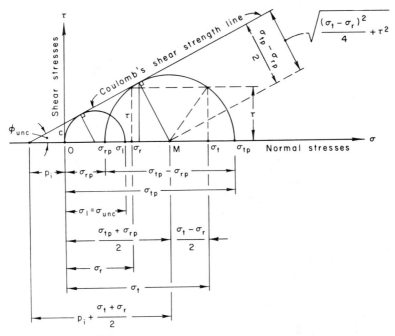

Fig. 8-3
Plasticity condition in COULOMB-MOHR presentation

the plastic zone, the condition of the internal resistance of the rock against shear at the ultimate limit (MOHR's theory of failure) is utilized. This is the condition when COULOMB's shear strength line tangents MOHR's stress circles (Fig. 8-3). This line depends upon the uniaxial compressive strength σ_{unc} of the rock mass and the test parameter φ_{unc}, popularly known as the angle of internal friction. The radial and tangential stresses, σ_{rp} and σ_{tp}, respectively, in the plastic zone can be determined by means of MOHR's theory of failure. Here the index p refers to stresses in the plastic condition.

The plasticity condition can now be written from MOHR's stress circle diagram, as in Fig. 8-3 (3, 4):

$$\sin \varphi_{unc} = \frac{\sigma_{tp} - \sigma_{rp}}{2p_i + \sigma_{rp} + \sigma_{tp}} \tag{8-4}$$

where φ_{unc} = stress parameter (angle of internal friction as obtained in the unconfined compression test (unc),

$p_i = c \cdot \cot \varphi_{unc}$ = initial stress in rock, and

c = cohesion of rock.

On the other hand:

$$\sin \varphi_{unc} = \frac{\sigma_{unc}}{2p_i + \sigma_{unc}} \tag{8-5}$$

and

$$p_i = c \cdot \cot \varphi_{unc} = \frac{\sigma_{unc}}{2} \cdot \frac{1 - \sin \varphi_{unc}}{\sin \varphi_{unc}} \tag{8-6}$$

where σ_{unc} = unconfined compressive strength of rock (see Fig. 8-3).

Substitution of (8-6) into (8-4) renders:

$$\sigma_{tp} - \frac{1 + \sin \varphi_{unc}}{1 - \sin \varphi_{unc}} \cdot \sigma_{rp} - \sigma_{unc} = 0 \tag{8-7}$$

With

$$(1 + \sin \varphi_{unc})/(1 - \sin \varphi_{unc}) = \lambda_{rp} = \zeta \tag{8-8}$$

where $\lambda_{rp} = \zeta$ is the coefficient at σ_{rp} in Eq. (8-7), the plasticity condition takes the form:

$$\sigma_{tp} - \zeta \cdot \sigma_{rp} - \sigma_{unc} = 0 \tag{8-9}$$

Introducing the stresses from the equilibrium condition — Eqs. (8-1) and (8-2) — into the plasticity condition — Eq. (8-9) — obtain a differential equation of the AIRY's stress function Φ:

$$\frac{d^2\Phi}{dr^2} - \zeta \cdot \frac{1}{r} \cdot \frac{d\Phi}{dr} - \sigma_{unc} = 0 \tag{8-10}$$

As shown by KASTNER (4), integration of Eq. (8-10) renders the sought AIRY's stress function ϕ:

$$\Phi = C_1 \cdot \frac{r^{\zeta+1}}{\zeta+1} - \frac{\sigma_{unc}}{\zeta-1} \cdot \frac{r^2}{2} + C_2 \tag{8-11}$$

$$\frac{d\Phi}{dr} = C_1 \cdot r^\zeta - \frac{\sigma_{unc}}{\zeta-1} \cdot r \tag{8-12}$$

$$\frac{d^2\Phi}{dr^2} = C_1 \cdot r^{\zeta-1} - \frac{\sigma_{unc}}{\zeta-1} \tag{8-13}$$

The boundary conditions are:

at $r = r_i$, $\sigma_{rp} = 0$, yielding the value of the integration constant C_1 as:

$$C_1 = \frac{1}{r_i^{\zeta-1}} \cdot \frac{\sigma_{unc}}{\zeta-1} \tag{8-14}$$

The integration constant C_2 does not enter into the equation of the plasticity condition.

With C_1, C_2, Φ, $d\Phi/dr$, and $d^2\Phi/dr^2$ now known, the stress components of the plastic zone are:

$$\sigma_{rp} = \frac{\sigma_{unc}}{\zeta-1} \cdot \left[\left(\frac{r}{r_i}\right)^{\zeta-1} - 1\right] \tag{8-15}$$

$$\sigma_{tp} = \frac{\sigma_{unc}}{\zeta-1} \cdot \left[\zeta \cdot \left(\frac{r}{r_i}\right)^{\zeta-1} - 1\right] \tag{8-16}$$

$$\tau_p = 0 \tag{8-17}$$

8-3. Extent of Plastic Zone

The plastic zone is delineated from the elastic zone by a circle whose radius is r_o (Fig. 8-1).

By Eqs. (7-28), (7-56), and (7-57), the elastic stresses in the elastic zone induced by the all-sided, equal, primary, compressive stress $\sigma = \sigma_v = \sigma_h$ are:

$$\sigma_{rel} = \sigma(1 - \alpha^2) \tag{8-18}$$

$$\sigma_{tel} = \sigma(1 + \alpha^2) \tag{8-19}$$

$$\tau_{el} = 0 \tag{8-20}$$

Herein, the index e pertains to the elastic stress; the index '1' refers to the primary (initial) stress condition, and $\alpha = r_o/r$ [Eq. (7-28)].

Besides primary stresses in the initial stress condition, there also prevails in the elastic zone a yet unknown axial-symmetrical radial stress σ_{r_o}. Thus, along the circular boundary of radius r_o between the elastic and plastic zones, these additional stresses are:

$$\sigma_{re2} = \sigma_{r_o} \cdot \frac{r_o^2}{r^2} \tag{8-21}$$

$$\sigma_{te2} = -\sigma_{r_o} \frac{r_o^2}{r^2} \tag{8-22}$$

$$\tau_{e2} = 0 \tag{8-23}$$

Summation of stresses (8-18) and (8-21), (8-19) and (8-22), and (8-20) and (8-23) renders the total elastic stresses on the elastic-plastic boundary:

$$\sigma_{re} = \sigma_{re1} + \sigma_{re2} = \sigma\left(1 - \frac{r_o^2}{r^2}\right) + \sigma_{r_o} \cdot \frac{r_o^2}{r^2} \tag{8-24}$$

$$\sigma_{te} = \sigma_{te1} + \sigma_{re2} = \sigma\left(1 + \frac{r_o^2}{r^2}\right) - \sigma_{r_o} \cdot \frac{r_o^2}{r^2} \tag{8-25}$$

$$\tau_e = \tau_{e1} + \tau_{e2} = 0 \tag{8-26}$$

At the elastic-plastic boundary zone $\left(r = r_o \text{ or } \frac{r_o}{r} = 1\right)$ the prevailing stresses should be equal, such as:

$$\sigma_{r_o} = \sigma_{rp} = \sigma_{re} \tag{8-27}$$

and

$$\sigma_{tp} = \sigma_{te} \tag{8-28}$$

Introducing Eq. (8-27) into Eq. (8-15), obtain

$$\frac{\sigma_{unc}}{\zeta - 1} \cdot \left[\left(\frac{r_o}{r_i}\right)^{\zeta-1} - 1\right] = \sigma_{r_o} \tag{8-29}$$

$$\frac{\sigma_{unc}}{\zeta - 1} \cdot \left[\zeta \cdot \left(\frac{r_o}{r_i}\right)^{\zeta-1} - 1\right] = 2\sigma - \sigma_{r_o} \tag{8-30}$$

Herein:

$$\zeta = \frac{1 + \sin \varphi_{unc}}{1 - \sin \varphi_{unc}} \tag{8-8}$$

Upon adding Eqs. (8-29) and (8-30), the σ_{r_0} stress cancels out, resulting in

$$\left(\frac{r_o}{r_i}\right)^{\zeta-1} \cdot (\zeta + 1) = \frac{2\sigma(\zeta-1)}{\sigma_{unc}} + 2 \qquad (8\text{-}31)$$

From this equation, the boundary (radius r_o) of the extent of the plastic zone calculates as:

$$r_o = r_i \left[\frac{2}{\zeta + 1} \cdot \frac{\sigma(\zeta-1) + \sigma_{unc}}{\sigma_{unc}}\right]^{\frac{1}{\zeta-1}} \qquad (8\text{-}32)$$

This equation shows that:

$$r_o = f(r_i; \lambda_o = 1; \sigma = \sigma_v = \sigma_h = \gamma \cdot h; \sigma_{unc}; \varphi_{unc}) \qquad (8\text{-}33)$$

The thickness of the plastic zone is thus $\Delta r = r_o - r_i$.

Summary of stresses for $\lambda_o = 1$

$$\sigma_{re1} = \sigma(1 - \alpha^2) = \sigma\left(1 - \frac{r_o^2}{r^2}\right) \qquad (8\text{-}18)$$

$$\sigma_{te1} = \sigma(1 + \alpha^2) = \sigma\left(1 + \frac{r_o^2}{r^2}\right) \qquad (8\text{-}19)$$

$$\tau_{e1} = 0 \qquad (8\text{-}20)$$

$$\sigma_{re2} = \sigma_{r_o} \cdot \frac{r_o^2}{r^2} \qquad (8\text{-}21)$$

$$\sigma_{te2} = -\sigma_{r_o} \cdot \frac{r_o^2}{r^2} \qquad (8\text{-}22)$$

$$\tau_{e2} = 0 \qquad (8\text{-}23)$$

At the plastic-elastic boundary $(r = r_o)$

$$\sigma_{r_o} = \sigma_{re} = \sigma_{rp}; \quad \sigma_{tp} = \sigma_{te} \qquad (8\text{-}28)$$

$$\sigma_{r_o} = \sigma_{rp} = \frac{\sigma_{unc}}{\zeta - 1} \cdot \left[\left(\frac{r_o}{r_i}\right)^{\zeta-1} - 1\right] \qquad (8\text{-}29)$$

$$2\sigma - \sigma_{r_o} = \frac{\sigma_{unc}}{\zeta - 1}\left[\zeta\left(\frac{r_o}{r_i}\right)^{\zeta-1} - 1\right] \qquad (8\text{-}30)$$

Stresses in the plastic zone

$$\sigma_{rp} = \frac{\sigma_{unc}}{\zeta - 1}\left[\left(\frac{r}{r_o}\right)^{\zeta-1} - 1\right] \tag{8-34}$$

$$\sigma_{tp} = \frac{\sigma_{unc}}{\zeta - 1}\left[\zeta\left(\frac{r}{r_o}\right)^{\zeta-1} - 1\right] \tag{8-35}$$

$$\tau_p = 0 \tag{8-36}$$

8-4. Discussion on Stresses in Elastic and Plastic Zones

For the sake of the discussion that follows, refer to Fig. 8-4 after KASTNER (4). In this figure are shown radial and tangential stress distributions in elastic and plastic zones in rock around a horizontal, circular underground opening. This figure pertains to the following example.

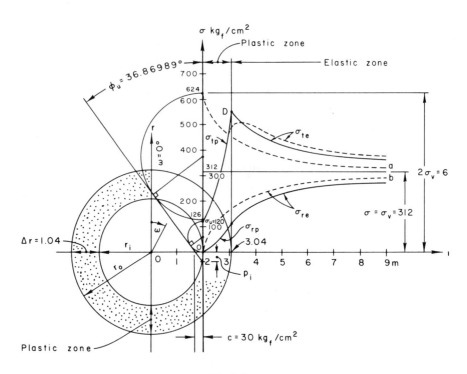

Fig. 8-4
Radial and tangential stresses around a horizontal, circular underground opening in elastic and plastic zones of sound rock

Assume a sound rock whose unit weight is $\gamma = 2.60$ t$_f$/m^3. Its unconfined compressive strength is $\sigma_1 = \sigma_u = 120$ kg$_f$/cm^2, and the test parameter $\varphi_u = 36.86989°$, resulting in a shear strength at zero pressure of $s = c = 30$ kg$_f$/cm^2, and in an initial stress of $p_i = c \cdot \cot \varphi_u = (30)(1.33333) = 40$ [kg$_f$/cm^2]. These strength data are shown on the COULOMB-MOHR strength diagram in Fig. 8-4.

The thickness of the overburden rock surcharge is given as $h = 1200$ m. Thus, the all-sided, uniform, overburden pressure for $\lambda_o = 1.0$ is:

$$\sigma = \sigma_v = \sigma_h = \gamma \cdot h = (2.6)(1200) = 3120 \ [t_f/m^2]$$

$$= 312 \ [kg_f/cm^2]$$

shown in Fig. 8-4.

The inside radius of the opening is given as $r_i = 2.00$ m. By Eq. (8-7), for $\xi = (1 + \sin 36.86989°)/(1 - \sin 36.86989°) = 4.0$, the outside radius r_o of the plastic zone calculates as $r_o = 304$ cm $= 3.04$ m. Thus, the thickness of the annular ring of the plastic zone is $\Delta r = r_o - r_i = 3.04 - 2.00 = 1.04$ [m].

For the plastic zone not to occur, $r_o = r_i$ (for $\lambda_o = 1$ and $\alpha = a = r_o/r\ 1.0$), the stresses at the contour for $\omega = 90°$ are, by Eqs. (7-56), (7-57), and (7-58):

$\sigma_r = 0$

$\sigma_t = 2\sigma_v = 2\sigma = (2)(312) = 624 \ [kg_f/cm^2] = \sigma_{te} = \sigma_{tp}$

$\tau = 0$

The stresses for the plastic zone and for the elastic zone are computed by Eqs. (8-15) and (8-18), respectively.

By Eq. (8-15), the radial plastic stress σ_{rp} corresponding to the tangential plastic stress σ_{tp} calculates as 126 kg$_f$/cm^2.

For the conditions as indicated above ($\lambda_o = 1.0$, $\alpha = 1.0$, and $\omega = 90°$), the distribution of tangential and radial stresses around a horizontal, circular opening in plastic and elastic zones at all-sided primary, uniform pressure is sketched in Fig. 8-4.

In Fig. 8-4, the dashed curve marked as "a" pertains to tangential stress distribution in a sound, elastic rock. Likewise, the dashed curve marked as "b" pertains to radial stress distribution in a sound, elastic rock.

The studies on plastic deformations of rock justify the following observations:

1. The problem of rock pressure on an underground opening represents a particular phase in the discipline of rock mechanics.

2. The determination of the magnitude of rock pressure is one of the most intricate problem in rock mechanics.

3. The kind of rock stress depends upon many factors, especially on the quality of the rock, and on the depth of the underground opening below ground surface.

4. In the aforegoing discussion, only the strength properties of the rock material are considered. They are unaffected by that rock pressure for which a geostatic pressure distribution ($\sigma = \sigma_v = \sigma_h = \gamma h$) is assumed.

5. Because of excavating of an underground opening in a rock massif, secondary stresses in rock are brought about, resulting in stress redistribution in the rock around the opening, and plastic deformation in the rock sets in. A plastic zone develops in the rock between the contour of the opening and the elastic zone. The extent of the plastic zone depends to a considerable degree on the ratio ζ of vertical to lateral pressures.

6. The boundary between the plastic and elastic zones is assumed to be a circular zone, labeled by r_o (outer radius of the plastic zone).

7. At the plastic and elastic zone boundary ($r = r_o$), the elastic and plastic stresses as computed by the theories of elasticity and plasticity must be equal.

8. Generally, near the contour of the opening, there prevail high stress peaks in the rock. Because of plastic deformations and stress redistribution, the high contour stresses have a tendency to decrease gradually.

9. Also, because of plastic deformations and consequent stress redistribution, wider rock zones around the opening become stressed.

10. From Fig. 8-4, it can be observed that at the plastic-elastic boundary ($r = r_o$) the tangential stress diagram has a discontinuity (point D), where it exhibits a peak value. This peak value is in excess of the strength of the rock, leading to a plastic stress condition. Actually, between the plastic and active zones, in the vicinity of the discontinuity at D, especially for the tangential stresses, there will be most likely a transition zone of rounding up these stresses *(natura non facit saltus)* thus shifting the tangential stress toward the interior of the elastic rock mass.

11. After the peak value, the magnitude of the tangential stress decreases in a certain proportion with distance away from the contour of the opening.

12. Depending upon the kind and quality of the rock, the tangential stress peak may attain a multiple of the magnitude of the average vertical stress. At some depth, such a tangential peak stress may even exceed the strength of the rock massif.

13. Within the plastic zone, the magnitude of the radial stress increases with distance toward the interior of the elastic rock massif. This may be explained as follows. The particles of the plastic rock material are subjected to the normal tangential stress as a normal force. Because of friction between the particles, these particles become more and more capable of resisting radial shear stresses.

14. It should not be forgotten that the assumption $\lambda_o = 1.0$ pertains to an ideal condition only. Nonetheless, the application of $\lambda_o = 1.0$ or nearly 1.0: (a) affords a rigorous theoretical plasticity analysis, rendering a good idea about the pressure condition within the rock massif, and (b) may be considered to be a good basis for designing the size of a tunnel lining.

8-5. Zone of Disturbance

In reality, upon making an underground opening in rock (say, by means of explosives), unfavorable stress conditions in the rock are introduced in the immediate proximity around the opening. Upon blasting out an opening, the rock here becomes disintegrated. As a consequence, a loose zone, or zone of disturbance between the contour of the opening and the plastic zone, is brought about (Fig. 8-5). Because of the disintegrated condition of the

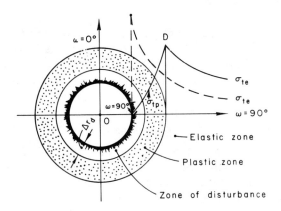

Fig. 8-5
Course of stress redistribution in rock brought about by a loose zone of rock disturbance

rock in this zone of disturbance, the rock here is not in a plastic state. It does not have any unconfined compressive strength, nor can it take and carry any stress.

Because of the loose zone, the stresses shift and redistribute farther away from the disturbed, loose zone into the rock medium. The value of the tangential stresses at the contour reduces to zero.

The thickness of the zone of blasting disturbance is designated as Δr_d. In order for this loose zone to exist, the opening must be braced or strutted.

Without support, the zone of disturbance may exist only by arching of the rock.

In Fig. (8-5), the stress distribution diagram shows an increase in the tangential stress from zero at the contour up to a peak (marked by D). This peak is attained at the boundary between the plastic and elastic zones. Farther away from the plastic-elastic boundary in the direction inward in the rock, the rock is affected only by elastic deformations. Thus, the rest of the course of the stress curve is similar to that shown in Fig. (8-4) for the case of plastic-elastic zones with no loose zone of disturbance.

The shift of the stresses displaces the centroid of the area of the stress diagram toward the interior of the rock, viz., away from the opening. This would mean nothing else than the natural self-aid of the rock material, in that the more remote ranges of the rock massif are being, so to say, mobilized to help to carry the stresses.

8-6. General Notes about Slip Lines

The phenomenon of slip, or displacement by translation of soil or rock particles, consists of a sliding (translocation) of particles along planes or surfaces of weakness relative to one another. This sliding of particles occurs in numerous planes, viz., surfaces forming a certain geometric slip line pattern. This type of deformation is often evidenced by numerous, remarkable parallel and mutually interesting markings on the mantle surface of soil specimens subjected to compression tests (Fig. 8-6). See also Reference 3.

Such markings or flow figures showing geometric patterns of families of mutually intersecting slip lines can also be observed on the cylindrical mantle surface of radially deformed rock specimens whenever the yield has reached in them under a radially symmetrical state of stress.

By the expression "geometric pattern of slip lines is the state of plane plastic strain" is understood two systems of plane curves in which the cylindrical surfaces of slip normal to the x,y plane intersect the latter plane.

PLASTIC ZONES AROUND OPENINGS

Fig. 8-6
Mutually intersecting markings on mantle surface of a soil specimen
in the state of plastic equilibrium

For a two-dimensional case, a curve whose tangent at every point is one of the shear directions at that point on the curve is variously called in the literature a slip line, or shear line, or line of rupture, or shear stress trajectory. Slip lines are the most useful means for representing the stress field.

In tackling rock mechanics problems as well as in dealing with pertinent problems in many other geotechnical engineering disciplines, in the evaluation of the shear stress conditions in or shear strength of the material, the engineer's interest centers about the determination of the shear stress in the rupture surface, and on the delineation of the shear stress lines and the rupture surface as well.

These shear stress lines are the subsequent slip lines, often called the lines of rupture, or rupture surface curves.

As a recourse from the theory of plasticity, it should be remembered that, in general, at limit equilibrium, there is a set of two impending rupture surface curves, or special slip lines, through every point in the material.

Knowledge of slip lines can be used to advantage for studying stress conditions and rupture surfaces in soil and rock materials brought about by structural loads.

Proceeding in the rock and soil mass from point to point, the direction of action of the principal stresses $\sigma_1 = \sigma_{tp}$ and $\sigma_3 = \sigma_{rp}$ and the course of slip-line curves in the material can be traced. Such a tracing results in a network or family of curves such as the slip lines or shear stress trajectories. Tangents drawn to the shear stress curves would then show the direction of action of the major principal stress σ_1 (for example, at the given point of tangency). The minor principal stress σ_3 then acts at this point in the direction normal to the tangent — that is, normal to σ_1.

8-7. Form of Slip Lines

For a rotary-symmetrical plastic zone such as one around a horizontal, circular underground opening, the slip line form is derived mathematically as set forth.

At primary, all-sided uniform pressure ($\lambda_o = 1.0$), the principal stresses, σ_{rp} and σ_{tp}, in the plastic zone are given by Eqs. (8-15) and (8-16) as:

$$\sigma_{rp} = \frac{\sigma_{unc}}{\zeta - 1} \cdot \left[\left(\frac{r}{r_i} \right)^{\zeta - 1} - 1 \right] \tag{8-15}$$

$$\sigma_{tp} = \frac{\sigma_{unc}}{\zeta - 1} \cdot \left[\zeta \cdot \left(\frac{r}{r_i} \right)^{\zeta - 1} - 1 \right] \tag{8-16}$$

The shear stress in this zone is zero: $\tau_p = 0$.

From MOHR's stress diagram in Fig. (8-7b), COULOMB's shear strength line OT — a tangent to the stress circle — as a boundary line represents the plastic flow condition, viz., failure of the rock material. The directions of rupture surfaces 1-1 and 2-2 form angles of rupture ($+ a$) and ($- a$) with the stress axis (σ), such that:

$$\tan \alpha = \frac{\left(\frac{\sigma_{tp} - \sigma_{rp}}{2} \right) \cdot \cos \varphi}{\left(\frac{\sigma_{tp} - \sigma_{rp}}{2} \right)(1 - \sin \varphi)} = \frac{\cos \varphi}{1 - \sin \varphi} = f(\varphi) = \text{const} \tag{8-37}$$

where φ = angle of internal friction of the in-situ rock massif. The principal stresses, σ_{tp} and σ_{rp}, are to be computed by Eqs. (8-15) and (8-16).

Equation (8-37) indicates that the form of the rupture surfaces is equilateral logarithmic spirals. Their assymptote is the pole (P) of the polar coordinate system, viz., the center M of the circular opening (Figs. 8-7a and

PLASTIC ZONES AROUND OPENINGS

a) Slip line tangents 1—1 and 2—2 at point S in the plastic zone

b) Mohr's stress circle for plastic equilibrium

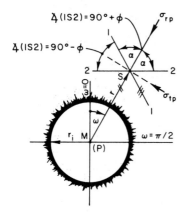
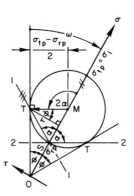

c) System of equilateral logarithmic spirals

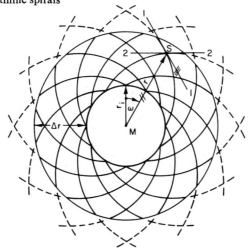

Δr = thickness of plastic zone

Fig. 8-7
Slip lines in the plastic zone

c). Therefore, the equation of this logarithmically spiraled rupture surface curve can be written at once as

$$r = r_i \cdot e^{\left(\frac{1}{\tan \alpha}\right) \cdot \omega} \tag{8-38}$$

$$= r_i \cdot e^{\left(\frac{1 - \sin \varphi}{\cos \varphi}\right) \cdot \omega}$$

or

$$r = r_i \cdot e^{\pm \omega \cdot \tan \varphi} \tag{8-39}$$

(see Refs. 1, 2),
where

$$\tan \varphi = \frac{1}{\tan \alpha} = \frac{1 - \sin \varphi}{\cos \varphi}, \tag{8-40}$$

r and ω = polar coordinates, and

r_i = initial radius vector = inner radius of the opening.

Thus, if r_i and φ are known, the course of the slip lines, and thus their pattern in the plastic zones around a circular opening at primary, all-sided uniform pressure, can be traced by Eq. (8-39). The form of the logarithmic spiral is thus governed by the value of $\tan \varphi$.

For the construction of the spiral, the magnitudes of the r-vectors are to be computed for each of the selected amplitudes ω with various lengths of radii starting with $r = r_i$ at the contour and ending up with $r = r_o$ at the plastic-elastic boundary.

The plotting of the logarithmically spiraled slip lines is greatly facilitated by means of the author's physical, dimensionless logarithmic spiral tables (2).

Figure (8-7c) shows a system of equilateral, logarithmically spiraled slip lines or shear stress trajectories in a plastic zone around a horizontal, circular underground opening for $\varphi = 30°$, and $\frac{\sigma_{tp} - \sigma_{rp}}{2} = 200 \text{ kg}_f/\text{cm}^2$, computed for every $\omega = 30°$.

8-8. Summary about Slip Lines

1. The slip lines are equilateral, logarithmic spirals.

2. The form of the logarithmic spiral depends upon the parameter φ, viz., value of tan φ.

3. The parameter r_i (initial radius) has no effect on the form of the spiral.

4. The family of slip lines intersects mutually.

5. No slip line in a homogeneous, isotropic rock mass has the advantage over the other in the same system of slip-line pattern.

6. The slip lines, viz., slip surfaces, terminate at the boundary between the plastic and elastic zones where $r = r_o$. Therefore, this zone boundary is not a slip surface.

7. The theory of plasticity has shown that the slip lines may be used with great advantage to represent the nature, condition, direction, and comparative amount of the shear stresses.

8. If the assumption of isotropy is not met, then preferred slip surfaces will prevail, dictated by the anisotropy.

REFERENCES

1. JUMIKIS, A. R., 1961, "The Shape of Rupture Surface in Dry Sand." *Proc. 5th International Conference on Soil Mechanics and Foundation Engineering*, held July 7—22, 1961. Paris: Dunod, Vol. 1, Paper 3 A / 23, pp. 693—698.
2. JUMIKIS, A. R., 1965, *Stability Analyses of Soil-Foundation Systems: A Design Manual* (Based on Logarithmically Spiralled Rupture Curves). New Brunswick, New Jersey: Rutgers University, College of Engineering, Bureau of Engineering Research. Engineering Research Publication 44 (contains 13 tables for the graphical construction of logarithmically spiralled rupture surface curves).
3. JUMIKIS, A. R. 1969, *Theoretical Soil Mechanics*. New York, N.Y.: Van Nostrand-Reinhold Company, pp. 41—55.
4. KASTNER, H., 1971, *Statik des Tunnel- und Stollenbaues auf der Grundlage geomechanischer Erkenntnisse*. Berlin: Springer-Verlag.
5. TIMOSHENKO, S., and J. N. GOODIER, 1951, *Theory of Elasticity*. McGraw-Hill Book Company, Inc., pp. 78—81.

CHAPTER 9

STRESSES IN ELASTIC ROCK AROUND VERTICAL SHAFTS

9-1. Primary Stress Conditions in the Elastic Zone

Prior to making a shaft, or drilling a borehole in a rock mass in situ, i.e., prior to disturbance, it is visualized that the rock mass is subjected to the effects of some of the rock-forming forces as well as to the force of gravity.

The rock-forming forces are difficult to comprehend. One generally assumes that the effects of the rock-forming forces (tectonic forces, for example) have already died or faded out, and that only the effect of the gravity force is usually considered in rock stress analysis of shafts, boreholes, galleries and tunnels made in rock.

In order to simplify calculations, one further assumes a homogeneous geologic material such as soil or rock the unit weight of which is γ. On a horizontal plane at depth z below the ground surface, there acts, then, a vertical compressive stress, $\sigma_v = \sigma_z = \gamma \cdot z$, known in rock mechanics as characterizing the primary stress field in the rock mass. Furthermore, as a consequence of the above assumption, it is assumed that at a given depth $z = h$, there acts a uniformly distributed lateral pressure from the rock mass:

$$\sigma_h = \sigma_x = \gamma \cdot z \tag{9-1}$$

around the outer circular contour of the shaft.

If there is a possibility for the rock mass to expand laterally then there occurs a corresponding lateral strain ε_x. However, the subsurface mass of rock, assumed in the theory of elasticity to be an infinitely extending hemispatial medium, cannot expand. Therefore, the lateral strain ε_x in HOOKE's stress-strain relationship is equal to zero, i.e.:

$$\varepsilon_x = \frac{1}{E} \cdot \left[\sigma_x - \frac{1}{m}(\sigma_y + \sigma_z)\right] = 0 \qquad (9\text{-}2)$$

where E = modulus of elasticity of rock, viz., soil;
m = POISSON's number; and
σ_x and σ_y = lateral stresses (σ_h) at depth z below ground surface.

The two lateral stresses, $\sigma_x = \sigma_y = \sigma_h$, are considered to be of equal magnitude. Therefore, by Eq. (9-2):

$$\sigma_x = \sigma_y = \sigma_h = \frac{\sigma_z}{m-1} = \frac{\gamma \cdot z}{m-1} \qquad (9\text{-}3)$$

Thus, assuming that the rock-forming forces are equal to zero, Eqs. (9-1) and (9-3) represent the undisturbed state of stress condition in a rock or soil formation at depth z prior to sinking a shaft or making a boring in the ground.

9-2. Secondary Elastic Stress Conditions

Upon sinking a shaft or drilling a borehole, the undisturbed stress condition in the rock or soil becomes considerably disturbed and altered. The new, or secondary elastic stress condition, for a central-symmetrical system, for any arbitrary depth z is calculated by means of AIRY's stress function, and is characterized in polar coordinates as set forth.

Radial stress:

$$\sigma_r = \frac{1}{2}\left(A - \frac{B}{r^2}\right) \qquad (9\text{-}4)$$

Tangential stress:

$$\sigma_t = \frac{1}{2}\left(A + \frac{B}{r^2}\right) \qquad (9\text{-}5)$$

Shear stress:

$$\tau = 0. \qquad (9\text{-}6)$$

Here r is a variable radius (Fig. 9-1).

STRESSES AROUND VERTICAL SHAFTS

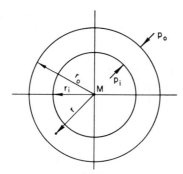

Fig. 9-1
Horizontal section of vertical, circular opening in the ground at depth z

The constants A and B are determined by means of the following boundary conditions:

1. When $r \to \infty$, then:

$$\sigma_r = \sigma_t = \sigma_x = \sigma_y = \sigma_h = \frac{\gamma \cdot z}{m-1} \quad (9\text{-}7)$$

2. When $r \to r_o$, then $\sigma_r = 0$

$$\therefore \quad \sigma_r = \frac{A}{2} = \frac{\gamma z}{m-1} = \sigma_t \quad (9\text{-}8)$$

$$\therefore \quad A = \frac{2 \cdot \gamma \cdot z}{m-1} \quad (9\text{-}9)$$

$$\therefore \quad \sigma_r = 0 = \frac{1}{2}\left(\frac{2 \cdot \gamma \cdot z}{m-1} - \frac{B}{r_o^2}\right) \quad (9\text{-}10)$$

$$\therefore \quad B = \left(\frac{2\gamma z}{m-1}\right) \cdot r_o^2 \quad (9\text{-}11)$$

Hence:

$$\sigma_r = \frac{1}{2}\left(\frac{2\gamma z}{m-1} - \frac{2\gamma z}{m-1} \cdot \frac{r_o^2}{r^2}\right) \quad (9\text{-}12\,\text{a})$$

or

$$\sigma_r = \frac{\gamma z}{m-1}\left(1 - \frac{r_o^2}{r^2}\right) = \sigma_h\left(1 - \frac{r_o^2}{r^2}\right) \quad (9\text{-}12)$$

$$\sigma_t = \frac{\gamma z}{m-1}\left(1 + \frac{r_o^2}{r^2}\right) = \sigma_h\left(1 + \frac{r_o^2}{r^2}\right) \quad (9\text{-}13)$$

Equations (9-12) and (9-13) represent the new, or the secondary stress condition in a soil or a rock mass at a particular depth z upon making a vertical, circular underground opening.

Because the shear stresses are equal to zero for reasons of central symmetry, the tangential stress σ_t is the major principal stress, and the radial stress σ_r is the minor principal stress. The vertical stress $\sigma_z = \sigma_v = \gamma \cdot z$ is an intermediate principal stress.

The theoretical secondary elastic stress conditions σ_r and σ_t in an elastic rock or soil medium beyond the contour of a vertical shaft or a borehole are shown in Fig. 9-2. These stress conditions are good as long as the limit of elasticity of the material is not exceeded. The minute the limit of the elastic deformations of the rock is exceeded, the foregoing assumptions made for these elastic stress condition calculations are no more valid. Therefore, in such a case, our knowledge about the prevailing real stress conditions and behavior of the rock or soil is confined to one's qualitative reasoning.

9-3. Discussion

1. The assumption of homogeneity of rock does not correspond to the reality. The rock is usually fractured and fissured, and the bedding planes of the rock formations are not always horizontal, nor parallel. Recall rock defects.

2. The shaft is usually blasted out by means of explosives. Hence the inside wall around the contour (circumference) of the shaft is irregularly formed, the rock is broken, and fissures may be brought about. They may propagate far out radially as well as deep in depth of the rock.

3. Equations (9-1), (9-12) and (9-13) show that all stress components, σ_r, σ_t and $\sigma_z = \sigma_v$, are compressive stresses.

4. The stress components σ_r and σ_t do not depend on the diameter of the shaft or the borehole, but on the (r_i/r)-ratio.

5. With increase in distance r from the vertical axis of the shaft, the radial stress σ_r increases, and the tangential stress σ_t decreases (Fig. 9-2).

6. At the contour, $r = r_i = r_o$, therefore, by Eqs. (9-12) and (9-13):

$$\sigma_r = 0$$

and

$$\sigma_t = \frac{2\gamma z}{m-1} = 2\sigma_h = \frac{2\sigma_v}{m-1} = 2 \cdot \sigma_x = \text{const} \qquad (9\text{-}14)$$

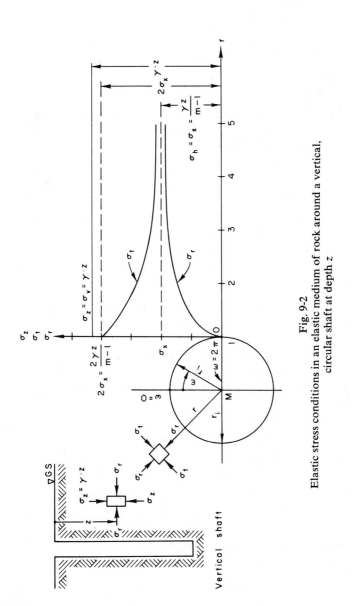

Fig. 9-2
Elastic stress conditions in an elastic medium of rock around a vertical, circular shaft at depth z

Hence, at the contour of the shaft, the radial and tangential stresses σ_r and σ_t, respectively, do not depend on the (r_i/r)-ratio of the shaft, but for a definite depth z below ground surface, and for a given POISSON's number m on hand, are constant quantities:

$$\sigma_r = 0; \quad \sigma_t = (2 \cdot \sigma_v)/(m-1) = 2\sigma_h; \quad \sigma_v = \sigma_z = \gamma \cdot z. \tag{9-15}$$

Notice that in a sound rock, the tangential stress σ_t has the double value of the horizontal stress $\sigma_h = \sigma_x$.

7. At the contour, the increase and decrease in σ_r and σ_t, respectively, is such that the sum of these two stresses at each point on the contour for a certain depth z is equal to $2\sigma_h = 2\sigma_x$, i.e.,

$$\sigma_t + \sigma_r = \frac{2\gamma z}{m-1} = 2\sigma_x \tag{9-16}$$

8. Because the displacements of the rock are directed radially toward the shaft, and because simultaneously no relative displacement between the individual particles of soil or rock mass can take place, physically the shear stress is equal to zero. Therefore, the radial and tangential directions represent the direction of action of principal stresses σ_r and σ_t.

9. The directions of sliding planes intersect the lines of action of principal stresses, viz., principal planes at a constant angle.

10. The shear stress is the same along the contour of the shaft. In general, the maximum shear stress τ_{max} is equal to one-half of the difference between the major and minor principal stress. In this particular case:

$$\tau = \frac{\sigma_1 - \sigma_3}{2} = \frac{\sigma_t - \sigma_r}{2} \tag{9-17a}$$

$$= \frac{\gamma z}{2(m-1)} \left[\left(1 + \frac{r_i^2}{r^2}\right) - \left(1 - \frac{r_i^2}{r^2}\right) \right]$$

$$= \frac{\gamma z}{2(m-1)} \left(2 \frac{r_i^2}{r^2}\right) = \sigma_x \cdot \frac{r_i^2}{r^2} \tag{9-17}$$

Its maximum value τ_{max} is attained (as long as $\sigma_t > \sigma_r$) at the contour when r is at a minimum, i. e., when $r_{min} = r = r_i$:

$$\tau_{max} = \sigma_x = \frac{2\sigma_x}{2} = \frac{\sigma_{t\,max}}{2} \tag{9-18}$$

9-4. Secondary Stress Conditions in the Plastic Zone

In that instance when the elastic deformations of a rock are exceeded, the foregoing assumptions do not apply any more; our knowledge about the real behavior of the rock is confined almost only on qualitative reasoning. In such a case, one usually makes use, then, of the theory of plasticity, especially of the Coulomb-Mohr failure condition. The reasoning may be such as set forth. If, in the rock, rupture or fracture occurs forming cracks and fissures, then in the ruptured rock around the shaft there sets in a plastic zone, or zone of stress release, called also the zone of *stress relaxation*. The relaxation of stresses decreases with increase in distance away from the shaft.

The stress release brings about a loosening of the rock, whereby the loosened rock material tends to intrude freely radially inward into the interior of the shaft. Because of this stress relaxation effect, the cross-section of the shaft is reduced. A lining of the wall of the shaft takes up the rock pressure, thus establishing a new condition of equilibrium, and the stress-relaxed zone extends out radially and terminates at a distance r_0 from the vertical axis of the shaft (Fig. 9-3).

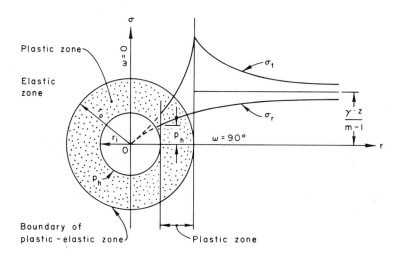

Fig. 9-3
Theoretical course of stress distribution in plastic and elastic zones of rock around a vertical shaft

In this figure, p_h is the lateral pressure exerted on the outside mantle surface of the vertical, cylindrical lining of the shaft or borehole, whichever the case is.

The pressure p_h on the lining from granular material is given as:

$$p_h = \frac{\gamma \cdot z}{m-1}(1-\sin\varphi)\left(\frac{r_i}{r_o}\right)^{\frac{2\sin\varphi}{1-\sin\varphi}} \quad (9\text{-}19)$$

This pressure attains a maximum when the plastic zone reaches the lining of the shaft, i.e., when $r_o = r_i$. Then the pressure is:

$$p_{h_{max}} = \frac{\gamma \cdot z}{m-1}(1-\sin\varphi) \quad (9\text{-}20)$$

The pressure decreases with increase in r_o. Equation (9-20) also shows that pressure decreases with increase in φ.

The stress components are:

$$\sigma_r = \left[\frac{\gamma \cdot z}{m-1}(1-\sin\varphi)\right]\left(\frac{r}{r_o}\right)^{\frac{2\sin\varphi}{1-\sin\varphi}} \quad (9\text{-}21)$$

$$\sigma_z = \gamma \cdot z \quad (9\text{-}22)$$

$$\sigma_t = \left[\frac{\gamma z}{m-1}(1+\sin\varphi)\right]\left(\frac{r}{r_o}\right)^{\frac{2\sin\varphi}{1-\sin\varphi}} \quad (9\text{-}23)$$

These equations are valid only when in the plastic zone the value of the third principal stress σ_z lies between the values of the other two principal stresses σ_r and σ_t.

To satisfy these equations, POISSON's number should be in the order of magnitude of 2. This means that the validity of such computations is severely limited, because in practice there are no rock materials which would deform without change in volume.

For solid rocks in the plastic zone, the following relationship can be written:

$$\sigma_t - \sigma_r = \sigma_{unc} \quad (9\text{-}24)$$

where σ_{unc} = unconfined strength of the rock.

The lateral pressure p_h in the plastic zone for solid rock calculates as:

$$p_h = \frac{\gamma \cdot z}{m-1} - \frac{\sigma_{unc}}{2}\left(2\ln\frac{r_o}{r_i}+1\right) \quad (9\text{-}25)$$

Within the region of $r_i < r < r_o$ at a distance r from the axis of the shaft, the stress components are:

STRESSES AROUND VERTICAL SHAFTS

$$\sigma_r = \frac{\gamma \cdot z}{m-1} + \frac{\sigma_{unc}}{2}\left[2\ln\left(\frac{r}{r_0}\right) - 1\right] \tag{9-26}$$

$$\sigma_z = \gamma \cdot z \tag{9-27}$$

$$\sigma_t = \frac{\gamma \cdot z}{m-1} + \frac{\sigma_{unc}}{2}\left[2\ln\left(\frac{r}{r_0}\right) + 1\right] \tag{9-28}$$

These three equations describe the stress conditions in the plastic zone.

At the lining or casing, by Eq. (9-25),

$$p_{h_{max}} = \frac{\gamma \cdot z}{m-1} - \frac{\sigma_{unc}}{2} \tag{9-29}$$

In contradistinction to granular material, the plastic zone in solid rock is limited. The larger the cohesion, the smaller is the zone where the material is plastic.

All these calculations show that the prevailing stresses in the individual strata are considerably altered when sinking a shaft, and that these changes in stress conditions are independent of the radius of the shaft or borehole.

Borings are usually penetrated into rock to a much greater depth than tunnels are driven.

This favorable circumstance (the depth) is not merely because the boring diameter is much less than that of a tunnel, but also because the horizontal tunnel, or gallery, is subjected to the vertical surcharge load of the superimposed rock, whereas in the case of the borehole the lesser horizontal stresses bring about a rotational-symmetrical stressing of the rock around the borehole.

9-5. Other Forms of Openings and Modes of Stress Distribution in Rock

For analyses of other forms than circular openings — such as elliptic, parabolic, horseshoe-like, square, and rectangular ones — refer to FENNER (1), TERZAGHI and RICHART, JR. (15), TERZAGHI (14), KASTNER (12), SZÉCHY (13), for example.

For surface loadings of an elastic medium by concentrated single and line loads; uniformly distributed; triangular and circular flexible surface loading giving influence values for stresses in the elastic medium at any point; and for triaxial, central-symmetrical vertical stress from rigid circular surface loading, refer to the author's publications, References 2-11.

REFERENCES

1. FENNER, R., 1938, "Untersuchungen zur Erkenntnis des Gebirgsdrucks," Glückauf, 1932, No. 32.
2. JUMIKIS, A. R., 1956, "Rupture Surface in Sand Under Oblique Loads." *Proc. ASCE, Journal of the Soil Mechanics and Foundations Division,* 1956, Vol. 82, Paper 861, pp. SM-1 to SM-26.
3. JUMIKIS, A. R., 1961, "The Shape of Rupture Surface in Dry Sand," *Proc. 5th International Conference on Soil Mechanics and Foundation Engineering,* held July 7—22, 1961. Paris: Dunod, 1961, Vol. 1, Paper 3A/23, pp. 693—698.
4. JUMIKIS, A. R., 1962, *Soil Mechnics.* Princeton, N. J.: D. Van Nostrand Company, Inc., pp. 641—651.
5. JUMIKIS, A. R., 1965, *Stability Analyses of Soil-Foundation Systems:* A Design Manual (Based on Logarithmically Spiralled Rupture Curves). New Brunswick, New Jersey: Rutgers University, College of Engineering, Bureau of Engineering Research, Engineering Research Publication 44 (contains 13 tables for the construction of logarithmically spiralled rupture surface curves).
6. JUMIKIS, A. R., 1969, *Stress Distribution Tables for Soil Under Concentrated Loads.* New Brunswick, New Jersey: Rutgers University, College of Engineering, Bureau of Engineering Research. Engineering Research Publication No. 48, 1969 (233 pages).
7. JUMIKIS, A. R., 1969, *Theoretical Soil Mechanics.* New York, N. Y.: Van Nostrand Reinhold Company, pp. 41—55.
8. JUMIKIS, A. R., 1971, *Vertical Stress Tables for Uniformly Distributed Loads on Soil* (for any point under square, rectangular, strip and circular bearing areas on the boundary surface of an elastic, homogeneous, semi-infinite medium). New Brunswick, New Jersey: Rutgers University, College of Engineering, Bureau of Engineering Research. Engineering Research Publication No. 52, 1971 (495 pages).
9. JUMIKIS, A. R., 1973, "Settlement Influence-Value Chart for Rigid Circular Foundations." *Highway Research Record No. 457.* Washington, D. C.: National Research Council-National Academy of Sciences — National Academy of Engineering, 1973, pp. 27—38.
9a. JUMIKIS, A. R., 1973, "Vertical Stress Chart for Rigid Circular Foundations." *Proc. ASCE, Journal of the Soil Mechanics and Foundations Division,* 1973, Vol. 99, No. SM 12, Paper 10189, pp. 1196—1201.
10. JUMIKIS, A. R., 1973, *Settlement Tables for Centrally Loaded, Rigid Circular Footings on Multilayered Soil Systems.* New Brunswick, New Jersey: Rutgers University, College of Engineering, Bureau of Engineering Research. Engineering Research Bulletin 54, 1973.
11. JUMIKIS, A. R., and A. A. JUMIKIS, 1975, *Red Brunswick Shale and its Engineering Aspects.* New Brunswick, New Jersey: Rutgers University, College of Engineering. Engineering Research Bulletin 55, 1975.
12. KASTNER, H., 1971, *Statik des Tunnel- und Stollenbaues auf der Grundlage geomechanischer Erkenntnisse.* Berlin: Springer-Verlag.
13. SZÉCHY, K., 1966, *The Art of Tunneling.* Budapest: Académiai Kiadó, pp. 83—84.
14. TERZAGHI, K., 1946, "Tunnel Hazards," in *Rock Tunneling with Steel Supports,* by R. V. Proctor and T. L. White. Youngstown, Ohio: The Commercial Shearing and Stamping Company, pp. 96—99.
15. TERZAGHI, K., and F. E. RICHART, Jr., 1952, "Stresses in Rock about Cavities." *Géotechnique* (London), Vol. 3, No. 2, p. 57.

CHAPTER 10

SOME ENGINEERING PROBLEMS ASSOCIATED WITH WORK IN ROCK

10-1. Kinds of Problems in Rock Engineering

Design in rock commonly comprehends structures as:
1. Foundations and anchorages;
2. Rock slopes; and
3. Various subsurface underground excavations: tunnels, shafts, mines, adits, and the like.

In underground rock engineering, the activities involved are:
1. Breaking and removing rock, and
2. Maintaining control over surrounding rock.

Therefore, the basic engineering science for underground as well as surface rock engineering is the discipline of rock mechanics.

Rock pressure, mechanical and chemical action of the underground water, pore-water pressure, temperature variations, frost action, gases, prestressing forces, and tectonic forces—they all impart their effects on underground structures, as well as on tunnel linings (17, 35, 62, 71, 77, 78, 82, 91—93, 102).

In rock engineering, one encounters several possible problems called by TERZAGHI (92) hazards.

Economically, hazards in underground opening works constitute unanticipated contingencies such as expenses and delay in such works, loss of materials, and sometimes also loss of life.

Usually hazards are caused by the departure in performance of the rock from the statistical average of similar materials, affecting seriously the state of its structure.

Generally, the causes of hazards are:

1. Abnormally high rock loads and thus high natural, primary stresses in rock;
2. Unfavorable rock conditions, such as rock defects, and orientation of stratification, joints, fissures, and underground cavities;
3. Excessive influx of water into an underground opening;
4. Temperature conditions in the interior of the rock massif;
5. Presence of harmful gases;
6. Seismic activity; and
7. Other possible unforeseen causes, among them chemical reaction of pore water on rock minerals, and radioactivity, for example.

Obviously, such hazards depend on the regional geology and the *en masse* properties of rock materials under consideration. Thus, in a way, in geotechnical engineering such hazards mean a financial risk.

Rock Pressure

Rock and soil pressures, and water pressure on underground structures at great depths, are the most important potential loads. The determination of the magnitude of rock pressure is one of the most intricate problem in rock mechanics.

There are forces prevailing in the rock, the causes of which reside in the deeper formations of the earth. These forces have brought about stresses in the earth's crust, bringing about plastic deformations as well as fractures in the rock. But even now, the earth's crust has not come to rest, but is always in motion. Tectonic and orogenic rock-forming processes are in progress even today.

It is, therefore, impossible to comprehend the rock-forming forces. In order to determine them, measuring instruments would have to be installed underground. However, to do this, it would be necessary to make a hole, an opening, or a borehole, or a shaft in the ground. To do this, the stress conditions in the rock would be readily disturbed. The results of in-situ measurements are also very much affected by tectonics of the location and by the method of measurement.

Therefore, in analyzing stress conditions in rock around a vertical, circular shaft, one usually assumes that the effects of rock-forming forces on rock have already faded out, and only the effect of the gravity force is considered.

The rock structure, depending upon its age and position of formation, is not homogeneous; that is, the composition and therefore, also its mechanical properties are not the same in all directions. The transformation of the original deposition which the rock has gone through—such as folding, faulting, and the natural inferiorities of the rock—all these, too, influence its mechanical behavior.

In rock engineering, many cases are known where the occurrence of plastic phenomena in, and fracturing of rock around a relatively narrow zone of underground openings sometimes cannot be satisfactorily explained by the prevailing overburden surcharge load—whether at great depth, at the ground surface or, especially, when near the ground surface.

10-2. Inferiorities of Rocks

In rock massifs, seldom, if ever, can ideal conditions be expected. Because of their presence, rock joints are important factors in excavating rock. Open, continuous joints form passages for groundwater circulation in rock.

Closely spaced joints impair stability of underground structures and foundation excavations in rock and rock slopes. In tunneling work, joints contribute to overbreak.

Working below the groundwater table, open joints are usually the source of water seepage into the underground opening, thus contributing to difficulties in work, as well as to added expenses for coping with the seepage by pumping and/or drainage facilities. Also, rock weathering, because of water, is a very undesirable factor.

Knowledge of fault and joint locations and their orientation is also of particular importance in rock engineering. Their extent also can lead to water problems, because joints can act as channels for conveying water through relatively dense strata.

Gouge is of the utmost concern in rock engineering. In underground openings, gouge in faults and joints may create disastrous hydrostatic head, viz., pressure.

Defects which intersect any kind of rock in situ may cause a very large scatter of strength values depending upon the number, character, distribution, and location of the various kinds of rock defects, and on the volume of the domain involved.

Natural and man-made activities in rock can develop into large-scale surface settlements or local sinkholes.

10-3. Subsurface Water

There is more than one effect of subsurface water on an underground structure. Groundwater finds its way toward a tunnel, bringing about erosion, washouts, and/or silting of tunnel drainage facilities (82).

Water dissolves certain chemicals, and some of them may be deleterious to concrete. Also, frost and its bursting action may contribute damage to a structure. Serious water problems may arise in the construction and maintenance of tunnels intersecting synclines containing water-bearing strata. Therefore, determination of physical characteristics of the water-bearing materials and their structure is of paramount importance.

Driving of tunnels in built-up urban areas may bring about dangerous lowering of the groundwater table, causing subsidence of streets and basement floors as well as rotting of timber piles, thus imparting distress to buildings. Drying-out of wells near caverns and tunnels is a well-known phenomenon. Underground openings are drainage centers. Relatively small water seepage into underground openings cause considerable subsidence, viz., settlement.

10-4. Temperature

From measurements made in underground openings and deep borings it is known that there exists a geothermal gradient in the earth's crust from the interior to the exterior of the earth. This means that there exists a terrestrial heat flux from the hot interior of the earth toward the cooler ground surface. This also means that the thermal regimen in the upper courses of the earth depends to a certain degree upon the geothermal gradient of the rock and/or soil materials.

The geothermal gradient varies greatly from place to place, depending upon the geologic history of the region, the kind of material, and stratification of the rock. Underneath mountain ranges, the average geothermal depth step is 35 to 60 m/°C, and underneath valleys it is 20 to 25 m/°C.

Some extreme geothermal depth steps are:

                Comstock mine (U.S.A.)   17 m/°C

Canadian Shield	125 m/°C
Gold ore mines, Johannesburg	110-130 m/°C

The thickness of the overburden cover of geological materials above a tunnel likewise influences the temperature in tunnel tubes already in service. The temperatures in some tunnels are compiled in Table 4-4.

High temperatures in a tunnel cause uncomfortable, ineffective work, and need air-conditioning and ventilation.

Heat, positive and negative (freezing), also induces thermal stresses in the rock. Thermal stresses occur where a body (say, rock) at its various points is warmed or cooled at various heat intensities. The cause of thermal stresses is the cooling of the magmatic rocks, or warming of rock which formerly was covered with a sheet of ice.

If underground caverns in rock are designed for storage of liquefied gas, great problems arise at very low temperatures (< -100 °C). Upon freezing of the groundwater, the low temperature causes the opening up of new cracks in the rock. The ice formed in the joints and cracks has no sealing effect because the rock contracts as the cooling proceeds, causing opening up of the ice-filled joints. According to MORFELDT (62), the contractions are considerable, more than 0.1 m of cavern.

An increase in temperature lowers rock strength and increases ductility. In compression tests of granite, GRIGGS et al. (35) found nearly a threefold decrease in strength as temperature was raised from 25° to 800 °C. Basalt showed nearly an eightfold decrease over the same temperature range. Dolomite showed a twofold decrease and marble an eightfold decrease in strength.

10-5. Hazards in Various Kinds of Rocks

As to geology, underground hazards can be encountered in igneous rocks, sedimentary rocks, and metamorphic rocks (17, 92, 102).

Thus, a thorough rock exploration is imperative before any rock engineering design and construction are contemplated.

Igneous Intrusive Rocks

Generally, the probability of underground hazards in igneous instrusive rocks, such as sound granite, is relatively small and of rare occurrence. However, deviations from the normal are always possible (102). Even in the best granites in Maine, plastic flow has been found in the rock at some 150-ft depths. "In one case, the solid sheets buckled upward with serious pop-outs and the fractures made further excavation wasteful since so little usable stone was recovered," according to FELD (33).

The possibility of treacherous conditions that might exist in chemically altered igneous rocks is difficult to ascertain at depths from the surface.

Also, flaking, popping, and spalling of rock exposures in deep tunnel construction should be anticipated (93). In the French part of the Mont Blanc tunnel, over 100,000 rock anchors were inserted to prevent rock bursts in the overstressed granite.

Igneous Extrusive Rocks

Breccia and tuff formations may be found to be in a state of decomposition. Because of fault zones, water finds its way into the underground openings made in such rocks. The presence of soft breccia may cause movement in a tunnel that intersects a fault. In porous basalt, large quantities of water may be present.

Because of the porosity of extrusive rocks, harmful gas movement is quite common. Hence, such rocks present themselves as hazardous.

Sedimentary Rocks

Limestone

Usually, limestone and sandstone above a water table present relatively few problems. However, deformations and cracks caused by actual overburden pressure are most likely to occur in limestones.

Limestones as a foundation for a dam fail because of hydration along leaking cracks.

Frequently, in limestone, solution cavities are encountered. These, and also fissures, may be fully or partly filled with fine-particled soil or they may be empty.

Considerable leakage through the limestone can be expected, especially if a cavernous limestone is situated under a dam or in a water-impounded reservoir. Also, limestone may contain thin layers of sandstone. These sandstone layers are permeable, and are weaker than limestone. A large-pore or cavern-containing limestone has a very low compressive strength (< 211 kg$_f$/cm² $\approx$ 3000 psi).

Any man-made change introduced in underground conditions in the limestone-dolomite rocks should be taken as a warning of possible subsidence of such rocks, as well as change-of-water circulation that will dissolve the rock.

Sometimes limestone grains are cemented together with clayey material. The cementing strength may become reduced upon wetting.

Shale

The predominant characteristic of a shale is the high percentage of fine clay particles, and also silt. The strength of shales varies from that of a sound shale rock to that of a swelling clay. Where porous, water-bearing formations rest on shale, a large influx of water from the porous formation into the shale can take place.

Shales disintegrate easily, and so do sandstones and many slates.

Buried seams in rock filled with saturated gouge may become liquefied even by a minor earth tremor and may bring about a major rock slide. Shales may also contain thin seams of plastic clay. Slides on buried seams in the rock may be the cause of catastrophic dam failures. Elastic rebound and plastic deformations in shale are frequently observed phenomena, also.

With large overburden pressure, shale is to be regarded as hazardous.

Schist

The hazards in sound schist are generally rather mild. The tall PAN-AM Building over the Grand Central Railroad Station in New York City is founded on what is called Manhattan schist (a mica schist). In altered schists, however, a large influx of water into underground excavations may take place.

Gneiss

Fine-grained gneiss may have open internal seams, frequently filled with water-deposited soft mud, seriously affecting the gneiss' engineering properties.

Requirements in the Design in Rock

Just as other engineering design must meet certain design requirements,

design of structures on and in rock is also subject to the basic requirements, namely:
1. Structures in rock should be of safe design.
2. They should be purposely and safely constructed.
3. They should be esthetical in appearance.
4. They should be economical to build.

10-6. Factors Contributing to Possible Hazards in Rock Engineering

Some of the natural and man-induced factors in rock and foundation engineering contributing to instability of structures built on and in rock, or to hazardous working conditions, are:

1. creep
2. erosion
3. expansive soils (swelling of clay minerals in discontinuities of rock)
4. fire
5. flooding and inundation
6. frost action
7. gases
8. groundbreak (including hydraulic groundbreak)
9. groundwater regimen (fluctuation, saturation, aquifers; glaciers; seeps, springs, artesian water)
10. human error
11. incompetent rocks and soils
12. karst phenomenon (solution cavities in limestone)
13. landslides of soils and rocks
14. liquefaction of soils and gouge in discontinuities of rocks
15. meteorologic conditions (heavy, prolonged torrential rainstorms; snow loads; melting snow; thawing soil)
16. negligent acts of third parties
17. overdraft of groundwater (subsidence problems)
18. overloading of rocks and soils (exceeding of bearing capacity and shear strength)

ENGINEERING PROBLEMS

19. rock defects (bedding planes, faults, folds and their unfavorable attitude; insufficient friction and cohesion of gouge in discontinuities against sliding)
20. rock pressures and deformations (stress relaxation; initial residual pressures and rock bursts)
21. seismic activity and tsunamis
22. solubility of rock materials
23. tunneling beneath glaciers
24. urbanization in and around fault zones
25. vibration, explosion shocks
26. volcanisms, and

other possible factors.

10-7. Stress Relaxation

If an excavation is made in a stressed rock mass, the stresses, viz., stored internal energy, in the rock become relaxed, and displacement or movement of the rock material is initiated. The process of relaxation takes place for some period of time. It ceases as soon as a new state of equilibrium is attained. The deformation of the cross-section of an opening and its surrounding area depends upon the strength and deformation behavior of the rock mass, the state of stress prevailing in the yet unworked rock mass; the size and form of the cross-section; the time elapsed since the excavation was made, and also upon the tunnel lining (if any), and its properties.

When a rock mass loses its strength, an apparently stable rock wall can flow into an excavation.

Instantaneous and long-time deformations caused by the relief of internal energy in rocks resulting from natural and/or man-made disturbances may be hazardous to rock integrity and thus to stability of structures built on and in rock.

Some of the rock deformations are large enough to cause popouts; rock bursts; cracking of tunnel walls after their completion; lifting of bridges; buckling or heaving of rock in the floor of quarries.

Popout, or popping, is the sudden detachment of thin rock slabs from a tunnel or excavation face. Popping occurs only in hard and brittle rock. Spalling may occur with no great violence.

The phenomenon of a rock burst is a spontaneous expansion, resulting in a sudden and often violent failure or "explosion" of rock material occurring in excavations, quarries, tunnels, mines, and at the toes of high, steep slopes of open cuts (in the proximity to faults, and in places of rock heterogeneity). Rock bursts occur in overstressed rocks with high elastic and strength properties, and are believed to originate in large part from large internal stresses in the rock (9, 29, 69, 70a, 70b, 77, 78, 102).

When a free face of a rock slope is cut by excavation, the unbalanced stress in the rock becomes sufficiently large to bring about a sudden failure. Rock bursts may be accompanied by shocks, rock falls, rock slides, and air concussions.

Fig. 10-1 shows schematically a deformation of a rock slope of a cut before and after excavation.

The dominating factors controlling the stress-strain conditions in rocks are:

the topography of the dam foundation site

the type and cross-section of the massive dam

the properties of the rock material underneath and adjacent to the dam

the time-rate of construction of the dam, as well as

the rate of filling of the reservoir.

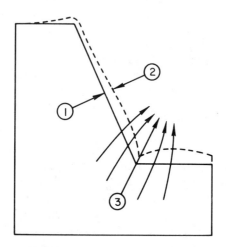

① SLOPE BEFORE EXCAVATION

② BULGED SLOPE AFTER EXCAVATION

③ ZONE OF ROCK BURST

Fig.10-1
Deformation of rock slope and excavation floor in a cut before and after excavation

The aforegoing discussion on stress relaxation in rock suggests that the correct evaluation of stress conditions in rock at the site of structures built in rock is of great importance. Regretfully, the complex make-up and the in-situ properties of the rock make it difficult to evaluate not only the correct stress regimen prevailing in the rock, but it is also difficult to determine the shape of the structure to be built in and/or on rock.

10-8. Hazards from Earthquakes and Faults

Nature of Earthquakes

A phenomenon that may constitute hazards to engineering works is the earthquake. Earthquakes have occurred in all parts of the globe. Earthquakes occur most commonly in volcanic regions, but often occur elsewhere. The West Coast of the United States is frequently affected by earthquakes of various intensity and magnitude.

Earthquakes and their related events are the most dramatic, terrifying and disastrous of all natural phenomena. They have been feared since ancient times because of their sudden, unpredictable occurrence and great capability of destruction. The toll of life and property taken by earthquakes within historical times is enormous. Therefore, knowledge about earthquakes and their effects on engineering structures is of paramount importance to geotechnical and structural engineers, and to engineering geologists who work in earthquake-prone regions where the occurrence of earthquakes is probable.

Earthquakes are vibrations or tremors of a part of the earth, particularly the vibration of the earth's surface. An earthquake is a perceptible trembling to violent shaking, or undulating of the ground, caused by a shock brought about by the sudden displacement of a slip of rocks over a fault surface (or possibly by a sudden slip in folding) below the earth's surface.

An earthquake zone (or seismic area) is an area of the earth's crust in which earth movements, sometimes associated with volcanism, occur.

Sometimes trembling or sudden shocks of the surface of the earth are accompanied by fissuring of the ground or by a permanent change of level. There have also been expressed opinions suggesting that earthquakes cause faulting rocks rather than the reverse, and that they themselves are due to sudden volume changes in the earth (32, 60).

Be that as it may, earthquakes are groups of elastic waves propagating in the earth, set up by a transient disturbance of the elastic equilibrium of a part of the earth.

As energy accumulates in rock during its deformation history on both sides of a fault, the rock is pressed tightly together and becomes gradually and slowly distorted until the elastic limit of the rock is attained. When the rock is stressed beyond its elastic limit, the rock ruptures by breaking or faulting suddenly and releases its stored-up internal energy with explosive suddenness in the form of earthquake shock waves through the rock as it snaps back or rebounds elastically, wholly, or partially into positions releasing the strain, the rock acting as an elastic body.

Natural and Artificially Induced Earthquakes

One distinguishes between natural earthquakes and artificially induced small quakes. As mentioned already, natural earthquakes result from a sudden release of energy. There are two basic genetic kinds of earthquakes, namely:

volcanic earthquakes and

tectonic earthquakes.

Volcanic earthquakes are vibrations generated by volcanic activity-volcanic movements below the earth's crust-resulting from explosions incident to the eruption of a volcano, or from violent subterranean movements of lava.

Tectonic earthquakes. The real significant earthquakes are all of *tectonic* or structural origin, being due to movements within the solid earth along fault planes. Specifically, earthquakes are most frequently caused by faulting associated with mountain building. These earthquakes are associated with the sudden release of stored energy and large strains in the earth's crust, resulting in sudden dislocations of large blocks of rock. Another prevailing theory holds that tectonic earthquakes are caused by slips as the causative mechanism of tectonic earthquakes. However, there is still doubt concerning the mechanisms that cause tectonic movements.

Artificially induced minor quakes are vibrations which may be brought about by heavy vehicular rolling stock on streets and highways; by train movement on railroads; by vibrating machinery at construction sites, and in factories; by pile driving operations; minor earthquakes may result also from explosions; in blasting operations, and any other possible man-induced disturbances.

The cause-and-effect relationship among the various categories of the natural and artificial earthquakes and man-made structures are generally complexly interrelated.

Landfills in water, imparted even by a moderate seismic shock, or mechanical vibration may liquefy the soil causing buildings, which are erected on such water-soaked fills, to crack and even collapse. Loose, water-saturated natural soil is also dangerous.

Hazards of Earthquake Destruction

Effects of earthquakes on engineering structures are of prime importance to engineers. Earthquakes may affect adversely the stability of, and work in, tunnels and other kinds of underground-openings, and damage foundation caissons at work and in service. Upon seismic vibration, a rock structure may be shaken loose, and water may find its way into the openings. Also, earthquakes may disrupt groundwater circulation, displace wells, springs, lakes and rivers.

Also, earthquakes may be catastrophic to densely populated areas and residential developments. Here buildings may collapse, power-, communication-, and gas pipelines break, and fires get out of hand because water systems may become disrupted. Also, the ground may become permanently displaced along faults, up, down, and shifted laterally. Unfortunately, it is seldom possible to site an engineering structure on and/or in the rock such as to avoid all major fault zones.

Perceptible, naturally induced earth tremors within shallow levels of the earth's crust may also be caused by an avalanche; rockfall; a landslide, by the slumping of submarine sediments; seismically generated water waves or tsunamis (23, 41, 54, 57, 64, 67); the collapse of a cavern roof, and by other possible causes.

Some damage is caused by the effect of shaking, not directly on the structure, but on its foundations. The effect of shaking, slumping, settling, and faulting have to be considered in relation to railways, highways, bridges, ditches, canals, tunnels, dams, tall structures, and other engineering works.

It has been experienced that earth bodies consisting of loose, relatively fine, uniform, submerged sand are susceptible to "liquefaction" during an earthquake (25, 80, 81). It has also been observed that buildings founded on deep alluvial deposits, or on "man-made" land filled in water, or on any other unconsolidated water-saturated sediments, or filled over swamps, are affected more severely from seismic shaking than similar structures founded on bedrock. Even a minor earth tremor can liquefy a saturated fill of sand, or a mud seam, and cause a major rock slide.

In soil mechanics usage, the liquefaction phenomenon is the sudden, large decrease in the shearing resistance τ of a cohesionless soil. Liquefaction is caused by the collapse of the soil structure by a sudden shock or other types of small strains, such as vibration, associated with sudden but temporary increase in pore water pressure u. It involves a temporary transformation of the soil material, dry or saturated, into a state of a fluid mass.

During liquefaction, an increase in pore-water pressure transforms a sand sediment into a suspension with a minimal shear strength (51-53).

The sudden shock means a suddenly applied shearing stress to the soil mass. Upon receiving the shock, the water (= pore-water) in the voids of the sand receives a suddenly applied pressure or neutral stress u, decreasing the intergranular, normal stress σ_n entering into the shearing process of the soil:

$$\tau = (\sigma_n - u) \cdot \tan \varphi, \tag{10-1}$$

where

τ = shear stress
$\sigma_n - u = \sigma_{n_{eff}}$ = normal effective intergranular stress
$\tan \varphi$ = coefficient of internal friction of noncohesive soil, and
φ = angle of internal friction of sand.

The decrease in shear strength τ means decrease in, or even exhaustion of, the bearing capacity of the soil, afflicting distress to a building, or bringing about its collapse.

Spectacular effects of soil liquefaction occurred during the earthquake of 1964 at Niigata, Japan: many buildings which were built on landfills in water were subjected to a "quicksand" condition, and only needed to be raised erect before they could be used again.

Responsibility for keeping track of earthquakes in the U.S.A. rests with the U.S. Geological Survey, and the Geological Surveys of each of the states in the confederation (see References 98—100).

Earthquake Intensity

The size or violence of an earthquake is usually measured in terms of either actual effects — intensity, — or magnitude (energy). While an earthquake can have only one magnitude, it can have several intensities. The intensity of the seismic shock is highest at the epicenter, and the intensity and damage gradually decreases as the distance from the epicenter increases. The intensity of an earthquake is rated in terms of the physical damage or geologic change it brings about, or both.

The intensity rating is based on personal, subjective observations, whereas the magnitude of an earthquake is based on instrument records. In the absence of any instrument readings of the ground motion, seismologists describe and classify the severity of the ground shaking by means of the so-called earthquake intensity scales. These scales have been devised based on the effects on people, objects, and visible damage to structures. Thus, earthquake intensity is usually measured qualitatively by estimating its visible geologic effects and destruction of life and property.

The scale most commonly used to rate the intensity of earthquakes is the popular Modified MERCALLI Intensity Scale as modified by WOOD and NEUMANN (31, 38, 40, 45, 74, 84, 105), see Columns 1, 3 and 4 in Table 10-1.

TABLE 10-1

Modified MERCALLI-CANCANI-SIEBERG Earthquake Intensity Scale (Condensed)

Rating of Intensity of Earthquake	Cancani's Maximum Acceleration a_{max} cm/s²	Mercalli Scale in Degrees of Intensity	Mercalli-Sieberg-Wood-Neumann Description of Earthquake Effects
1	2	3	4
Instrumental	< 0.25	I	Almost imperceptable. Microseismic tremors and shocks detectable by instruments only.
Very light	0.25 to 0.5	II	Felt only by a few sensitive people at rest. Recorded by seismographs. Bodies of water may sway gently. Doors may swing very slowly.
Light	0.5 to 1.0	III	Felt by a number of people indoors at rest like passing of light trucks. Hanging objects swing slightly. Motor vehicles may rock slightly.
Perceptible or Moderate	1.0 to 2.5	IV	Felt indoors by many. Generally perceptible by few people in motion outdoors. Loose, movable objects, dishes, windows, doors disturbed. Glassware trembles. Ceilings and walls crack. Hanging objects swing. Liquids in open vessels disturbed slightly. Sensation of a jolt like a heavy body striking building. Vibration sensation like passing of heavy trucks. Standing motor cars rock noticeably.
Strong	2.5 to 5.0	V	Generally felt by nearly everyone outdoors. Many sleepers awakened. Dishes, windows and doors rattle. Glassware broken. Ringing of bells. Cracked windows in some cases. Pendulum clocks stoped. Doors and shutters swing, close, open abruptly. Walls and house frame crack. Buildings tremble throughout. Liquids disturbed; some spilled from well-filled open containers. Oscillations of suspended objects. Small or unstable objects overturn. Stationary automobiles rock noticeably.
Very strong	5.0 to 10.0	VI	Felt by all — indoors and outdoors. Awakened all. Many people frightened. General panic. Slight damage in poorly built buildings. Buildings tremble throughout. Some plaster cracks. Windows crack. Hanging objects swing considerably. Dishes, glassware break. Moving and

Rating of Intensity of Earthquake	Cancani's Maximum Acceleration a_{max} cm/s²	Mercalli Scale in Degrees of Intensity	Mercalli-Sieberg-Wood-Neumann Description of Earthquake Effects
1	2	3	4
			overthrow of furniture and movable objects. Liquid set in strong motion. Collapse of chimneys in some instances.
Strongest	10.0 to 25.0	VII	Everybody runs outdoors. People find it difficult to stand. General alarm. Noticed by people driving motor vehicles. Damage to poorly built buildings and adobe houses. Light damage to brick buildings and light structures. Light cracks in walls. Fall of plaster, loose bricks, ornaments, roof tiles, and chimneys. Damage negligible in buildings of good design and construction. Liquids are set in strong motion. Waves form on ponds, lakes and streams. Water turbid from stirred-up mud. Gravel and sand stream banks slide and cave in. Large bells ring. Concrete irrigation ditches considerably damaged.
Destructive	25.0 to 50.0	VIII	Frightens everyone. Slight damage to buildings of special, good design and construction. Chimneys, columns, monuments, smoke stacks, towers and elevated tanks twist and fall. Cracks in filled panel walls in steel lattice structures having no corner stiffeners. Much damage in ordinary substantial buildings. Collapse of more than $1/4$ of all buildings Heavy furniture overturned. Persons driving cars are disturbed. Steering of automobiles affected. Sand and mud erupts in small amounts. Flow of springs and wells is temporarily and sometimes permanently changed. Dry wells renew flow. Temperature of spring- and well-waters varies. Cracks in ground and on wet slopes. Shaking down of loosened brick-work and tiles. Considerable damage to concrete irrigation ditches.
Devastating	50.0 to 100.0	IX $a = (0.1)g$	General panic. Great damage to poorly built substantial structures and to lattice-type steel structures (not frames). Collapse of more than $1/2$ of all buildings. General damage to foundations. Ground cracked

Rating of Intensity of Earthquake	Cancani's Maximum Acceleration a_{max} cm/s²	Mercalli Scale in Degrees of Intensity	Mercalli-Sieberg-Wood-Neumann Description of Earthquake Effects
1	2	3	4
			conspicuously. Pipes broken. Underground reservoirs seriously damaged. Decayed piling broken off. Well-designed modern frame structures thrown out of plumb.
Disastrous	100.00 to 250.0	X	Panic in general. Crushing of column heads. Many buildings ($\sim$ 3/4) and their foundations are destroyed; most of them collapse. Light damage to aseismically designed steel and reinforced concrete structures and bridges. Crevices in soil. Ground and pavements crack up to widths of several inches. Large landslides from river banks and steep coasts. Shifting sand and mud. Water is thrown on canal banks. Dams, dikes, embankments are seriously damaged. Underground pipelines torn or crushed. Rails bent slightly.
Catastrophic	250.0 to 500.0	XI	Widespread general panic. Most buildings collapse. Few structures left standing. Great damage to bridges, embankments, and dams. Wide cracks in ground. Earth slumps. Rock and landslides. Tsunamis of significant magnitude. Water charged with sand and mud is ejected in large amounts. Rails bent greatly. Pipelines are put out of service completely.
Violent Catastrophe	> 500.0	XII	Panic is general. Great disaster. Total destruction of all structures, including foundations. Nothing stands which is created by man's hands. Ground deformations to a large extent. Disturbance of strata. Fissures in the earth's crust. Rockfalls from mountains. Landslides. Lakes run out; rivers are diverted. Large destructive tidal waves (tsunamis). Waves seen on ground surface. Large rock masses displaced. Objects are thrown upward into the air. Fault slips in firm rock, with noticeable horizontal and vertical offset displacements. Surface and underground water channels are disturbed and modified greatly. Lakes are dammed, new waterfalls are formed.

The MERCALLI-CANCANI-SIEBERG earthquake intensity scale contains grading or degrees of intensity of destructiveness by means of a single Roman number — the intensity — from I (not directly felt) to XII (nearly complete destruction of most manmade structures). However, this earthquake rating is too subjective and too crude a basis for most engineering purposes.

Because, for design purposes, a rating of earthquake severity based on maximum acceleration of the ground is of most interest to the engineer, the maximum acceleration values a_{max} (cm/s²) after CANCANI's scale (84) are added to the MERCALLI-SIEBERG earthquake intensity table in Column 2 of Table 10-1.

Hence, the title of Table 10-1 reads: Modified MERCALLI-CANCANI-SIEBERG Earthquake Intensity Scale.

CANCANI tried to evaluate the descriptive MERCALLI earthquake scale based on ranges of seismic acceleration. Thus, the CANCANI values in Col. 2 of Table 10-1 are actual measurements in physical terms (maximum acceleration, a_{max}, in cm/s²). However, as all earthquake intensity scales are in some way deficient, so also is the CANCANI scale, namely: in any one earthquake there are many different accelerations. Therefore, period must be considered together in defining intensity. Also, because of variations in geologic conditions from place to place, a scale which fits in one place does not fit in another. Therefore, correlation between the various earthquake intensity scales must be done with caution (74).

During an earthquake, the ground moves in a random manner in both horizontal and vertical directions. Therefore, in earthquake engineering, where appropriate, earthquake factors — both in horizontal and vertical directions should be used (see Chapter 12).

Earthquake Magnitude

Destructiveness brought about by an earthquake is directly related to the energy that the ground motion transmits to people and to man-made structures. Whereas "intensity" is a measure of the earthquake's local destructiveness, and refers to any particular point, the *"magnitude"* of an earthquake is a measure of the total energy released. Energy does not depend on maximum intensity alone, but, in the shaken area, it also depends upon the extent of faulting, depth range of displacement, and other factors.

An earthquake scale independent of human subjective estimates of damage is based on the *magnitude* of the effect of earthquake waves on a standard seismograph situated on a firm ground at a distance of 100 km from the epicenter. Specifically, magnitude is computed from the measurement of the maximum amplitude of the horizontal trace on the seismogram considering decrease in energy with increase of distance of the epicenter. For other distances the magnitude must be calculated (37—40).

A convenient means of rating severity of earthquakes according to magnitude is the so-called RICHTER Magnitude Scale (37—40; 74). The RICHTER Magnitude Scale is based on instrumental records. It measures the energy release of an earthquake at the epicenter, and is the most commonly used today. On the RICHTER Scale, the magnitude of the earthquake is expressed in whole numbers and decimals. The RICHTER Magnitude Scale has no fixed maximum or minimum. Observations have placed the smallest normally felt earthquakes by humans at about 3, and the largest recorded ones in the world at about 8.9. Earthquakes of magnitude 5 or greater generate ground motions sufficiently severe to be potentially damaging structures.

According to RICHTER (75), "Magnitude can be compared to the power output in kilowatts of a broadcasting station. Local intensity on the Mercalli Scale is then comparable to the signal strength on a receiver at a given locality; in effect, the quality of the signal."

On the RICHTER scale, the magnitude varies logarithmically (decimal logarithms) with the maximum wave amplitude of the quake recorded on the seismogram. This means that an increase of a whole number on the magnitude scale represents a tenfold increase in the size of the earthquake record. For example, the earthquake magnitude of 8.3 is not twice that of shock whose magnitude is 4.3, but 10,000 times as great.

The wave amplitude depends only on the quantity of energy released at the focus and distance from the focus. Because amplitude varies through a wide range, it is more convenient to plot the logarithm of amplitude A. Such curves $\log_{10} A = f$ (distance) form the basis for the RICHTER Scale of earthquake magnitude M (74).

The magnitude M is expressed as

$$M = \log_{10} A - \log_{10} A_o, \qquad (10\text{-}2)$$

where A = maximum recorded trace amplitude of any given earthquake recorded by a standard seismograph at a distance of 100 km from the center of disturbance, and

A_o = amplitude of a particular earthquake selected as a standard or the "zero" earthquake.

The zero level A_o of the scale is fixed by setting M = 3 when the amplitude A_o is one thousandth of a milimeter at the standard distance of 100 km from the epicenter.

It was mentioned earlier that the Richter earthquake scale relates to magnitude and energy. In an earthquake, the amount of energy E released and magnitude M on the RICHTER Scale are related empirically by

$$\log_{10} E = (1.5) M + 11.4, \tag{10-3}$$

where E = energy released in ergs (74).

Comparison of Magnitude and Intensity

It is difficult to compare magnitude and intensity because intensity is linked with the particular ground and structural conditions of a given area, as well as distance from the earthquakes epicenter, while magnitude is a measure of energy released in the bedrock at the focus (hypocenter) of the earthquake.

Table 10-2 shows an approximate comparison of magnitude and intensity after RICHTER (74).

Some representative magnitudes of California earthquakes are compiled in Table 10-3 after California Geology (18,96), IACOPI (45), and U.S. Geological Survey Earthquake Information Bulletins (98, 100).

It is believed that the San Francisco earthquake of 1906 was caused by a strike-slip displacement on the San Andreas fault. Great fires resulted from the quake. Railroad tracks were buckled and broken, and in one place a train was thrown over on its side.

TABLE 10-2

Approximate Comparison Between RICHTER MAGNITUDE and Expected Modified MERCALLI-SIEBERG Maximum Intensity Effects at Epicenter

Richter Magnitude Scale	Mercalli-Sieberg	
	Intensity Scale	Approximative description of intensity effects
1	2	3
2	I—II	Usually detected only by instruments
3	III	Felt indoors
4	V	Felt by most people. Slight damage
4.5	VII	May cause slight damage in small area
5	VI—VII	Felt by all. Many frightened and run outdoors. Damage minor to moderate
6+	VII—VIII	Everybody runs outdoors. Damage moderate to major
7+	IX—X	Major damage
8+	X—XII	Major and total damages

Thirty-three states in the U.S.A. had earthquakes in 1976. One hundred and eighty-eight earthquakes of M = 4 or larger occurred in California during the period 1975 through March 1979 (96), causing various kinds of damage.

Earthquakes do not only happen in California, but shake all countries. Most loss of life and property has been because of the collapse of antiquated and unsafe structures.

The 1964 Alaska earthquake (Anchorage, Seward, Valdez) was the most destructive in the area since 1899. It rocked the entire south-central Alaska; there was a loss of life; change of land level took place;, differential subsidence and landslides occurred; waves were generated by massive submarine landslides; harbor facilities, structures, water supply- and sewer systems were damaged and/or destroyed. At Seward, the railroad marshalling yard was left a tangled mass of wreckage by wave action and fire. Twisted rails and overturned locomotives at Seward attested to the great energy of waves (23, 41, 57, 64).

Because of the extensive earthquake damage at Valdez, a new site for relocating the town at Valdez has been designated. Valdez is an ice-free port at the southern end of the Alaska pipeline.

TABLE 10-3

Some Representative Magnitudes of California Earthquakes

Richter Magnitude M	Place	Date	Approximate Intensity
1	2	3	4
5.2	Homestead Valley	March 15, 1979	—
5.3	Daly City	March 22, 1957	VII
6.3	Long Beach	March 10, 1933	VII—VIII
6.3	Santa Barbara	June 29, 1925	VIII—IX
6.6	San Fernando	February 9, 1971	VIII—XI
6.6	Imperial Valley	October 15, 1979	—
7.1	Imperial Valley	May 18, 1940	X
7.7	Kern County	July 21, 1952	—
8.3	San Francisco	April 18, 1906	X—XI
8.3—8.6	Alaska: Anchorage (8.5) Seward (8.3—8.4), Valdez (8.4—8.6)	March 27, 1964	

Figure 10-2 shows bent rails caused by faulting on February 4, 1976 in Guatemala (31).

Some other destructive earthquakes of recorded history, with magnitudes and brief remarks about the casualties and damage, are compiled in chronological order in Table 10-4 after HECK (42), KENDRICK (54), BÅTH (11), U.S. Geological Survey Earthquake Information Bulletin publications (98, 100, 103), and other sources.

According to HECK (42), in the Lisbon earthquake of 1755, first

> "there was a great shock accompanied by a roar like thunder, with great clouds of dust that were almost suffocating. There were three shocks in all, over a period of nine minutes, at the end of which the city lay in ruins. Immediately, fire driven by strong winds swept the ruins and then one of the greatest seismic seawaves of history rolled over the low lying portions of the city to complete the disaster. Loss of life and property was very great."

Fig. 10-2
Bent rails caused by faulting in Guatemala, February 4, 1976.
(Photo by A. F. ESPINOSA. Photo courtesy of U.S. Geological Survey, Denver, Colorado, No. 144. Reference: U.S. Geological Survey Earthquake Information Bulletin, 1977, Vol. 9, No. 2, p. 8)

TABLE 10-4

Some Selected Destructive Earthquakes

Date	Place	Magnitude M	Loss of Lives	Remarks
1	2	3	4	5
1755, Nov. 1	Lisbon, Portugal	8.75	60 000	Perhaps the strongest quake in history. Seismic sea waves
1811, Dec. 16 and 1812, Jan.23 and 1812, Feb. 7	New Madrid, Missouri, U.S.A.			Thousands of shocks, some felt over most of United States. Created Reelfoot Lake in Tennessee. Changed topography and land boundaries
1897, June 12	Assam, India	8.70		Very severe. Scarp raised 10.5 m. Flooding. Almost complete destruction of buildings. Aftershocks continued 2 years. $a = 0.36$ g and $a = 0.43$ g
1899, Sept. 3 and Sept. 10	Yakutat Bay, Alaska, U.S.A.	8.6		Seacoast uplifted more than 14 m. Glaciers shattered. Landslides.
1905, Sept. 8	Calabria, Italy	7.9	2 500	—
1906, Aug. 16	Chile, Colombia, Ecuador			Very severe; $a = 0.2$ g. Very severe
1906, April 18	San Francisco region California, U.S.A.	8.3	700	Horizontal movement along San Andreas fault. Very severe quake. Devastating fire. Landslides. Some heavy aftershocks continued for more than a year.
1908, Dec. 28	Messina and Reggio, Sicily, Italy	7.5	100 000	Both cities were razed. Coast depressed ~ 1 m. Tsunamis
1920, Dec. 16 1929	Kansu Province, China	8.6	180 000	Most people buried in caves by landslides of loess.
1923, Sept. 1	Tokyo, Yokohama, Japan	8.3	250 000	Tokyo and Yokohama destroyed. Tsunamis. Destructive fires. Max. displacement 4.5 m. $a_{max} = 0.49$ g.
1931, Feb. 3	Hawkes Bay, Napier, Hastings and Wairea, New Zealand	7.9	225	Great earthquake disaster in New Zealand. Uplift. Displacement 1 to 2 m. Landslides and fires.
1957, July 27	Mexico	7.8	55	Intense quake.

TABLE 10-4 Continued

Date	Place	Magnitude M	Loss of Lives	Remarks
1	2	3	4	5
1959	Hebgen Lake, Montana			Landslide. New lake formed. Damage in Yellowstone Park.
1960, Feb. 29	Agadir, Morocco	5.8	10 000 to 15 000	Maximum observed intensity was XI. The greatest number of deaths ever for such a small shock. City extensively destroyed.
1963, July 26	Skopje, Yugoslavia	6.0	1 100	Shock occurred right under the city. 80% of the buildings destroyed; 295 aftershocks felt up to August 15, 1963. Intensity: VIII.
1964, March 27	Anchorage, Seward, Valdez, Wittier, Alaska, U.S.A.	8.5	114	Very severe, widespread, most destructive earthquake in this area since 1899. Widespread damage and tsunamis. Ground motion. Landslides. Fissures 10 m wide and 15 m deep. Horizontal and vertical displacements of 6 m. Fires. Tsunamis. Liquefaction of loose sand layers. Intensity: XI.
1971, July 9	Valparaiso, Chile	7.7	110	Great fires.
1971, Feb. 9	San Fernando, Cal., U.S.A.	6.6	65	Collapse of buildings; 30 000 buildings damaged. $a = 0.5\,g$ to $a = 0.75\,g$. Slope failures of 2 dams. $500 million property damage. Rails bent. VII—XI.
1976, Feb. 4	Guatemala City, Guatemala	7.5	23 000	Extensive damage. Houses collapsed. Landslides. Surface faulting. Irrigation canals offset.
1976, Nov. 24	Turkish-Iranian border, Turkey	7.4	5 000	Many villages completely destroyed.
1976, July 28	Tangshan, at Peking	7.8		XI. Loss of life. Serious damage. Liquefaction. Rails bent. The city of Tangshan is being rebuilt.

TABLE 10-4 Continued

Date	Place	Magnitude M	Loss of Lives	Remarks
1	2	3	4	5
1977, March 4	Vrancea region, Romania	7.2	1 578	11,221 injured. Considerable damage in Bucharest; $a_h = 0.2\,g$; $a_v = 0.12\,g$. 33 000 housing units destroyed or seriously damaged; > 200 000 homeless. Destroyed schools, hospitals, university buildings, and damaged 400 cultural institutions and 763 factroies.
1978, July	Hawaii Island	4.1	—	No damage reported.
Aug. 9		4.0	—	No damage reported.
1979, Jan. 16	Iran, northwestern	6.7	200	Many injured. Considerable damage.
Sept. 16	Iran, Tabas		15 000	Many injured.
Dec. 14	Iran, Izeh	6.3	76	Destructive earthquake.
1979, April 15	Yugoslavia, southwestern coast	7.0	121	Extensive damage; 1000 injured; 100 000 homeless
1979, July 1	Panama-Costa Rica border region	6.4		Damaging quake. $2 million damage to petroleum-loading facilities.
1980, Oct. 10	El Asnam, Algeria	7.5	~25,000	Violent quake. The earth moved 20 times (aftershocks). 80 % of El Asnam destroyed. ~200,000 injured or left homeless. Hotels, apartment buildings, city hall, a hospital, police station, a highschool, and a mosque were virtually demolished.

A description of the Skopje, Yugoslavia 1963 earthquake damage may be found in References 12 and 87, and that of the Romania earthquake of 1977 in Reference 68.

Hazard from Faults

Faulting has been observed to occur together with large earthquakes, and most important earthquakes are believed to originate this way. Earthquake energy does not depend on maximum intensity or magnitude alone, but, in the shaken area, on the extent of faulting, depth range of displacement, and other seismic and geologic factors. Fault movement is the result of elastic rebound — the slow build-up and sudden release of strain within the rock mass. Even little fault movements may cause severe local earthquakes capable of damaging engineering structures.

When faulting breaks the rocks, waves of elastic compression or distortion spread out through the solid earth as waves spread over a pond, or as quivering spreads through jelly. These waves arriving at the surface of the earth shake it — violently near the source of a great eartquake, mildly at distant points. This shaking is responsible for most of the ordinary earthquake damage, and for nearly all minor earthquake phenomena (74).

Fault movement may shear and/or offset a structure built across a fault by several meters. Just as earthquakes can destroy man-made structures, they can also shift rock formations. A primary fault movement ruptures and destroys everything that crosses its path. According to STEINBRUGGE (89), the extent of damage to a building depends more on the strength of the building and the supporting ground beneath it than it does on the strength of the shear waves.

A structure founded partly on soil or fill and partly on hard rock may undergo differential settlement. This may impose great strain on the structure that may result in a distress during an earthquake.

Faults may also disrupt and offset subterranean water courses. Hot springs are common along the San Andreas fault in California, and other faults.

Frequently, when two rock faces along a fault line rub one against each other by tremendous mechanical forces during an earthquake, the rock crushes and grinds, becoming physically and chemically altered, whereby some of the rock minerals are transformed to clay. The abraded, soft, pulverized material in a fault zone, bedding planes, and other kinds of geological discontinuity in rock produced by movement of rock across the shear plane is called *gouge*.

Gouge affects the water regimen in rock, and may form an impermeable barrier in the faults thus blocking the normal groundwater flow through rock discontinuities. The gouge may also back-up water on the uphill side of the fault forming springs, ponds, and lakes, and a relatively arid zone on the downhill side.

Faults constitute potential rock-slide hazards and create water influx problems in tunnels and other kinds of underground excavations.

When montmorillonite clay gouge in a dipped discontinuity acquires moisture, it swells, becomes slippery, thus presenting a potential hazard relative to stability of a rock mass, namely: the plastic, swelling gouge causes rock slides.

Also, if sufficiently wet and open-packed, coarse-material gouge in outwardly dipped joints, faults and bedding planes may also function as sliding surfaces.

San Andreas Fault

To appreciate the seriousness and geologic hazards from faults that may be encountered in site exploration, urban planning, foundation-, tunnel-, and hydraulic structures engineering, it is desirable to familiarize with at least one regional example of locations of significant faults. As such an example the California region is chosen here.

The most conspicuous, significant, and advertised rift — a major strike-slip fault — is the great San Andreas fault. In historical time, along this fault, considerable movement has occurred. The San Andreas fault is a dominant part of California's extensive fault system. This fault annually causes dozens of earthquakes, and is also one of the contributing factors to the many landslides occurring along the California coast ranges.

California is broken up into a series of crustal blocks that are separated by faults — great fractures that form lines and/or zones of weakness in the rock masses at the ground surface. The great San Andreas shear zone extends roughly parallel to the Pacific Ocean coast, and in part is hidden under water.

The San Andreas fault is not a single break in the rock, but a wide zone of varying width of up to several hundred meters, consisting of several fault lines that are approximately parallel, and being interlaced with many other faults.

According to Geologist D. P. SMITH (86),

> "The San Andreas fault is generally defined as the trace of the most recent surface rupture. However, through geologic time, the fault has shifted position sideways, back and forth across a zone that may be as much as several miles wide. The entire zone is termed the fault zone."

Figure 10-3 shows locations of significant faults of the California region. They are the

Garlock fault	—the second largest fault in California
Hayward fault	—a branch of the San Andreas fault
Imperial fault	—another branch of the San Andreas zone
Newport-Inglewood fault	—
San Andreas fault	—
San Fernando fault	—
San Jacinto fault	—part of the San Andreas zone
Santa Ynez fault	—
Sierra Nevada fault	—
White Wolf fault.	—

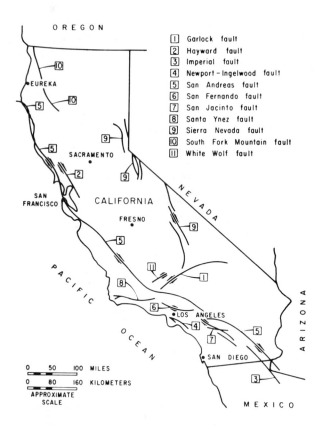

Fig. 10-3
Location of significant faults of the California region

ENGINEERING PROBLEMS 393

Figure 10-4 shows a trace of the San Andreas fault in Choia Valley, California. Notice the streams that have been offset by strike slip on the fault. Bulging along the fault trace is caused by compression. The San Andreas fault is heavily instrumented as part of a program for earthquake prediction (Ref. 100).

An example of disregard for existing geologic hazards is shown by an aerial photograph from California in Fig. 10-5. It is an aerial view of the San Andreas fault zone south of San Francisco. In this photo, the fault zone extends northwestward from San Andreas reservoir (lower left) and out to the coast (upper center). Since 1956, most of the tremendous urban growth in the vicinity of the San Andreas fault has taken place with little regard for hazards created by existing and potential geological processes (Ref. 18).

One of the best exposures of a part of the San Andreas fault zone is the large roadcut where the Antelope Valley Freeway (State Route 14) passes

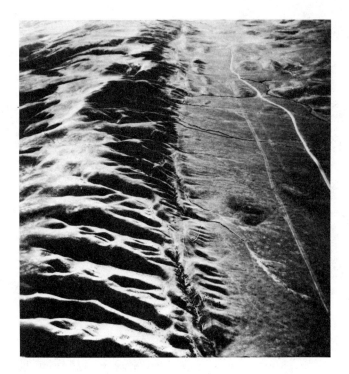

Fig. 10-4
Trace of San Andreas fault in Choia Valley, California
(Photo by R. E. WALLACE. Photo courtesy of the U.S. Geological Survey, Ref. 100)

through the San Andreas fault zone near Palmdale, California. The roadcut is about 93.6 km (60 miles) north of Los Angeles. This 27.4 m (90 feet) deep road cut lies between the San Andreas fault and the Little Rock fault (Fig. 10-6). This road cut provides a sloped profile in which to study the fault zone in the highly folded and faulted Tertiary Anaverde formation.

The Anaverde formation exposed in the cut is mainly nonmarine arkosic sandstone (the thicker light-toned beds) and gypsiferous shale, strongly folded and faulted (86). Notice the deformed, faulted syncline in Fig. 10-6 by compressive and shear forces within the San Andreas fault zone.

Codes

To safeguard life and property, earthquake, or aseismic, design codes are put into effect (2, 12, 13, 14, 16, 30, 36, 47, 65, 66, 67, 74, 76, 87, 89, 90, 97, 103).

Fig. 10-5
Oblique aerial photograph showing the trace of the San Andreas fault zone and urban development south of San Francisco. The fault zone extends northwestward from San Andreas Reservoir (lower left) and out to the coast (upper center). Most of this residential development in the vicinity of the San Andreas fault has taken place since 1956.
(Photo by M. Moxum. Photo courtesy of Mary C. Woods, Geologist, California Division of Mines and Geology)

ENGINEERING PROBLEMS 395

For aseismic design of structures, an earthquake resistant design code should give guidance on three main topics, namely:

1. loading and risk
2. overall performance criteria, and
3. aseismic detailing.

Aseismic design codes provide adequate strength for the structures and their foundations to resist ground motion. Also, these codes focus increased attention on the vulnerability of essential lifetime services, both within the building and throughout urban centers. Included are indications for the design of power lines, water supply, sewage disposal, communications, oil and gas pipelines, and transportation facilities.

During the earthquake the ground moves in a random manner in both horizontal and vertical directions. Therefore, where appropriate, earthquake

Fig. 10-6
View of the deformed, faulted syncline in a roadcut made in the San Andreas fault zone in southern California (Ref. 86).
(Photo courtesy of Geologist MARY C. WOODS, Editor-in-Chief of *California Geology*, California Division of Mines and Geology)

factors — both in horizontal and vertical directions — should be used (see Chapter 12).

The so-called "Field Act" of California provides for improvements in existing building codes. The "Field Act" gave the California State Division of Architecture authority and responsibility for approving designs and supervising construction of public schools and establishing severe penalties for violations.

10-9. Hazards from Water
Water Movement Through Rock

Water is the main hazard in geotechnical engineering. Many engineers have had the experience, at least once, that the "opponent" water has nullified the results of all their efforts. From the sequence of the matter as discussed in the previous sections in this book, it can be understood that the "opponent" water in many cases appears especially troublesome in all construction work concerning foundation engineering and earthwork- and rock engineering in general, and hydraulic structures engineering in particular. Many failures with structures on and in rock are caused by the *unanticipated action of water*, the regimen of which may change as a result of interference by man's activities.

Water is one of the most important independent variables governing work in soil- and rock engineering.

Basically, water movement in rocks takes place along bedding planes, cracks, joints, and numerous small adjustment faults and fractures. In these rock defects impermeable gouge, if present, impedes the free movement of groundwater affecting adversely the stability of dam-supporting rock foundations and tunnels (13, 59).

One distinguishes between two kinds of water movement in the ground, namely: seepage and leakage.

Seepage is the slow movement of water or any other liquid through a porous medium or interstices of soil or rock.

Leakage is the uncontrolled escape of water through bedding planes, cracks, joints, faults, holes, crevices, and other kinds of discontinuities in the rock.

Seepage may result in the loss of some of the stored water in the reservoir. Of course, some seepage through the rock foundation is practically unavoidable because bedrock is seldom, if ever, entirely unbroken. Therefore, before attempting any study of seepage flow, it is necessary to know the geology in depth.

Excess leakage from an impounded reservoir through a dam-supporting rock foundation, if uncontrolled, may take a dramatic course for the worst.

Water Control

Leakage testing of rock is usually done as a matter of course in most underground rock engineering works. For this reason it is seldom reported in the technical literature. Leakage management involves channeling the water which leaks and drains into the underground opening to a point where the amount of ingressed leak water can be measured.

Water control is usually accomplished by appropriate surface drainage, subsurface drainage, and rock-foundation drainage systems. Drainage adds to the increase in safety of foundations in tight rocks with fine fissures which are difficult or even useless to grout, but it is possible to drain. Here drainage is the only means of controlling seepage pressure.

Water tightness of rocks also improve their load bearing capacity.

Permeable rocks with wide, open fissures are possible to grout. Grouting reduces leakage and also seepage pressures downstream of the grout curtain.

A watertight curtain will reduce effectively the leakage- and seepage pressure downstream only if the curtain is thick enough and if its permeability is much less than that of the adjacent, surrounding rock mass. Grouting of a watertight curtain in a rock mass of low permeability is a difficult problem because grout only reduces the natural permeability. Besides, a grout curtain is very sensitive to local defects in the curtain. Thus, there is the problem of water flow through grout curtains. These opinions have been accepted by the rock engineering profession since the First RANKINE Lecture delivered by A. CASAGRANDE in 1961 (20).

Drainage is the most practical, effective and economical method to cope with seepage through earth dams, under massive dams, with soil and rock slides, mud flows, and to reduce uplift pressures on structures and rocks to acceptable values.

Hydraulic Fracturing

Sometimes, as reported at the 13th International Congress of Large Dams held 1979 at New Delhi, India (5), leakage through rocks may occur as a result of hydraulic fracturing or splitting of the rock effected by a high hydrostatic head. The excess leakage through cracks brought about by hydraulic fracturing is thought to depend upon the stress-strain condition in the dam supporting rock.

Solution Cavities

During excavation for the Hale's Bar Dam on the Tennessee River, many *solution cavities* were discovered. Coping with these cavities long delayed the completion of the dam. The grouting of the open cavities required about 5000 tons of cement. After the dam was put into service, the dam leaked enormously. Ten years after completion, and after extended remedial measures, leakage was cut down to an acceptable rate by introducing hot asphalt into numerous drill holes in the rock. More than 10 000 barrels of asphalt were used (3).

Other Water Problems

Then some other water problems may arise. For example, with the rising of the water level in a reservoir upon its filling, or because of heavy precipitation in the vicinity of the reservoir, the discharge of existing springs can increase, and some new springs may appear nearby the dam (see section on Vaiont reservoir disaster in this chapter further on).

Geological Hazards from Rock

To many people, the term "rock" means a very stable and durable material, and brings to the mind the concept of strength, reliability, and permanence (recall the saying "as safe as the Rock of Gibraltar," and the saying which comes down to us from Biblical times that "a house must be built on a rock foundation"). Only a few have an idea about the mechanical defects and weaknesses of rocks, and the perception of the real, natural, geological hazards and problems involved in rock engineering an engineer may encounter in his work with rock.

To the layman it might appear that a strong rock such as granite, may in general always be a reliable and suitable material for almost any engineering purposes. However, under certain special conditions this may be quite the contrary. Concerning granite, its soundness, strength, and quality depends very much on factors such as degree of weathering; its composition and structure; water regimen and permeability; details of faults, joints, fissures, and other hidden rock defects which have been encountered at many construction projects. The water hazard occurring in granite adjacent to faults and fault zones should especially be emphasized.

Also, water flowing from seams, joints, and faults in granite may bring about relief of pressure (i.e., decompression), and lift seams (open joints) because of elastic rebound. Or, hydrous solutions rising from depth may alter the feldspar of granite to kaolin.

A case in point are the geological difficulties encountered during the driving of the Mont Blanc tunnel, 11.6 km long. It was driven from 1959 to 1962 between Chamonix, France and Courmayeur, Italy (RICHEY, 1963, Ref. 73, and ZIGNOLI, 1965, Ref. 106).

ENGINEERING PROBLEMS

On the Italian side of the Mont Blanc tunnel (Fig. 10-7), an enormous quantity (86372 ℓ/min) of cold water rushed into the tunnel from fractures in the granite face and roof. The water, supposedly, came from a fault zone that extended upward to the ground surface, viz., the Toula glacier several hundred meters overhead.

Notice from the rock temperature graph that upon approaching the site of the influx of the cold water, the rock temperature drops sharply. The decrease in rock temperature is related to the cold melt-water descending through the fault zone from the Toula glacier. Thus, a decrease in rock temperature during tunnel driving may indicate that a site is being approached where a large influx of water from an overhead glacier can be expected.

Examples of underground powerplants built in granitic rocks are described by RICHEY (1963) in Ref. 73.

The national monument Valle de la Caidos near Madrid, Spain, was built underground in a weathered granite rock having many open seams and cracks. To cope with leakage into the cript, 50 m in diameter, the zone around the dome had to be injected.

In his book entitled "Barrages et Géologie," Ref. 61, Professor MAURICE LUGEON of Lausanne, Switzerland, the great elucidator of the Alps and of

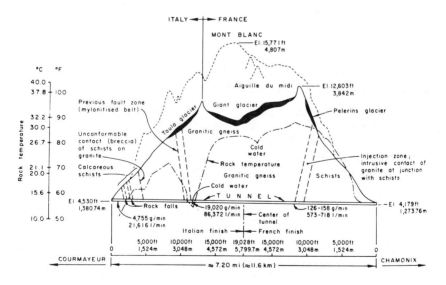

Fig. 10-7
Geological section, with temperature graph, through Mont Blanc vehicular tunnel.

many dam sites, wrote much concerning granite both in praise and otherwise, remarking:

> "Combien il faut se méfier au granite." This LUGEON's dictum may be freely translated as — how much one should distrust granite.

But LUGEON goes on to say that this dictum is not an absolute rule, and that the two-mica granites, in general, are sounder rocks than the others, and less liable to weather to granite sand.

About hazards in granitic rock TERZAGHI wrote:

> "...General experience with tunnelling through these rocks [intrusive igneous rocks such as granite] indicates that the probability of hazards is very low but the deviations from normal can be important to upset completely the original estimate of cost. Furthermore, in many instances the existence of abnormal conditions at depth cannot be predicted from surface evidence. Hence, the intrusive rocks should be considered decidedly treacherous."

Hydrostatic Uplift Pressure

If water gains access under the base of a foundation, or a massive dam, or under a rock stratum, it exerts an upward pressure on the base known as *uplift*. In general, uplift pressure acts against the weight of a submerged structure, thus increasing the possibility of danger of the structure sliding on its base, as well as a probable overturning of the structure.

Uplift pressure is usually reckoned with on the full base area of a structure. This pressure is a function of the pressure head, the perviousness of the soil or rock, and the effectiveness of the natural and/or artificial drainage system installed in the ground.

In practice, it is assumed that the uplift pressure varies linearly from the full pressure of the headwater at the heel of the structure to full pressure of the tailwater at the toe of the structure. By reducing uplift pressure, seepage can be reduced. This may be attained by the installation of grout curtains and/or appropriate drainage systems in the rock (refer to Canelles dam in this chapter).

Assuming in the simple case a linear uplift pressure distribution, and referring to Fig. 10-8, the uplift pressure p_w under the edge of the dam on the upstream side (heel) has its full value of $p_w = \gamma_w \cdot h$ (because of seams, joints and fissures in the rock), but on the downstream side (at the toe) it is $p_w = 0$. Here γ_w is the unit weight of water, and h is the full hydrostatic pressure head on the upstream side of the dam. The uplift pressure diagram in this case is a triangle, Δ (123), and the total uplift pressure U is

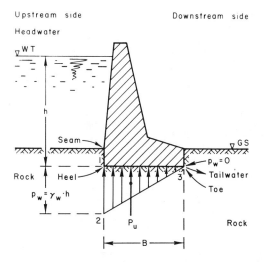

Fig. 10-8
Uplift pressure diagram under a massive water-impounding dam

$$U = (1/2)(p_w)(B)(1.0) = (1/2)(\gamma_w)(h)(B) \tag{10-4}$$

According to CRIEGER and McCOY (24), because of the grout cutoff at the heel of a homogeneous dam (refer to Fig. 10-9), the pressure ordinate p_{w1} is calculated as

$$p_{w1} = \gamma_w \left[h_2 + \zeta(h_1 - h_2) \right] \tag{10-5}$$

where ζ = uplift intensity factor depending upon the effectiveness of the grouted cutoff and drains.

However, the uplift intensity at the downstream face remains constant at $p_{w2} = \gamma_w \cdot h_2$, and the uplift pressure varies linearly as the line 1—2 shown in Fig. 10-9. The total uplift pressure U is the equal to

$$U = (\gamma_w)(B)\left[h_2 + \frac{\zeta}{2}(h_1 - h_2) \right] \tag{10-6}$$

The intensity factor ζ varies for well-grouted and drained foundations from close to 0 to not more than 0.5 (8). This uplift intensity is considered to be exerted on 100 percent of the total area of the base of the structure.

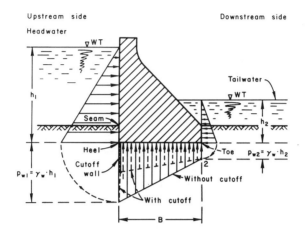

Fig. 10-9
Uplift pressure diagrams under a massive, homogeneous dam with and without cutoff

The practices of assessing the magnitude of the uplift pressure vary from agency to agency, and from country to country.

Uplift pressure can be reduced by means of cutoff walls, seepage-retarding aprons, high-pressure grouting of rocks, supporting dams or structural foundations by installing of drainage wells. These methods reduce the length of the seepage path, thus reducing the hydraulic gradient, and hence the seepage velocity, viz., discharge, and subsequently resulting in a lessened danger of sliding and overturning of the structure.

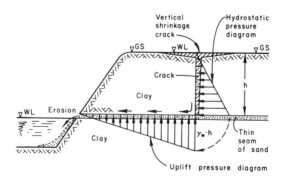

Fig. 10-10
Uplift in earth bank

Uplift in an earth bank may occur under conditions as sketched in Fig. 10-10. Here a clay bank contains a thin seam of permeable sand. Water permeates in this seam under pressure head from a vertical water-filled shrinkage crack out into the open. The seeping water may undermine the stability of the earth bank and result in underground erosion. Because of the uplift pressure, sliding of the entire mass of earth above the thin seam of wet sand can take place, or the sand may become eroded by the seeping water at the face of the slope.

For rock slides over gouge-filled discontinuities refer to Chapter 12.

Hoover Dam

Hoover dam, Fig. 10-11, formerly known as the Boulder dam, is an arch-type dam on the Colorado River between Arizona and Nevada southeast of Las Vegas, Nevada. The dam was built 1931—1936 as a massive curved gravity structure. The dam is 221.85 m (726 feet) high, 388.62 m (1244 feet) long, and has a bottom and top thickness of 200.86 m (659 feet) and 13.72 m (45 feet), respectively.

Fig. 10-11
Hoover dam.

At Hoover Dam, an extensive program of grouting was pursued to control seepage and leaks, and to reduce the hydrostatic uplift pressure on the base of the structure (85).

The treated rocks are of volcanic origin, namely breccias.

The possible sources of seepage and uplift pressure on the base of the concrete dam during the initial filling of the reservoir were rock faults and contributing geologic features at the dam site: two distinct minor faults crossed the right abutment and manganese seams in the foundation rock. The increasing uplift pressure on the base of the dam was of the greatest concern to the engineers responsible for the safety of this hydraulic structure. Thus, an additional program of flushing, grouting and draining of inadequate areas was necessary.

The purpose of additional grouting and a drainage program was to correct geological defects of rock in the foundation and abutments to

1. reduce the uplift pressure acting on the base of the dam
2. reduce seepage through the abutments of the dam
3. eliminate seepage of hot alkaline water through the concrete lining of the penstock tunnel, and
4. eliminate seepage of cold water into some parts of the penstock tunnels.

Grout cutoff curtains were installed beneath the dam, consisting of three or more rows of grout holes. They were spaced from 6.10 m by 6.10 m (20 feet by 20 feet) on centers in each direction to an irregular pattern in which no particular spacing predominated. The grout holes were 38.2 mm to 50.8 mm ($1^1/_2$ to 2.0 inches) in diameter. A total of 90947.14 m (298,383 linear feet) of drilling had to be made, of which 10398.25 m (34,115 feet) were for drainage purposes. 251,115 sacks of cement were injected into the rock foundation and abutments of Hoover Dam under difficult and severe conditions. In this work, a combination of stage (or successive) grouting with packer grouting was the best method used. In some instances, an extremely thin grout mixture had to be used.

The drainage system beneath the dam varied from gravel-filled trenches in the foundation rock below the masonry, to a regular pattern of drainage holes drilled into the bedrock slightly downstream from the grouted cutoff curtain.

Canelles Dam

More recently, important and difficult grouting work was executed during 1971—1974 by the Rodio-Kollbrunner organization in the rock foundation of the Canelles arch dam (151 m high; crest length 203 m) near Lerida, Spain (DIMAS et al., 1978, Ref. 26).

Figure 10-12 shows the main cross-section of the Canelles dam. The grout curtain A was designed to shut off leakage through the complex network of the cavernous karst-like limestone representing the dam-supporting foundation. The karstic caves here were very irregular, sometimes ~ 5 m × ~ 6 m in section. Sometimes at their bottoms ~ 1 m thick layers of mud deposits were encountered. This curtain (A) was installed during the course of construction of the dam. During the first filling of the reservoir up to elevation 470, leakage discharge was approximately 6000 ℓ/s. Many springs (s_2 — s_3 in Fig. 10-12) appeared on the left flank. The springs were located mostly underneath the marl layer first thought to be impervious.

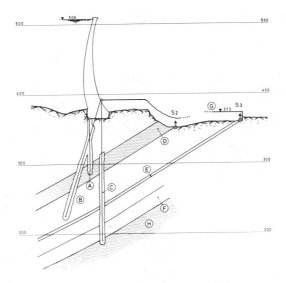

Fig. 10-12
Main cross-section of Canelles Dam (Ref. 26).
A. Grout curtain performed during dam construction
B. Grout curtain inclined 35 %
C. Auxiliary boreholes for drainage and grouting
D. Marl layer named "Capa Negra"
E. Another marl layer
F. Sandstone with red and yellow veins
G. Concrete of the spillway stilling basin
H. Thick deep marls named "Margos hondas"
S_2 — S_3 Springs

Figure 10-13 shows the toe of the Canelles Dam as seen from its crest. In the pool, with dirty water on the left, the clear zones (dark) in the center of the photograph is the water that rises from the springs in the bedrock. At the downstream end of the stilling pool to the right, the seepage discharge gauging station is seen between the last two energy dissipation blocks.

Figure 10-14 shows vertical drainage at the toe of the dam. The leakage discharge water spurts up from a depth of more than 100 m.

Figure 10-15 is a striking illustration of the placing of a packer: when the drilling bit intersects a rock fissure, water gushes up under pressure.

The numerous unexpected difficulties in treating the karstic rock formation were successfully met by using various kinds of grout mixes. Normally clay-cement mixes were injected. These were the most economical ones in carrying out water-proofing operations in rocks with large voids. Bentonite-cement, cement-sand, and neat cement grouts were also used for this grouting job. The use of special materials such as polyurethane foams and acrylic asphalt-resin mixes afforded the karstic areas to be quickly and economically sealed up to depths of 200 m in the presence of water at a pressure of 6 at and a flow rate as high as 16 m/s.

Fig. 10-13
The toe of Canelles dam, Spain, as seen from its crest. In the pool with dirty water on the left, the clear zone in the center is the water that rises from the springs in the bedrock.
(Photo courtesy of the Kollbrunner / Rodio Foundation, Ref. 26)

ENGINEERING PROBLEMS 407

Fig. 10-14
Canelles dam, Spain: vertical drainage boreholes at the toe of the dam; discharge leakage water coming from depth of more than 100 m.
(Photo courtesy of the Kollbrunner/Rodio Foundation, Ref. 26)

Fig. 10-15
Canelles dam, Spain. Placing the packer in a borehole from which water is flowing from a water-bearing crevice.
(Photo courtesy of the Kollbrunner/Rodio Foundation, Ref. 26)

When the grouting and drainage works were completed, the seepage was reduced from an initial discharge of 6000 l/s to 68 l/s.

10-10. Failure of the Malpasset Dam

The construction of the extremely thin Malpasset arch dam on the Reyan River at Malpasset near Fréjus, France, began on April 1, 1952, and was completed in 1954. The collapse of that dam occurred on December 2, 1959 without warning following five days of heavy rainstorms. Upon failure, the dam released the impounded water in a single wave that swept down the narrow river valley towards the Mediterranean Sea. In this disaster, more than 400 people lost their lives, and part of the town Fréjus (located ~ 7.2 km downstream) was destroyed (Refs. 4, 6, 7, 19, 44, 46, 48, 50, 72, and M. MARY, 1968, pp. 36—77).

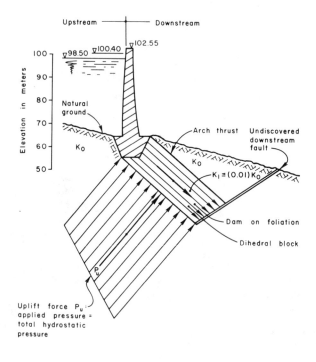

Fig. 10-16
Malpasset dam: pressures in rock at the left abutment. After Ref. 44.
k_0 and k_1 are cofficients of permeability

The Malpasset Dam was one of the most advanced examples of arched concrete dam engineering. It was the world's thinnest, double-curved arch dam. The cross-section of the Malpasset dam is shown in Fig. 10-16 after Reference 44. The thickness of the dam at its base was 6.91 m; at its top — 1.5 m. The height of the dam was 60.55 m.

The official report about the collapse of the dam states that the quality of the concrete was very good, the dam design was correct, and that the construction of the dam showed no trace of fault. Additional investigation also revealed that the concrete abutment of the dam was safely anchored in the rock. So the cause of collapse had to be in the rock.

The geology at the site was mainly carboniferous with underlying gneiss emerging in the area where the dam was built. In the foundation area of the dam some fine fissuration of the gneiss was present. The microfissured gneiss was considered to be the main weakness of the foundation rock.

Gneiss, which is moderately impervious under normal conditions, may become completely impervious when compressed perpendicularly to its foliation.

As to the causes of the failure of the dam, the ICOLD Committee on Failures and Accidents to Large Dams wrote in 1974 (Ref. 44) that

> "the Malpasset failure was the result of an unexpected combination of causes, some of which had not previously been envisaged."

The dam site had an impervious fault dipping obliquely downstream under the left bank of the dam, and another going obliquely upstream (Fig. 10-16). These faults formed a dihedral block of rock. Because of lack of drainage, the dihedral block was violently lifted by the large uplift pressure U (corresponding to the level of water in the reservoir) as if the rock foundation had exploded. It was said that this double fault had presented conditions not met with up to then.

The force of the arch thrust of the dam due to impounded hydrostatic pressure and selfweight of the dam, instead of spreading out into the foundation rock, remained concentrated to a very great depth in a prism of constant width, bringing about high compressive stresses on the foliated gneiss, rendering the gneiss a sort of watertight diaphragm in the foundation rock stopping seepage from the reservoir, and thus mobilizing a large uplift pressure. Thus it appears as if the left abutment of the Malpasset dam failed by sliding along a continuous seam of weak material covering a large area.

Based on the findings from investigation and tests by many consultants, experts, and various investigating agencies, many opinions and theories evolved, among those by CH. JAEGER (50). JAEGER put forward the following theory of the failure of the Malpasset dam. In summing up his theory, JAEGER suggested that basically the failure must have occurred in a series of several phases, namely (50):

1. A slow build-up of water pressure in the mass of the fissured and fractured gneiss upstream of the dam.

   Because of the large area over which the uplift acted, the sliding force was able to rupture a section of the foundation along one of the major rock faults (Fig. 10-16).
2. The slow progressive opening of a 10 mm to 20 mm gap along the dam heel..., and a progressive displacement of the dam foot and rotation of the dam shell about its crest.
3. A blow-out occurred of the rock mass on the left bank and the left concrete abutment slid causing collapse of the shell.

However, the real, plausible causes of failure of the Malpasset dam could not be determined with absolute certainty. The ICOLD review (44) concludes that "under such circumstances, it would have been of no use in grouting a cement curtain, upstream, during construction, and drains would have had to have been directed upstream to be efficient, as is often done now, since the Malpasset failure."

10-11. Landslides

Nature of Landslides

A landslide is the perceptible downward sliding over a slope or falling of a mass of earth, rock, or a mixture of the two under the influence of gravity.

The principal kinds of landslides are slides en masse, mud flows, creep, solifluction, rock falls, and rock slides.

Landslides can be classed as bedrock slides along planes of weakness in the rock (bedding planes, faults, joints), and slumps of unconsolidated or weakly consolidated surficial materials (natural deposits and man-made cuts and fills).

Some of the causes of landslides are high, steep slopes; weathering of rock and soil materials; heavy rain (the shear weight of water soaking into the upper soil may be partly responsible for the slides); water-filled rock cracks; hydrostatic and uplift pressures acting on loosened rock; temperature variations (freezing and thawing); seismic activities; artificially and naturally induced shocks and vibrations; loads, and other possible causes.

Landslides are almost exclusively conditioned by discontinuities in geologic materials such as rocks (inclined bedding planes, joints, cracks, faults, and zones of weathering).

Gravity and water are the main villains in promoting slides. Water adds weight to the soil and rock masses; imparts uplift pressure; induces seepage force, and pore-water pressure in a permeating mass of soil and rock. Upon freezing, ice exerts an expansive force loosening rock fragments.

Construction operations in highway, railway and canal engineering, cuts on outwardly dipped soil and rock slopes, and undercutting rock slopes at their toes have brought about many landslides. Sliding is especially troublesome if the bedding planes of the earth formations dip toward the free face of the slope (as in excavation) or valley (95).

The process of sliding is basically a continuous series of events from simultaneous causes — natural, man-induced, and a combination thereof — to the end effects. Hence, by their nature, most slides are complex phenomena.

In true slides, the movement of the sliding mass of earth or blocks of rock results from failure in shear along one or several sliding surfaces, or from shear failure of gouge filling the discontinuities of the rock.

A gouge is a geologic filling material that consists of an abraded, soft, pulverized mixture of rock and mineral materials found along a shear (fault) zone, in joints, cracks, and other kinds of mechanical defects and discontinuities in rock. Gouge is produced by differential movement of rock across the plane of slippage. If the gouge consists of montmorillonite, variation in its moisture content may bring about great catastrophic instability to the rock slope. Any clay gouge in a sloped discontinuity makes the rock mass more slippery. When such a gouge becomes wet, it promotes sliding of the rock blocks.

Landslides of natural and man-made earth slopes, and slopes of earth dams, may become activated also by natural and/or artificial, man-induced rapid draw-down of groundwater table (34, 53).

Extensive slides of slopes of the Panama Canal cuts (Fig. 10-17) are described in the Report of the Committee of the National Academy of Sciences (63).

Figure 10-18 shows a rupture of earth slope of the Delaware-Chesapeake Canal bank at the north pier of the Summit Bridge, Delaware. The pier had to be underpinned, and the embankment dewatered by a permanent groundwater lowering system.

The railway embankment slide shown in Figure 10-19 at Vita Sikudden, Sweden, occurred on October 1, 1918, at 19:00 h. Just after the slide occurred, a passenger train plunged into the slide; 41 people lost their lives, and many were injured. The upper part of the slope consists of sand and silt, whereas the lower part consists of loose clay.

The cause of the slide appears to have been an exceptionally great water infiltration into the embankment fill during the unusually heavy rains before

Fig. 10-17
Rupture of slopes in the Panama Canal cuts.
(By permission of the National Academy of Sciences. Ref. 63)

Fig. 10-18
Rupture of slope of the Delaware-Chesapeake Canal bank at the north pier of the Summit Bridge, Delaware.
(Photo courtesy of the U.S. Corps of Army Engineers, Philadelphia District)

the slide occurred. In Fig. 10-19, which is an aerial photograph of the slide, notice the earth mass of the slide in the Bråviken Bay, the torn, interrupted railroad line, the highway buried by the sliding earth mass, the overturned locomotive and cars in the slide, and some cars still on the track.

Conditions for sliding are many and too complex to be dealt with in detail in this volume. About these matters, the reader is referred to References 10, 21, 22, 27, 43, 56, 63, 83, 88, 94.

In recent years, destructive landslides have been plaguing in California from Seattle to San Diego as urban construction has spread out of the valleys and up steep slopes, especially in clastic sediments and in near-coastal environment. In regions of seismic activity, the problems associated with landsliding become worse.

Some of the causes of a "dip-slope" terraced hillside rupture and sliding could be a heavy, prolonged torrential rain; water ingressed into the slide mass from leaky cesspools; upper swimming pools; lawns, and gardens. The leaking water may find its way into the bedrock, thus promoting further sliding.

Fig. 10-19
Aerial view of a railway embankment slide at Vita Sikudden, Sweden.
(Photo courtesy of the Swedish State Railways' Geotechnical Commission)

Mud or Debris Flows

Mud flows almost invariably result from unusually intensive and prolonged precipitation, or from a sudden thawing of frozen soil wreaking havoc in areas where men had built precariously on steep hillsides. Vibrations of all sorts, too, may trigger mud flows.

Some mud-slides flow with a high speed, whereas others creep along at very low speed.

Creep

Mechanically, creep is the phenomenon of a continuously increasing deformation under constant load. Although less obvious than landsliding, creep is an imperceptibly slow, more or less continuous downslope movement of a slope-forming earth mass and/or rock under a sustained load.

Creep is also facilitated by the presence of water. Any process which causes a volume change in an earth mass on a slope, for example, alternate wetting and drying, freezing and thawing, results in a downslope creep.

Solifluction

Solifluction is a phenomenon occurring because of a sudden excess of water caused by the thawing of icy soil. Upon thawing, the soil becomes plastic or semi-liquid.

Though probably applicable to all kinds of creep downhill of weathered material, the term solifluction is particularly used to indicate the movement in cold regions of superficial material for a considerable distance. Here an upper saturated (from rain, melted snow, or thawing) and undrained layer of weathered material, lying on the frozen ground — level or sloped — underneath, readily slides downhill.

Quick-Clay Slides

The term *quick clay* refers to a natural mixture of water-soaked, fine-particled, glacial clay deposit the consistency of which sometimes, under certain conditions, can change suddenly from a solid to a rapidly flowing liquefied mud, causing catastrophic landslides. Quick clay that has lain undisturbed for thousands of years can be jarred into motion by any sudden shock. Quick clay is by far the most mobile of all common solid materials on the surface of the earth. Quick-clay slides can be triggered by an earthquake; a blast; vibration from rolling stock, or machinery, or even by the jar of a pile driver. Also, rainstorms may be responsible for many landslides.

Quick clay is made up largely of flaky mineral particles less than about two microns in size. The particles are mostly silicate minerals such as illite, montmorillonite, chlorite and kaolinite. It is the high consistency and the mineral texture that furthers the quick clay to flow with the utmost ease. The so triggered quick clay then flows rapidly like an avalanche along any slope of the ground or over a flat land with a slope of less than one degree, rafting along buildings, bridges and other structures in its path, and destroying roads, property, and lives.

The regions where quick-clay slide phenomena occurred most frequently are Norway (15, 28); Sweden (101); the valleys of St. Lawrence, Ottawa, and Saguenay River in Canada.

One of the latest extensive landslides occurred on November 30, 1977, at Tuve — a suburb of Göteborg, Sweden. Tuve is a new development since about 1964. The quick, progressive mass wasting landslide of Tuve destroyed 67 buildings completely; made 30 more buildings inhabitable; killed eight people, and injured 73 inhabitants (101). This practically liquid landslide is attributed to failure of the sensitive "quick clay." The slide moved downward toward the river. Fig. 10-20 is an aerial photograph showing the extent of the landslide. With reference to WESSELOH (1978, Ref. 101), the Tuve landslide was supposedly induced by blasting of a tunnel a few kilometers away from the site of the slide.

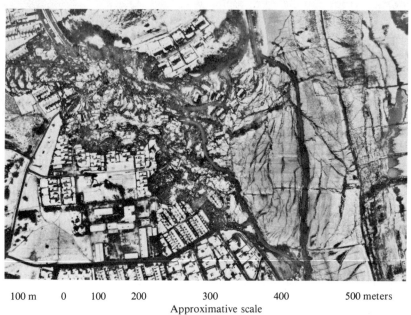

100 m   0   100   200   300   400   500 meters
Approximative scale

Fig. 10-20
Landslide at Tuve, Göteborg, Sweden.
a) Terrestrial view of slide area
(Photo courtesy of the Swedish Geotechnical Institute)
b) Aerial view of the extent of the landslide
(Admitted for publication by the National Land Survey of Sweden)
(Photo courtesy of the Swedish Geotechnical Institute)

The frequency of occurrence of large clay slides in the Göta Valley has lately increased because of the extensive construction activity in that region.

The "quick clay" phenomenon should be of practical interest to those concerned with the selection of location, and planning of sites of towns, housing developments, roads, airfields, farms and other construction endeavors.

Rock Falls

The term "rock fall" refers to the relatively free falling of blocks of rocks of newly detached fragments of rock of any size, or individual boulders, down a steep slope such as a mountain side, cliff, or from a ceiling of an underground opening, tunnel, or overhanging rock. Rock falls are also common along steep shorelines and in the higher mountain regions. Ocean, sea, lake and river shore vertical or overhung rock cliffs can be undermined by erosion. Under the influence of gravity, the loosened segments of blocks fall and/or slide down. Spring and fall are the most dangerous seasons of rock falls.

Niagara Waterfalls

The Niagara Falls is one of the most spectacular, magnificent natural seismic phenomena in North America, covering a story of 500 millions years of geological history. Fig. 10-21 is a view of the Horseshoe Waterfall from the Canadian side of the Niagara River. With the passing of time the tremendous force of rushing water has gradually cut back crestlines of the Falls. The immediate results of this erosion are most evident at the American Falls where a great amount of fallen rock has accumulated as talus at the base of the American Falls (Fig. 10-22).

Niagara Falls is formed in nearly horizontal beds by resistant rock strata (dolomite) about 24 m thick capping the weaker strata (shale) about 18 m thick. The falling water eroded the softer, less durable shale gradually undermining the dolomite cap.

The Horseshoe Falls have a crest length of 750 m and a height of 52 m. The American Falls have a crest length of 330 m and a height of 55 m.

To eliminate hazards to persons, property, or to scenic beauty in the region, and to preserve and enhance the beauty of this natural, scenic wonder of the American Falls for all people, the U.S. Army Corps of Engineers, Buffalo District, was assigned to undertake engineering geology and rock mechanics investigations and studies in order to obtain the necessary data for rock stabilization and corrective measures. The geotechnical work involved the following studies:

undermining

stratigraphy

jointing

talus

permeability

joint water pressure (believed to have played a major role in the occurrence of past rockfalls at Niagara)

Fig. 10-21
Horseshoe Falls.
(Photo courtesy of the Ontario Ministry of Industry and Tourism, Toronto, Ontario, Canada)

ENGINEERING PROBLEMS 419

rock stress measurements (the rock stresses may have influenced the rates of recession of both the American and Horseshoe Falls)

laboratory testing of rock chemical, physical, and strength properties

behavior of the rock at low temperatures, and

public safety at the flanks of the American Falls and at the Goat Island flank of the Horseshoe Falls.

For the purpose of these studies the American Falls was dewatered in June 1969, see Fig. 10-22. Figure 10-23 shows the Niagara Falls during the winter of 1964.

"During the dewatered period, the Corps of Engineers did not perform grouting or rock stabilization at the American Falls nor at the Goat Island flank of the Horseshoe Falls. However, in 1973 the Niagara Frontier State Parks and Recreation Commission performed some surficial stabilization work in the vicinity of Luna Island and the Bridal Veil Falls. There has been no stabilization work performed at the Horseshoe Falls. However, some

Fig. 10-22
Dewatered American Falls, June 1969.
(Photo courtesy of U.S. Corps of Engineers, Buffalo District, New York)

Fig. 10-23
The Niagara Falls during the winter of 1964.
(Photo courtesy of the Canadian Consulate General, New York, N.Y.)

Fig. 10-24
Rock fall
(Photo courtesy of Ohio Department of Highways)

ENGINEERING PROBLEMS   421

stabilization work has been performed along the gorge wall immediately above the Ontario Hydro plant which is located downstream of the Horseshoe Falls'' (Ref. 1, 58, 104).

Figure 10-24 shows a geological hazard in terms of a huge boulder that slid down a slope, landed in the middle of a road, and blocked traffic.

Rock Slides

A rock slope slide is the downward and outward, and usually rapid movement of essentially consolidated slope-forming rock material of newly detached segments of bedrock sliding on geologically predetermined sliding surfaces such as bedding planes, joint or fault surfaces, or any other plane of separation or discontinuity.

There are natural rock slopes and man-made slopes in rock. Rock slides occur when the equilibrium of a rock mass is disturbed.

Rock slides are common in cuts made through hard rock if the bedding planes, fault planes and other kinds of potential outward sliding surfaces prevail in the rock. A rock slope is especially prone to sliding if the rock strata are dipping toward the excavation or valley.

Fig. 10-25
Rock slide
(Photo courtesy of Ohio Department of Highways)

Figure 10-25 shows a rock slide endangering the road. Figure 10-26 shows a granitic rock slide on inclined, geologically predetermined planar discontinuities at Emerald Bay, Lake Tahoe, California.

Rock slides may be brought about by natural causes, or by man-induced activities. Some of the natural causes of rock slides are: earthquakes; heavy rain; change in physical, chemical and mechanical properties of rock, and gouge in its discontinuities. The so-called Good Friday earthquake (the great Alaska earthquake at Anchorage in 1964) triggered some 51 rock avalanches. In south-central Alaska, the land level during that quake was altered in an area of about 71,000 sq. miles ($\sim$ 179 200 km²).

Snow, too, may be a natural factor triggering a rock slide: snow adds weight on rock, and upon thawing, the meltwater surface runoff contributes to erosion and ingress into the openings of the rock. Upon sliding over rock, snow may bring about sliding of the rock blocks or slabs because of friction between the surface of the rock and the shifting snow.

Fig. 10-26
Slide of granitic rock. Emerald Bay, Lake Tahoe, California
(Photo courtesy of Transportation Laboratory, California Department of Transportation)

Naturally induced rock slides occur frequently in mountainous regions.

Man-induced rock slides may occur as a result of oversteepening of rock slopes in highway, railway and canal engineering. Oversteepening of a high rock slope brings about large stress concentrations around the toe of the slope, and a subsequent rock burst at the toe. Removal of support from a down-dip, down slope side at the toe of the slope may either start the slide or prepare the prerequisite condition for a slide.

Seismic effects on rock slopes from detonation of explosives, too, are examples of man-made, artificially induced rock slides. Blasting concussions tend to loosen the rock mass thus reducing the strength of the rock associated with such a loosening. Artificial inundation of, and a rapid drawdown of water from a fractured rock slope, as well as excessively loading up the top and/or face of a fractured, outward dipping stratification of rock may result in rock slides.

Depending upon the state of stress in the rock mass; fissures and degree of fissuring in rock; attitude of bedding planes of rock strata; crushed zones, as well as of the morphological shape of the rock mountain, rupture and sliding of a jointed hard-rock mass may occur in various ways in which the degree of participation of the mechanical properties of the rock material, such as compressive strength, friction, cohesion and shear strength, play a dominant role. All these and other factors pose problems of stability of fissured, blocky and slabby rock masses of various geological structures in rock foundations, rock slopes, underground openings, tunnel engineering, and hydraulic structures engineering, for rock engineers.

10-12. The Vaiont Reservoir Disaster

One of the landslides that has attracted the attention and interest of engineers and engineering geologists was that of the rock slide of the reservoir's left bank slope of the Mont Toc mountain in the valley of Vaiont, in north-eastern Italy (Fig. 10-27), Refs. 49, 50, 55, 79. It was one of the worst landslides in the course of reservoir history. The reservoir was associated with the world's second highest double-curved arch dam. The dam was completed in 1960.

This major, spontaneous landslide occurred on October 9, 1963, when an enormous mass of soil and rock on the upstream side of the dam broke away, slid into the water of the Vaiont reservoir, causing the water level to rise; spilled water over the top of the dam; flowed down the narrow valley, and damaged but did not break the dam. The sliding earth mass was pushed 140 m up the opposite slope of the reservoir.

The monstrous rock slide sent a surge of water of about 100 m high above the crest of the dam. A giant tidal wave flushed down the narrow Vaiont Valley. In this narrow gorge downstream of the dam, the height of the flood wave was given by various estimates over 70 m. About one mile away from the dam, the Vaiont Valley confluences with the Piave Valley. The flood

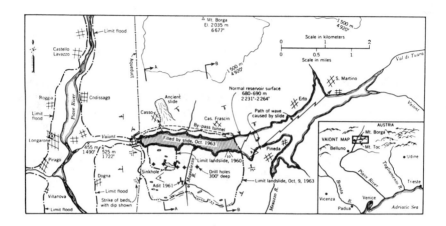

Fig. 10-27
Map of Vaiont reservoir site and Piave Valley, limits of slide and destructive flood waves (After G. A. KIERSCH, *Civil Engineering (ASCE)*, March 1964, Vol. 34, No. 3, p. 33, Ref. 55)

disaster continued for miles sweeping away and destroying several villages and a great part of the town of Longarone (Fig. 10-28), and killed 2117 people. The discharge shot down the narrow gorge at a fantastic speed, destroying everything that stood in its path. After razing Longarone, the waters spread down the Piave Valley with about the same effect as the flush flood. A rebound wave, however, raced back to hit the remants of Longarone once again. The time from the start of the slide to complete destruction downstream was seven minutes.

The greatest loss of life in any similar disaster was 2209 in the Johnstown flood in Pennsylvania, U.S.A., in 1889 (55).

The Vaiont slide created strong earth tremors, recorded as far away as Trieste, Vienna, Stuttgart, Basel, Brussels and Rome.

As to the geology, the Vaiont reservoir area is characterized by a thick deposit of fractured sedimentary rocks, primarily limestone of karst nature with sinkholes at the surface of the slope, solution cavities and openings in the rock, with frequent clayey interbeds and a series of alternating limey and marl layers. Marl denotes a deposit consisting of a mixture of calcareous sand, or clay, or loam. During the geologic past, the sedimentary formations have undergone folding and faulting, resulting in weak bonding between layers. Abundant fractures in the rock, bedding planes and jointing form a blocky mass of rock.

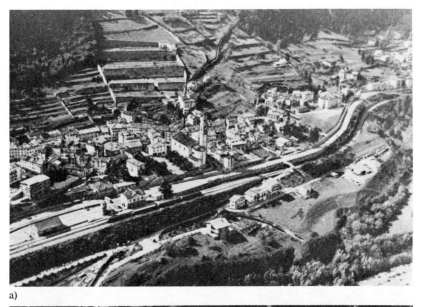
a)

b)

Fig. 10-28
Longarone in the Piave Valley
a) Before disaster
b) After disaster.
(Courtesy of Prof. G. SCHNITTER, Swiss Federal Institute of Technology, Zürich, Switzerland, Ref. 79)
(Reproduction permitted by "Wasser, Energie, Luft" of the Schweizerischer Wasserwirtschaftverband, Baden, Switzerland)

It is believed that the slide occurred over a geologically predetermined sliding surfaces — the bedding planes.

According to KIERSCH (1964, Ref. 55), the rock units involved were inherently weak. Further on, Professor KIERSCH indicates that the stability of the slope was adversely affected by a poor shear strength of the weak rock adjacent to the reservoir; the steeply dipping slope of the Mont Toc bank toward the reservoir; the weight of water infiltrating the rock formations from the raised water table after filling the reservoir; the hydrostatic uplift of the rock material, and creep prior to the actual landslide. The heavy rainfall two weeks prior to the disaster and thus the excessive runoff into the reservoir and as a recharge into the rock slopes, too, raised the water table in the reservoir and in the adjacent rock, thus having also contributed to the sudden landslide.

Professor KIERSCH (1964) concludes (55) that it is imperative that in rock exploration the following three factors must be recognized, namely:

"1. rock masses, under changed environmental conditions, can weaken within a short period of time — days, weeks, months.
 2. The strength of a rock mass can decrease very rapidly once creep is under way.
 3. Evidence of active creep should be considered as a warning that warrants immediate technical assessment, since acceleration to collapse can occur very quickly."

10-13. On the "Crushing Drama"

The "crushing drama" about failures of engineering structures by the sudden onslaught of force majeure brought about by natural causes and/or induced by human error shows how necessary it is in geological exploration and engineering to work hand in hand.

From the foregoing discussion it becomes apparent that rock engineering represents a complex field of engineering endeavor which discloses itself to the engineer only after many years of experience in which theory and practice are indivisible.

FROUDE once said: "Experience teaches slowly and at the cost of mistakes."

Earthquakes and their related events are among the most dramatic and disastrous of natural phenomena. The shaking permits gravity to displace large bodies of rock and soil material down slopes.

The variety and number of slide prevention methods are evidence that there can be no rule-of-thumb system of prescribing treatment. Also, for a particular landslide there is seldom one "correct" method of treatment. Rather, prevention may consist of a combination of several methods.

The discussion in this chapter on the "crushing drama" also reminds one forcefully not to ignore the destructive element of water which foundation engineers call the "public enemy No. 1." Once the negative effect of the factor of seepage on rock-foundation stability was fully realized after the Malpasset dam failure, drainage came to be considered one of the most effective means of seepage control in most cases.

Successful dam safety programs require more than the enactment of laws. They require money and knowledgeable professionals who understand the behavior of dams and are dedicated to their safety.

REFERENCES

1. American Falls International Board, 1974. A Final Report prepared for the International Joint Commission: "Preservation and Enhancement of the American Falls at Niagara."
2. ANDRUS, F. M., 1952, "Earthquake Design Requirements of the Uniform Building Code." Symposium on Earthquake and Blast Effects on Structures, held June 26—28, 1952 at Los Angeles. Berkeley, Cal.: Earthquake Research Institute and University of California, Berkeley, 1952, pp. 314—316.
3. Anonymous, 1926, "Asphalt Grouting Under Hales Bar Dam." *Engineering News Record*, 1926, May 20, Vol. 96, No. 20, pp. 798—802 (G. W. CHRISTIANS).
4. Anonymous, 1959, "French Dam Collapse." *Engineering News Record*, December 10, 1959, pp. 24—25.
5. Anonymous: "ICOLD'S XIII Congress, 1979." *International Water Power and Dam Construction*. January 1980, Vol. 32, No. 1, pp. 31—37.
6. Anonymous, 1959, "The Malpasset Dam Design." *Engineering*, December 18, pp. 642—643.
7. Anonymous, 1960, "The Malpasset Dam." *Engineering News Record*, January 21, 1960, Vol. 164, p. 26, and April 7, 1960, Vol. 164, p. 25.
8. ASCE Committee Report on "Uplift in Masonry Dams." *Trans. ASCE*, 1952, Vol.117, pp. 1218—1250.
9. BAIN, G. W., 1931, "Spontaneous Rock Expansion." *Journal of Geology*, 1931, Vol. 39, No. 8, pp. 715—735.
10. BARNEY, K. B., 1960, "Madison Canyon Slide," *Civil Engineering* (ASCE), August, 1960, pp. 72—75.
11. BÅTH, M., 1973, *Introduction to Seismology*. New York, N.Y.: John Wiley and Sons, pp. 110—119; 125—128.
12. BERG, G. V., 1963, *The Skopje, Yugoslavia Earthquake*. New York, N.Y.: American Iron and Steel Institute.

13. BERGMAN, M., Editor, 1978, *Storage in Excavated Rock Caverns*. Proceedings of the First International Symposium, held September 5—8, 1977 at Stockholm, Sweden. Oxford/New York, N.Y.: Pergamon Press, 1978 (three volumes).
14. BERKEY, C. P., 1929, "Responsibilities of the Geologist in Engineering Projects." *Technical Paper 215*, A.I.M.E., 1929, pp. 4—9.
15. BJERRUM, L., 1955, "Stability of Natural Slopes in Quick Clay." *Géotechnique* (London), 1955, Vol. 5, No. 1, pp. 101—119.
16. BOLNI, H. W., 1952, "The Field Act of the State of California." *Symposium on Earthquake and Blast Effects on Structures*, held June 26—28., 1952 at Los Angeles, Berkeley, Cal.: Earthquake Engineering Research Institute and University of California, Berkey, 1952, pp. 309—313.
17. BOLT, B. A., HORN, W. L., MACDONALD, G. A., and R. F. SCOTT, 1977, *Geological Hazards* (2nd ed.), New York, N.Y.: Springer-Verlag.
18. California Division of Mines and Geology, 1977. *California Geology,* January, 1977, p. 24, and July, 1979.
19. CAMBEFORT, H., 1963, "The Malpasset Report." *Water Power*, April 1963, Vol. 15, No. 4, pp. 137—138.
20. CASAGRANDE, A., 1961, "Control of Seepage Through Foundations and Abutments of Dams." (First Rankine Lecture). *Géotechnique* (London), 1961, Vol. 11, pp. 159—181.
21. CLEVELAND, G. B., 1971, *Regional Landslide Prediction*. Prepared by the California Division of Mines and Geology for the U.S. Department of Housing and Urban Development, Federal Insurance Administration, June 1971 (33 pages).
22. COLLIN, A., *Landslides in Clays*, translated from the French by N. R. Schriever, 1956. University of Toronto Press.
23. COULTER, H. W., and R. R. MIGLIACCIO, 1966. "Effects of the Earthquake of March 27, 1964 at Valdez, Alaska." U.S. Geological Survey Professional Paper 542-C. Washington, D.C.: U.S. Government Printing Office, 1966, pp. C1 to C36.
24. CRIEGER, W. P. and B. O. McCOY, 1956, "Dams," in R. W. Abbett (ed.), *American Civil Engineering Practice*. New York, N.Y.: Wiley and Sons, 1956, Vol. 2, pp. 14—19.
25. DONOVAN, N. C., and S. SINGH, 1978, "Liquefaction Criteria for Trans-Alaska Pipeline." *Proc. ASCE, Journal of the Geotechnical Engineering* Division. Paper 13666, April 1978, Vol. 104, No. GT 4, pp. 447—462.
26. DIMAS, J., SAVINI, T., and W. WEYERMANN, 1978, *Rock Treatment of the Canelles Dam Foundations*. Zürich: A publication by Rodio in collaboration with the Institute for Engineering Research Foundation Kollbrunner-Rodio. May, 1978, No. 42 (37 pages).
27. ECKEL, E. B. (Editor), 1958, *Landslides in Engineering Practice*. Washington, D.C.: Highway Research Board Special Report 29, 1958, NAS-NRC Publication 544.
28. EIDE, O., and L. BJERRUM, 1955, "The Slide at Bekkelaget." *Géotechnique* (London), March, 1955, Vol. 5, No. 1, pp. 88—100.
29. EMERY, C. L., 1964, "Strain Energy in Rocks," published in *State of Stress in the Earth's Crust*, Proceedings, International Conference held at Santa Monica, California, 1963. New York, N.Y.: American Elsevier Publishing Co., 1964, pp. 234—279.
30. ENGLE, H. M., 1953, "Earthquake Provisions in Building Codes." *Bulletin of the Seismological Society of America*, 1953, Vol. 43, pp. 233—237.
31. ESPINOSA, A. F., 1977, "Intensity Distribution of the Guatemala Earthquake," *U.S. Geological Survey Earthquake Information Bulletin*, March—April, 1977, Vol. 9, No. 2, p. 8.

32. EVISON, F. F., 1963, "Earthquakes and Faults." *Bulletin of the Seismological Society of America*, Vol. 53, No. 5, pp. 873—891.
33. FELD, J. 1966, *Rock as an Engineering Material.* Evanston, Illinois: Soiltest, Inc., p. 18.
34. GORDON, G., 1937, "Freezing Arch Across Toe of East Forebay Slide, Grand Coulee Dam." *Reclamation Era*, Vol. 17, 1937, pp. 12—16.
35. GRIGGS, D. T., et al., 1960, "Deformation of Rocks at 500° to 800°C," *American Geological Society Memoir 79.* Denver, Colorado: The Geological Society of America, Inc., pp. 39—104.
36. GUMENSKY, D. B., 1957, "Earthquake and Earthquake-Resistant Design." *American Civil Engineering Practice.* R. W. Abbett, Editor. New York, N.Y.: John Wiley and Sons, Vol. 3, Section 34, pp. 34-01—34-34.
37. GUTENBERG, B., and C. F. RICHTER, 1941, "Seismicity of the Earth." *Geological Society of America, Special Paper 34, 1941.*
38. GUTENBERG, B., and C. F. RICHTER, 1942, "Earthquake Magnitude, Intensity, Energy and Acceleration." *Bulletin of the Seismological Society of America.* July 1942, Vol. 32, No. 3, pp. 163—191.
39. GUTENBERG, B., and C. F. RICHTER, 1949, *Seismicity of the Earth and Associated Phenomena.* Princeton, N.J.: Princeton University Press, p. 10.
40. GUTENBERG, B., and C. F. RICHTER, 1956, "Earthquake Magnitude, Intensity, Energy, and Acceleration," (Second Paper). *Bulletin of the Seismological Society of America*, April 1956, Vol. 46, No. 2, pp. 105—145.
41. HANSEN, W. R., 1966, "Effects of the Earthquake of March 27, 1964 at Anchorage, Alaska." *U.S. Geological Survey Professional Paper 542-A.* Washington, D.C.: U.S. Government Printing Office, 1966, pp. A 1 to A 68.
42. HECK, N. H., 1936, *Earthquakes.* Princeton, New Jersey: Princeton University Press, pp. 54—56.
43. HOLMSEN, P., 1953, "Landslips in Norwegian Quick-Clays." *Géotechnique* (London), 1953, Vol. 3, No. 5.
44. ICOLD (CIGB), 1974, "Malpasset Dam, Lessons from Dam Incidents." Paris: International Commission on Large Dams, Complete Edition 1974, pp. 33—39.
45. IACOPI, R., 1973, *Earthquake Country.* Menlo Park, California: Lane Magazine and Book Co. a) Significant Faults in California, p. 15; b) Representative California Earthquakes, p. 32; c) Modified Mercalli Scale of Earthquake Shock Intensities, pp. 34—35.
46. International Committee on Large Dams (ICOLD), 1974, *Lessons from Dam Incidents*, Complete Edition, 1974, pp. 33—39.
47. International Conference of Building Officials, *Uniform Building Code,* 1976 ed. Whittier, California: June 1976.
48. JAEGER, C., 1963, "The Malpasset Report," *Water Power*, February 1963, Vol. 15, No. 2, pp. 55—61.
49. JAEGER, C., 1965, "The Vaiont Rock Slide." *Water Power*, March 1965, Part I, pp. 110—111, and April 1965, Part 2, pp. 142—144.
50. JAEGER, C., 1972, *Rock Mechanics and Engineering.* Cambridge at the University Press, pp. 325—339.
51. JUMIKIS, A. R., 1962, "Theory of Quicksand Condition." *Soil Mechanics.* Princeton, New Jersey: D. Van Nostrand Company, Inc., pp. 333—345.
52. JUMIKIS, A. R., "Theory of Quicksand Condition." *Introduction to Soil Mechanics.* New York, N.Y.: D. Van Nostrand Company, pp. 189—195.

53. JUMIKIS, A. R., 1982, *Soil Mechanics*. Melbourne, Florida, R. E. Krieger Publishing Company, Inc. (in press).
54. KENDRICK, T. D., 1955, *The Lisbon Earthquake*. Philadelphia, Pa.: Lippincott.
55. KIERSCH, G. A., 1964, "Vaiont Reservoir Disaster." *Civil Engineering* (ASCE), March 1964, Vol. 34, No. 3, pp. 32—39.
56. LADD, G. E., 1935, *Landslides, Subsidences and Rock-Falls*. Proceedings of the American Railway Engineering Association, 1935, Vol. 36, pp. 1091—1162.
57. LEMKE, R. W., 1966, "Effects of the Earthquake of March 27, 1964 at Seward, Alaska." *U.S. Geological Survey Professional Paper 542-E.* Washington, D.C.: U.S. Government Printing Office, 1966, pp. E1 to E43.
58. LO, K. Y., LUKAJIC, B., YUOEN, C. M. K., and M. HORI, 1979, "In Situ Stresses in a Rock Overhang at the Ontario Power Generation Station, Niagara Falls." *Fourth International Congress on Rock Mechanics*, held September 2—9, 1979 at Montreux, Switzerland.
59. LONDE, P., 1973, "Water Seepage in Rock Slopes." *Quarterly Journal of Engineering Geology*, 1973, Vol. 6, pp. 75—92.
60. LOUDERBACK, G. D., 1942, "Faults and Earthquakes." *Bulletin of the Seismological Society of America*, 1942, Vol. 32, pp. 305—330.
61. LUGEON, M., 1933, *Barrages et Géologie*. Methodes des recherches, terrassement et imperméabilization. Lausanne: Librarie de l'université F. Rouge et Cie. S.A., p. 38.
62. MORFELDT, C. O., 1970, "Significance of Groundwater at Rock Construction of Different Types," in *Large Permanent Underground Openings*. Oslo: Universitetsforlaget, pp. 305—317.
63. National Academy of Sciences, 1924, *Memoirs of the National Academy of Sciences*. Washington, D.C.: Government Printing Office. Vol 18, Report of the Committee of the NAS on Panama Canal Slides.
64. National Academy of Sciences, 1973, *The Great Alaskan Earthquake of 1964*. Washington, D.C.: NAS.
65. National Academy of Sciences, 1975. *Earthquake Predictions and Public Policy*. Washington, D.C.: NAS.
66. *National Building Code of Canada, 1977*. Ottawa, Canada: Associate Committee on the National Building Code; National Research Council of Canada, 1977 (374 pages).
67. National Bureau of Standards, 1972, *Engineering Aspects of the 1971 San Fernando Earthquake*. Washington, D.C. NBS Building Science Series 40, 1972.
68. National Bureau of Standards, 1977, "Observation on the Behavior of Buildings in the Romania Earthquake of March 4, 1977." *National Bureau of Standards Special Publication 490*. Washington, D.C.: U.S. Government Printing Office (160 pages).
69. NICHOLS, T. C., Jr., and J. R. ABEL, Jr., 1975, "Mobilized Residual Energy — a Factor in Rock Deformation." *Association of Engineering Geologists Bulletin*, 1975, Vol. 12, No. 3, pp. 213—225.
70a. NICHOLS, T. C., Jr., 1975, "Deformations Associated with Relaxation of Residual Stresses in a Sample of Barre Granite from Vermont." *U.S. Geological Survey Professional Paper 875*, 1975 (32 pages).
70b. NICHOLS, T. C., Jr., and W. Z. SAVAGE, 1976, "Rock Strain Recovery-Factors in Foundation Design." Published in *Rock Engineering for Foundations and Slopes*. New York, N.Y.: American Society of Civil Engineers, Vol. 1, pp. 34—54.
71. OBERT, L., and W. I. DUVALL, 1967, *Rock Mechanics and the Design of Structures in Rock*. New York., N.Y.: John Wiley and Sons, Inc., p. 495.

72. PULS, L. G., 1963, "The Malpasset Report." *Water Power*, June 1963, Vol. 15, No. 6, pp. 228—230.
73. RICHEY, J. E., 1963, "Granite." *Water Power.* June, 1963, Vol. 15, No. 6, pp. 237—242; July, 1963, Vol. 15, No. 7, pp. 303—307; Aug., 1963, Vol. 15, No. 8, pp. 326—332; Sept., 1963, Vol. 15, No. 9, pp. 374—382; Oct., 1963, Vol. 15, No. 10, pp. 409—417; Nov., 1963, Vol. 15, No. 11, pp. 475—481; Dec., 1963, Vol. 15, No. 12, pp. 505—509.
74. RICHTER, C. F., 1958, *Elementary Seismology.* San Francisco, Cal.: "Earthquake Effects," p. 5; "The Lisbon Earthquake — 1755," pp. 104—105; "Modified Mercalli Intensity Scale of 1931," pp. 136—139; "Magnitude and Energy," pp. 364—367; "Earthquake Risk and Protective Measure," pp. 379—390; "San Andreas Fault — 1838 and 1857," pp. 473—476.
75. RICHTER, C. F., 1980, an interview about magnitude, in *U.S. Earthquake Information Bulletin*, January—February, 1980, Vol. 12, No. 1, pp. 5—8.
76. RINNE, J. E., 1952, "Building Code Provisions for Aseismic Design." *Symposium on Earthquake and Blast Effects on Structures,* held June 26—28 at Los Angeles, Berkeley, Cal.: Earthquake Engineering Institute and University of California, Berkeley, 1952, pp. 291—308.
77. ROCHA, M., 1965, *Some Problems on Failure of Rock Masses, Memória No. 258.* Lisbon: Laboratório Nacional de Engenharia Civil.
78. ROŠ, M., and A. EICHINGER, 1949, *Die Bruchgefahr fester Körper.* Zürich: Eidgenössische Materialprüfungs Anstalt — EMPA Report No. 172.
79. SCHNITTER, G., 1964, "Die Katastrophe von Vaiont in Oberitalien." *Wasser- und Energiewirtschaft* (Baden), 1964, No. 2/3, pp. 1—7.
80. SEED, H. B., and I. M. IDRISS, 1967, "Analysis of Soil Liquefaction; Niigata Earthquake." *Proc. ASCE, Journal of the Soil Mechanics and Foundations Division*, May 1967, Vol. 93, No. SM 3, pp. 83—108.
81. SEED, H. B., and I. M. IDRISS, 1971, "Simplified Procedure for Evaluating Soil Liquefaction Potential." *Proc. ASCE, Journal of the Soil Mechanics and Foundations Division,* September, 1971, Vol. 97, No. SM 9, pp. 1249—1273.
82. SERAFIM, J. L., 1968, "Influence of Interstitial Water on the Behavior of Rock Masses." *Rock Mechanics in Engineering Practice.* London: John Wiley and Sons, Inc., Chapter 3, pp. 55—97.
83. SHARPE, C. F. S., 1938, *Landslides and Related Phenomena.* New York, N.Y.: Columbia University Press.
84. SIEBERG, A., 1933, *Erdbebenforschung.* Jena: Gustav Fischer, pp. 26—28.
85. SIMMONDS, A. W., 1953, "Final Foundation Treatment at Hoover Dam." *Transactions ASCE*, 1953, Vol. 118, Paper 2537, pp. 78—79.
86. Smith, D. P., 1976, "Roadcut Geology in the San Andreas Fault Zone." *California Geology*, May 1976, Vol. 29, No. 5, pp. 99—104.
87. SOZEN, M. A. 1964, *Structural Damage Caused by Skopje Earthquake 1963.* Illinois University, Department of Civil Engineering Structural Research Series 279, January, 1964.
88. Statens Järnvägars Geotekniska Commission, 1922, *Slutbetänkande 1914—1922.* Stockholm: Statens Järnvägas, 1922.
89. STEINBRUGGE, K. V., 1968, *Earthquake Hazard in the San Francisco Bay Area: A Continuing Problem in Public Policy.* Published by the University of California Institute of Governmental Studies.

90. STEINBRUGGE, K. V., and H. J. DEGENKOLB, 1975, "Meeting the Earthquake Challenge: California's New Laws." *Civil Engineering* ASCE, February 1975, Vol. 45, No. 2, pp. 44—47.
91. SZÉCHY, K., 1966, *The Art of Tunnelling*. Budapest: Académiai Kiadó, pp. 83—84.
92. TERZAGHI, K., 1946, "Tunnel Hazards," in *Rock Tunneling with Steel Supports*, by R. V. Proctor and T. L. White. Youngstown, Ohio: The Commercial Shearing and Stamping Company, 1946, pp. 96—99.
93. TERZAGHI, K., 1962, "Dam Foundation on Sheeted Granite." *Géotechnique* (London), 1962, Vol. 12, No. 3, p. 199.
94. TERZAGHI, K., 1950, "Mechanics of Landslides." *Application of Geology to Engineering Practice* (Berkey Volume). Geological Society of America, Denver Colorado.
95. TERZAGHI, K., 1962, "Stability of Steep Slopes on Hard Unweathered Rock." *Géotechnique* (London), 1962, Vol. 12, No. 4, pp. 251—270.
96. TOPPOZODA, T. R., C. R. REAL, and D. C. PIERZINSKI, 1979, "Seismicity of California January 1975 Through March 1979." *California Geology*, July 1979, Vol. 32, No. 7, pp. 139—142.
97. *Uniform Building Code, 1976*, Whittier, California: California International Conference of Building Officials, 1976.
98. U.S. Geological Survey, 1977, *Earthquake Information Bulletin*, March—April, 1977, Vol. 9, No. 2, Fig. 2, p. 8.
99. U.S. Geological Survey, 1977, "Modified Mercalli Intensity Scale of 1931." *Earthquake Information Bulletin*, July—August, 1977, Vol. 9, No. 4, pp. 30—31 (Adapted from Sieberg's Mercalli-Cancani Scale, modified and condensed).
100. U.S. Geological Survey, 1977, *Earthquake Information Bulletin*, Vol. 9, No. 6, November—December, 1977.
101. WESSELOH, J., 1978, "Erdrutsch in Tuve." *Geotechnik*, September 1978, pp. 60—61.
102. WHITE, W. S., 1946, *Rock Bursts in the Granite Quarries at Barre, Vermont*. U.S. Geological Survey Circular No. 13 (15 pages).
103. WIEGEL, R. L. (Editor), 1970, *Earthquake Engineering*. Englewood Cliffs, New Jersey: Prentice-Hall, Inc.
104. WILKINSON, T. A., District Geologist, Buffalo District, personal correspondence of August 9, 1979.
105. WOOD, H. O., and F. NEUMANN, 1931, "Modified Mercalli Intensity Scale of 1931," *Bulletin of the Seismological Society of America*, 1931, Vol. 21, No. 4, pp. 277—283.
106. ZIGNOLI, V., 1965, *Il Traforo del Monte Bianco*, Estratto dalla Rivista "Autostrade", No. 6, Giugno, 1965 (44 pages).

OTHER RELATED REFERENCES

CARLSON, R. W., 1957, "Permeability, Pore Pressure and Uplift in Gravity Dams." *Transactions ASCE*, 1957, pp. 587—602.

CHAE, Y. S., 1978, "Design of Excavation Blasts to Prevent Damage." *Civil Engineering — ASCE*, April 4, 1978, Vol. 48, No. 4, pp. 77—79.

DOWRICK, D. J., 1977, *Earthquake Resistant Design*. New York, N.Y.: John Wiley and Sons.

EIDE, O., and L. BJERRUM, 1955, "The Slide at Bekkelaget." *Géotechnique* (London), March 1, 1955, Vol. 5, No. 1, pp. 88—100.

FELLENIUS, B., 1955, "The Landslide at Guntorp." *Géotechnique* (London), March, 1955, Vol. 5, No. 1, pp. 120—125, plus Discussion.

FOLBERTH, P. J., "Beitrag zur rechnerischen Ermittlung der Spannungszustände in Felsbauwerken." Wien and New York, N.Y.: Springer Verlag. *Safety in Rock Engineering*. 15th Symposium of the Austrian Regional Group of the International Society for Rock Mechanics, held September 24—25, 1964 at Salzburg. Supplementum II, 1965, pp. 25—33.

FRANKLIN, J. A., 1970, "Observation and Tests for Engineering Description and Mapping of Rocks." *Proceedings of the 2nd Congress of the International Society for Rock Mechanics*, held September 21—26 at Beograd, Jugoslavia. Vol. 1, pp. 11—16.

FRÖHLICH, O. K., 1955, "General Theory of Stability of Slopes." *Géotechnique* (London), March 1955, Vol. 5, No. 1, pp. 37—47.

GAZIEV, E. G., and RECHITSKI, V. I., 1974, "Stability of Stratified Rock Slopes." *Advances in Rock Mechanics*. Proceedings of the 3rd Congress of the International Society for Rock Mechanics, held September 1—7, 1974 at Denver, Colorado. Washington, D.C.: National Academy of Sciences, Vol. II, Part B, pp. 786—791.

GUTENBERG, B., 1941, "Mechanism of Faulting in Southern California Indicated by Seismograms." *Bulletin of the Seismological Society of America*. October 1941, Vol. 31, No. 4, pp. 263—302.

GUTENBERG, B., 1956, "Effects of Ground on Earthquake Motion." *Bulletin of the Seismological Society of America*. July, 1957, Vol. 47, No. 3, pp. 221—225.

HARZA, L. F., 1953, "Uplift and Seepage Under Dams on Sand." *Transactions ASCE*, 1953, Vol. 61, No. 8, Part 2, pp. 1352—1385. Discussions on pp. 1386—1406.

HINO, K., 1956, "Fragmentation of Rock Through Blasting and Shock Wave Theory of Blasting," Quarterly School of Mines, Special Paper.

HOEK, E., and J. W. BRAY, 1981 (Revised 3rd edition), *Rock Slope Engineering*. London: The Institution of Mining and Metallurgy.

KENNEY, N. T., 1969, "Southern California's Trial by Mud and Water." *National Geographic*, October 1969, Vol. 136, No. 4, pp. 552—573.

KJELLMAN, W., 1955, "Mechanics of Large Swedish Landslips." *Géotechnique* (London), 1955, Vol. 5, No. 1, pp. 74—78.

LONDE, P., 1973, "The Role of Rock Mechanics in the Reconnaissance of Rock Foundations, Water Seepage in Rock Slopes and the Analysis of the Stability of Rock Slopes." *Quarterly Journal of Engineering Geology*, 1973, Vol. 6, pp. 57—127.

MARY, M., 1968, *Barrages-voûtes: Historique, Accidents et Incidents*. Paris: Dunod (159 pages).

MARCELLO, C., 1964, "Some Considerations on Accidents which have Happened to Water-Storage Dams." *Transactions of the 8th International Congress on Large Dams*, held May 4—8, 1964 at Edinburgh, Vol. 5, pp. 573—579.

Mencl, V., 1966, "Mechanics of Landslides with Non-Circular Slip Surfaces with Special Reference to the Vaiont Slide." *Géotechnique* (London), 1966, Vol. 16, No. 4, pp. 329—337.

Middlebrooks, T. A., 1942, "Fort Peck Slide." *Transactions ASCE*, 1942, pp. 723—764.

Müller, L., 1964, "The Rock Slide in the Vaiont Valley." *Rock Mechanics and Engineering Geology*, 1964, Vol. 2, pp. 148—212.

Newmark, N. M., J. A. Blume, and K. K. Kapur, 1973, "Seismic Design Spectra for Nuclear Power Plants." *Proc. ASCE, Journal of Power Divison*, November, 1973.

Oakeshott, G. B., 1976, *Volcanoes and Earthquakes*. New York, N.Y.: McGraw-Hill Book Company.

Palmer, J. H. L., and K. Y. Lo, 1976, "In Situ Stress Measurement in Some Nearsurface Rock Formations — Thorold, Ontario." *Canadian Geotechnical Journal*, Vol. 13, No. 1, pp. 1—7.

Rainer, J. H., and T. D. Northwood, 1979, "Earthquakes and Buildings in Canada." *Canadian Building Digest*, Division of Building Research, National Research Council of Canada. Ottawa, Canada; October 1979, pp. 208-1 to 208-4.

Schroter, G. A., and R. D. Maurseth, 1960, "Hillside Stability — the Modern Approach." *Civil Engineering*, June 1960, Vol. 30, No. 6, pp. 66—69.

Voight, B. (editor), 1979, *Rockslides and Avalanches*, 2. Amsterdam: Elsevier, Vol. 14B, Engineering Sites. Developments in Geotechnical Engineering.

Wood, A. M., and Viner-Brady, N. E. V., 1955, "Folkstone Warren Landslip Investigations, 1948—1950." and "Folkstone Warren Landslip Remedial Measures, 1948—1950." *Proceedings Institution of Civil Engineers* (London), 1955, Vol. 4.

Series on *Rock Mechanics and Engineering Geology*, Vienna and New York, N.Y.: Springer-Verlag.

Supplementum I. *Principles in the Field of Geomechanics*, 14th Symposium, held Sept. 27—28, 1963 at Salzburg. Published in 1964; Supplementum II. *Safety in Rock Engineering*, 15th Symposium, held Sept. 24—25, 1964 at Salzburg. Published in 1965; Supplementum III. *Rock Engineering in Theory and Practice*, 16th Symposium, held Sept. 30 to Oct. 1, 1965 at Salzburg. Published in 1967.

Taschenbuch für Tunnelbau, 1980. Essen: Glückauf.

Panel on Rock-Mechanics Research Requiremants. U.S. National Committee for Rock Mechanics Assembly of Mathematical and Physical Sciences, National Research Council, 1981, *Rock Mechanis Research Requirements for Resource Recovery, Construction, and Earthquake-Hazard Reduction*. Washington, D.C.: National Academy Press, 1981 (222 pages).

PART 4
STABILIZATION OF ROCK

CHAPTER 11

ROCK REINFORCEMENT

11-1. Anchoring

In general, rock bolt support or rock anchorage is replacing or supplementing conventional timber and steel supports for the roofs and sides of rock in underground openings (12, 21). Bolt support or rock anchorage is also used for stabilizing rock slopes and mines. Rock anchoring is used, for example, for reinforcement of arch-dam rock abutments, caverns, and other rock support in foundation, tunnel, and hydraulic structures engineering (Figs. 11-1 through 11-4). Whenever a naturally solid, self-supporting roof is encountered above a false hanging wall or weak layer, the wall can be held up by a bearing plate and bolts anchored in the solid rock formation above (Fig. 11-1).

11-2. Rock Bolting

To avoid failure of structures made in rock and rock slopes, rock reinforcement is used. Rock reinforcement is a means by which strength is added to the rock to avoid its failure. The reinforcing elements are steel rods, rock bolts, or stressed steel cables. Rock bolting, or rock anchor, is used for reinforcing rock slopes to prevent rock blocks from falling away from the main rock mass when isolated by joints and faults, or sliding down along an inclined plane of rock. Hence, the reinforcement has for its purpose to load the rock in such a manner that the resistance to shear along planes of weakness is increased.

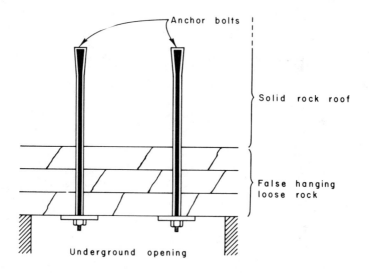

Fig. 11-1
Anchoring a weak zone of rock to a stronger zone with rock bolts and bearing plates

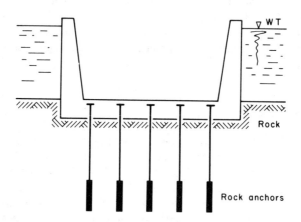

Fig. 11-2
Anchoring of a U-shaped structure against hydrostatic uplift

ROCK REINFORCEMENT

a) Failure in shear of sand under an inclined load, $\varphi = 30°$
   (Author's study, Refs. 14, 16, 17a, 17b)

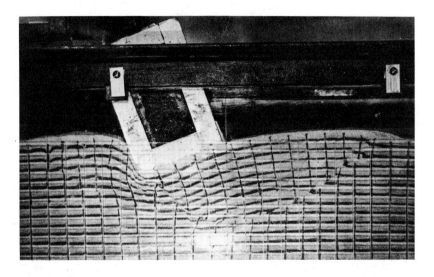

b) Anchoring of a potential rupture wedge in rock against shear and subsequent lateral expulsion from undernearth the base of a foundation

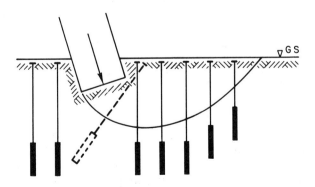

Fig. 11-3
Curved rupture surfaces in soil and rock

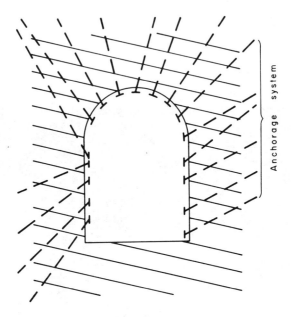

Fig. 11-4
Rock bolting in fissured rock around an underground opening

Requirements of Anchorage

1. The anchorage must be strong.

2. The rock must be *strong* and *continuous* on a limited scale to provide a firm anchorage for the bolts. However, anchorages may also be provided by cement grouting.

3. The bolts should be of adequate length to create a precompression zone of sufficient extent around, or along, a structure in rock to cope with stresses of failure. Rock bolts should be anchored beyond the tension zone.

11-3. Effect of Rock Bolting on Shear Stress

In reinforcement of jointed rock, the effect of shear strength of the joint on the design of rock bolts and anchors enters into consideration. Upon introduction of the reinforcement, precompression is applied to the solid rock through tensioning, viz., prestressing of the anchor bolt. Before prestressing, the ends of the rock bolts, or anchors, should be firmly concreted into the borehole of the anchor.

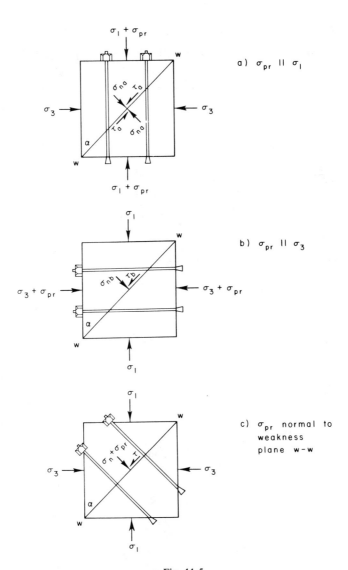

Fig. 11-5
Imparting precompression stress σ_{pr} on bolted rock elements
(After FARMER, Ref. 6)

As shown by FARMER (6), a precompression stress, σ_{pr}, can be imparted to the element of a rock, separated by a diagonal plane of weakness, in one of the three different ways as sketched in Figs. 11-5a, b, and c:

1. Precompression σ_{pr} applied parallel to the major principal stress σ_1;
2. Precompression σ_{pr} applied parallel to the minor principal stress σ_3, and
3. Precompression stress σ_{pr} applied normal to the shear plane, viz., plane of weakness.

The maximum load P per bolt is computed as

$$P = \sigma_{ult} \cdot a \tag{11-1}$$

where σ_{ult} = ultimate compressive stress of the rock, and
a = cross-sectional area of the anchorage.

The applied precompressive stress σ_{pr} to rock is:

$$\sigma_{pr} = \frac{n \cdot P}{A} \tag{11-2}$$

where
n = number of bolts, and
$A = \sum_{1}^{n} (a)$ = total area of the bolted rock.

Using the normal- and shear-stress equations

$$\sigma_n = \frac{\sigma_1 + \sigma_3}{2} + \frac{\sigma_1 - \sigma_3}{2} \cos 2\alpha \tag{11-3}$$

$$\tau = \frac{\sigma_1 - \sigma_3}{2} \sin 2\alpha \tag{11-4}$$

and adjusting them for the prestressing condition, obtain normal and shear stresses on the shear plane:

Case (a), $\sigma_{pr} \parallel \sigma_1$:

$$\sigma_{na} = \sigma_n + \frac{\sigma_{pr}}{2} \cdot (1 + \cos 2\alpha) \tag{11-5}$$

$$\tau_a = \tau + \frac{\sigma_{pr}}{2} \cdot \sin 2\alpha \tag{11-6}$$

Case (b), $\sigma_{pr} \parallel \sigma_3$:

$$\sigma_{nb} = \sigma_n + \frac{\sigma_{pr}}{2} \cdot (1 - \cos 2\alpha) \tag{11-7}$$

$$\tau_b = \tau - \frac{\sigma_{pr}}{2} \cdot \sin 2\alpha \tag{11-8}$$

Case (c), σ_{pr} normal to weakness plane $w - w$:

σ_{nc} and τ_c to be evaluated by Eqs. (11-3) and (11-4).

Furthermore, FARMER (6) shows that the COULOMB shear strength line or envelope for $\alpha = 45°$ for the above mentioned three cases are:

$$\text{a)} \quad \tau = \sigma_n \cdot \tan \varphi - \frac{\sigma_{pr}}{2}(1 - \tan \varphi) + c \tag{11-9}$$

$$\text{b)} \quad \tau = \sigma_n \cdot \tan \varphi + \frac{\sigma_{pr}}{2}(1 + \tan \varphi) + c \tag{11-10}$$

$$\text{c)} \quad \tau = \sigma_n \cdot \tan \varphi + \sigma_{pr} \cdot \tan \varphi + c \tag{11-11}$$

Observation of these equations shows that there will be a weakening in the shear strength τ of the rock structure in case a) and strengthening in cases b) and c). The maximum strengthening is obtained in the case (b) when precompression aids the minor principal stress σ_3, which is the lateral confining stress.

11-4. Rock Bolt Support

After excavation of an underground opening in rock, the stresses of a rock mass around and near the opening and/or excavation tend to become relaxed. To maintain the initial, natural, tightly packed state of the rock mass, rock bolts are used.

Rock bolts not only take care of excessive relaxation of the rock mass around the contour of the opening, but they can also prevent a sudden decompression of the rock material brought about by rock spalling and rock bursts. To avoid this, the bolts must be installed immediately upon and after making the excavation of the opening, viz., driving of the tunnel. Rock bolts prevent massive falls of loose rock fragments, thus supporting roofs and ceilings of underground openings in rock, and slope faces of rock in open excavation.

The performance of rock-bolted layers of rock formations resembles that of a bolted beam. The rock layers are usually tied together by means of pretensioned rock bolts. On the average, the maximum spacing of the bolts is taken as approximately one-half of the length of the bolt. Figure 11-6 is a schematic representation of some rock-bolt support systems.

In principle, a rock bolt is installed as follows.

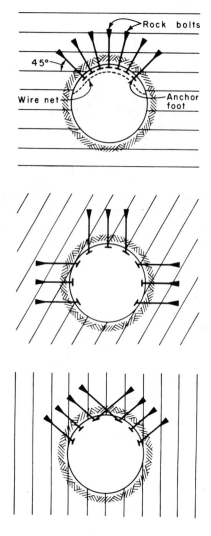

Fig. 11-6
Some rock-bolt support systems

ROCK REINFORCEMENT 445

A hole is drilled through the unstable rock into a stable one, into which hole a long steel bolt is inserted. With the appropriate head-wedge assembly employed, the bolt is securely fixed and keyed in the hole. In this way, the outer layer of a rock slab is bolted to the stable layer in which the bolt is keyed.

To protect against falling of smaller rock fragments from the ceiling, the feet of the anchors or bolts around and along the vault-like ceiling are covered with a wire net.

The most suitable form of rock bolting, bolt tensions, and rock displacements can be determined by appropriate tests.

As to the types of rock bolts and anchors, there are several types of them on the market. Almost every rock engineering firm which is in the business of rock anchorage has its own trademarked and/or patented anchor design. The type of rock bolts and rock anchors to use depends upon the nature of the construction site, the geology of the rock and its properties, the load to cope with, the method of installation of the anchors, and possibly other factors. Each anchor type has its place and use in foundation, hydraulic structures, an rock engineering. To choose the type of anchor to use, one would refer to anchor manufacturers' catalogs.

11-5. Determining Rock Bolt Support

The analytical delineation, viz., thickness of the tensile stress zone in rock above the crest of a circular opening, is discussed in Section 7-8. This delineation enables one to calculate the weight of the loosened rock, also called the gravity or overburden load of the rock mass in the upper part of an underground opening. The thickness of this loose, fragmented rock mass depends upon the quality of the rock, of course, as well as upon the construction method in the rock and the kind and method of rock support used against caving-in.

One method to stabilize the loose rock above the crest of the circular opening, and thus to safeguard against rock fragments falling into the opening, is the one known by the term "rock-bolt support."

As shown by CORDING (4, 5), the thickness t of the zone of the loosened rock above the crest of the circular opening in rock comprising the gravity load W (weight of the loosened rock) may be determined empirically as

$$t = n \cdot D \tag{11-12}$$

where t = thickness of the rock load on the crest of the opening (Fig. 11-7),
  n = a function of support rigidity and strength of the rock mass (10), and
  D = diameter of the circular underground opening in rock.

Equation (11-12) thus indicates that the thickness ($n \cdot D$) of the rock load depends upon (1) the rock quality, (2) the kind of support and construction method used, and (3) the width, viz., diameter, of the opening. As indicated by CORDING, the value of n will be small for an excellent-quality rock if the rock is rapidly reinforced and supported with rock bolts. In such a case, the bolt pressures p_i on rock necessary to support the gravity load W will be relatively small. For example, if bolt pressures in the dome of a cavity were 20 psi = 137.9 kN/m², this pressure would support a 20-foot-thick (6.096 m) gravity-loaded rock (n = 20 feet/100 feet = 0.2 = 6.096 m/30.48 m for this case). The bolt pressures p_i needed for stability increase in proportion to the width (diameter) of the opening.

If the cohesion of the rock mass is small (visualize a heavily jointed rock mass around a large opening), then the pressure p_i required for stability, as given empirically by CORDING (4), is

$$p_i = n \cdot D \cdot \gamma \tag{11-13}$$

where γ = unit weight of the rock mass.

If the size of the underground opening is small with respect to the fractures and cohesion must be considered, the bolt pressure p_i on rock is

$$p_i = n \cdot D \cdot \gamma - n' \cdot c \tag{11-14}$$

where n' = a function of the rock mass quality and the kind of support method, and
  c = cohesion per unit area of the rock mass.

In a small tunnel driven through a sound, excellent-quality rock, the cohesion of rock is usually large enough so that no support (= no p_i) is needed.

The opening must also be safeguarded against movement of the most critical rock wedge into the opening (Fig. 11-8). The normal pressure P_N tends to drive the wedge into the opening (4). The necessary bolt pressure p_i on the rock for preventing failure of the wedge depends upon the total shear resistance S along the failure plane. Low shear resistance may be incurred because of a high hydrostatic pressure U on the failure plane, or because there is present, on the potential plane of failure, a gouge material with a low frictional strength.

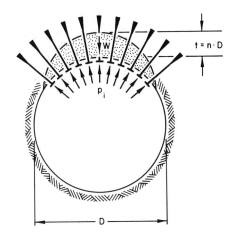

Fig. 11-7
Support of gravity load

W = weight of loosened rock
p_i = internal pressure applied by bolts
n = function of support rigidity and strength of rock mass

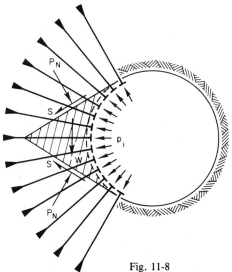

Fig. 11-8
Support of critical wedge
(After CORDING, Ref. 4)

$S = (P_N - U) \cdot \tan \varphi + C$
U = groundwater pressure
φ = angle of friction
C = cohesion
W = weight of wedge

The shear resistance, or shear strength S, is written as

$$S = (P_N - U) \cdot \tan \varphi + C \qquad (11\text{-}15)$$

where U = hydrostatic or neutral pressure,
 φ = angle of friction, and
 C = total amount of cohesion in the shear plane.

The internal pressure p_i to support the worst rock wedge is thus a function of the normal force P_N and the size of the opening.

The lengths of the rock bolts should be long enough to extend them through the worst rock wedge. Usually bolt lengths are one-fourth to one-third of the span of the underground opening.

11-6. Some Advantages and Disadvantages of Rock Bolting

Although rock-bolting is not a cure-all and will not work in all cases where applied, it has several advantages over timber support methods for stabilizing rock masses. Some of the rock-bolting advantages are:

1. A better underground-opening roof, and an open-excavation slope support where other methods fail;
2. Economy;
3. Relatively short installation time of anchors and/or bolts;
4. Uncluttered storage space for materials;
5. Unobstructed working space (there are no "forests" of supporting timber; this facilitates a convenient use of mechanical equipment);
6. Relatively long duration in service, outlasting timber;
7. Improved ventilation in underground openings; and
8. Choice in direction and magnitude of the retaining forces, thus taking care of the various attitudes of the rock and the different orientations and magnitudes of shear strength of the various-quality joints.

Some of the disadvantages of rock-bolting are:

1. Poorer performance in extremely blocky and loose rock formations where no solid anchoring medium can be found, and
2. Impossibility of ascertaining the exact condition of the bolt in the bore-hole.

11-7. The Williams Rock Bolt

During the course of time, as power plants moved underground and tunnels and underground storage spaces became more prevalent, methods of

ROCK REINFORCEMENT

bolting rock for reinforcement were developed.

Estimates of rock reinforcement are made in the light of detailed engineering-geological investigations.

In the rock, bolting provides the most satisfactory means of successfully securing the loosened near-surface rock, providing that the bolt type, method of installation and layout are designed with reference to the rock structure.

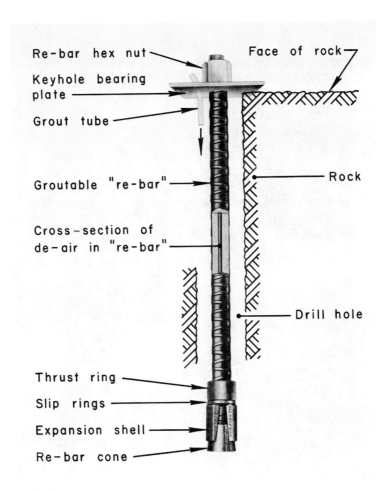

Fig. 11-9
The Williams rock bolt.
(Photo courtesy of Williams Form Engineering Corporation, Grand Rapids, Michigan)

450   ROCK MECHANICS

Rock bolts are now used extensively to prevent movement of rock. There are several makes of trademarked rock bolts on the market. One of the popular kind of rock bolts invented and patented by C. I. WILLIAMS is the so-called Williams prestressable, hollow core groutable re-bar rock bolt (Fig. 11-9), Ref. 27. This bolt consists of a hollow re-bar high-tension steel reinforcing bar with an expansion shell anchoring device at one end of the bolt shaft, and on the other end of which a large washer or retainer plate — keyhole bearing plate — is provided under the hex nut at the face of the rock. The tightening of the hex nut actuates the expansion device of the bolt placed in the drillhole.

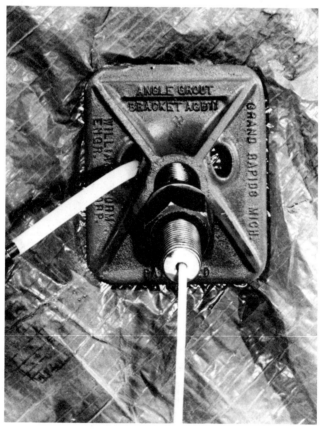

Fig. 11-10
Pressure grouting hollow core rock bolt. Notice return flow of grout through hollow core of bolt giving positive proof of the grouting operation.
(Photo courtesy of Williams Form Engineering Corporation,
Grand Rapids, Michigan)

After the bolt is set into a flushed, cleaned drillhole, the steel keyhole bearing plate is applied to the face of the hole (Fig. 11-10). Then the nut is torqued to stress the rock bolt to the desired loading of the bolt, consistent with the compressive strength of the rock, but not to exceed the allowable load of the bolt. The high-tensile steel permits high torquing of long anchor bolts in deep holes without any substantial loss of power that would result through the use of mild steel rods.

After prestressing the bolt, grout is pumped into the drill hole through the plastic grout tube until the entire hole is filled, and the grout returns through the hollow bar (Figs. 11-9 and 11-10). Visible return of grout completes inspection. The bolt is permanent, completely engulfed in grout, and is protected from corrosion (27).

According to WILLIAMS, it is one of "the most efficient methods of rock bolting in the world because it guarantees complete grouting by returning grout through the center of the rock bolt itself." (27)

11-8. Rock Caverns

Tunnel Beneath Oroville Dam

Figure 11-11 shows rock bolting at water tunnel portal beneath Oroville Dam, Northern California. The rock is fine- to coarse grained, massive to schistose amphibolite (metavolcanic).

Washington, D.C.'s Metro

A study of the rocks encountered along the Metro tunnel routes showed crystalline metamorphic schists and gneiss of Precambrian age. The main rock types encountered were mapped under the broad categories of schistose gneiss, chlorite schists, and quartz-diorite gneiss (9). The schistose gneiss unit, as broadly identified, includes complex interfingerings of varying amounts of quartz-hornblende gneiss, and quartz-biotite gneiss.

The quartz-diorite gneiss unit here is the most structurally favorable bedrock encountered and has, in general, the least support requirement.

Figures 11-12a and 11-12b show the construction of the rock tunnel for the Rosslyn subway passenger station of the Washington, D.C. Metro system. The rock face was lined with steel sets, reinforced with rock bolts, and shotcrete which provided almost immediate support for the blocky rock.

The heavy shotcrete lining is 10 cm (4 inches) thick over the steel ribs and 45.7 cm (18 inches) between the ribs.

Fig. 11-11
Rock bolting at water tunnel portal beneath Oroville Dam, Northern California — December 1965.
Rock: fine-to coarse-grained, massive to schistose amphibolite (metavolcanic).
(Photo courtesy of W. W. PEAK, California Department of Water Resources)

Because of Washington's geographic position on the "fall line", a wide variety of tunneling techniques were used, ranging from work in hard rock to tunneling in marine clay.

Wire Net

To protect workers against falling rock fragments, the roofs of underground openings are draped with a wire net (Fig. 11-13).

Underground Locomotive Maintenance Workshop of the Norwegian State Railways

This underground workshop was built in Oslo in 1963—1965. Figs. 2-14 and 11-14 show the round-house as part of the locomotive maintenance workshop. These figures convey to one an idea of the size of that rock cavern.

Fig. 11-12a
A drilled-and-shot pilot bore. Although quite foliated and rather jointed, the rock in this small tunnel is standing with only rock bolts for support.
Washington Metro tunnel.
(Photo courtesy of ROBERT S. O'NEIL, P. E., Senior Vice President of De Leuw, Cather and Company, Washington, D.C.)

Fig. 11-12b
Rock tunneling for the Rosslyn subway station of the Washington, D.C. Metro system. (WMTA)
(Photo courtesy of PAUL J. MYATT, Washington Metropolitan Area Transit Authority)

454 ROCK MECHANICS

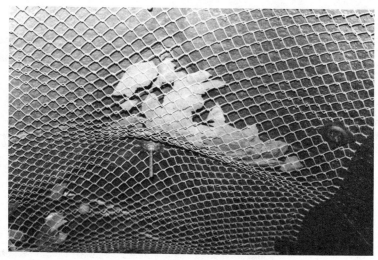

Fig. 11-13
Wire mesh draped under the roof of an underground opening for protection of workers from loose, falling rock fragments.
(Photo by ANDRIS A. JUMIKIS)

Fig. 11-14
Underground locomotive maintenance workshop in rock in Oslo, Norway.
(Photo by ARNE SVENDSEN. Photo courtesy of H. HARTMAN, Executive Offices of the Norwegian State Railways in Oslo, Norway)

The rock quality here is not one of the best (8). The rock comprises sediments of the Silurian Age, consisting partly of clay schist and partly of layers of limestone. The attitude of the stratification here is almost horizontal. The schist is fissured, and splits easily into thin flakes.

For the purpose of safety against rockfall, 3 m long bolts were installed. In special cases, shotcrete with and without reinforcement was used. The lining of the cavern with a 30 cm thick concrete gives sufficient safety as a permanent precaution against rock pressure.

There was a considerable influx of groundwater through the fissures and joints into the cavern. This water was coped with by means of a proper drainage system (8).

Underground Oil Storage Plant in Göteborg

The rock cavern shown in Fig. 11-15 is part of an underground oil storage plant in Göteborg, Sweden. This automatically operated oil storage plant comprises four parallel caverns each 500 m long, 20 m wide, and 30 m high. The plant was commissioned and filled in 1978 with 1.2×60^6 m^3 of crude oil (10).

The rock is a pegmatite gneiss. The roof of the cavern was reinforced with rock bolts when required for the safety of the workers during the construction. The roof of the cavern was 100% shotcreted 2.5 cm (1 inch) thick. Shotcreting was also used on the walls as a reinforcement when required, as well as on protruding bolts, and on parts of the rock rich in quartzite. The latter mentioned was done in order to satisfy the requirements of the Safety Inspection Agency as a protection against ignition from falling of loose rock fragments (10, 23).

World Trade Center Basement Foundation

In connection with the preparation of the building site for the World Trade Center in New York City to prevent undermining of adjacent buildings, surrounding property and settlement of streets, the reinforced concrete slurry-walls surrounding the site were anchored in several tiers by tie-backs 9.14 m to 10.67 m (30 to 35 feet) into the Manhattan schist bedrock (Fig. 11-16), and the tie-backs were pretensioned. After that, unimpeded excavation within the site was started as soon as the watertight basement perimeter cut-off wall was completed (1, 20).

Fig. 11-15
An underground rock cavern for storage of oil in Göteborg, Sweden.
(Photo courtesy of Kurt Hellblom, NYA ASFALT A.B., Rock Division, Stockholm)
a) Installation stage.
b) Finished oil storage cavern.

Fig. 11-16
Reinforced concrete perimeter basement wall for the World Trade Center in New York City anchored in Manhattan schist.
(Photo courtesy of the Port Authority of New York and New Jersey)

11-9. Prestressing

The main function of rock bolting is to reinforce and support partially loosened, thinly laminated or incompetent near-surface rock which otherwise would be subject to failure. Bolts introduce additional stresses and strains in the rock mass, which should improve its general stability, providing that the bolt type, method of bolt installation and layout are designed with reference to the rock structure.

Estimates of rock reinforcement are made in the light of the detailed engineering-geological exploration.

For permanent rock stabilization, bolting must be carried out with the assurance that the anchorage in rock is secure, the tension in the bolt will not relax, and that the bolt itself will not deteriorate.

After the steel bolt has already been fixed into the drillhole in the rock, the nut is tightened against the bearing plate on the surface of the face of the rock. Upon tightening the nut, the bearing plate exerts a tightening or contact pressure σ_r on the surface of the face of the rock. Simultaneously, the shaft of the steel bolt becomes subjected to a tensile force T_{pr} (prestressing or pretensioning force), bringing about, upon tightening of the nut, elongation δ_s of the bolt.

The average tightening compressive stress σ_r on the rock from the bearing plate calculates as

$$\sigma_r = \frac{T_{pr}}{A_r} \qquad (11\text{-}16)$$

where A_r = net bearing area of rock, viz., the area of the anchor bearing plate [$= (a \times a)$ minus cross-sectional area of the shaft of the bolt], and

a = length of side of the steel bearing plate in contact with rock.

The compressive stress σ_r on rock causes an average indentation deformation δ_r in the surface of the rock (25):

$$\delta_r = \sigma_r \cdot \frac{a}{E_r} \qquad (11\text{-}17)$$

where E_r = modulus of elasticity of rock.

Usually, δ_r varies between 0.1 and 0.5 mm.

Upon prestressing, the steel bolt elongates by an amount of δ_s:

$$\delta_s = \frac{T_{pr}}{A_s} \cdot \frac{L}{E_s} = \sigma_t \cdot \frac{L}{E_s} \qquad (11\text{-}18)$$

where A_s = net cross-sectional area of steel rock bolt

L = length of shaft of steel rock bolt

E_s = modulus of elasticity of steel bolt material

$\sigma_t = T_{pr}/A_s$ = tensile stress in the steel bolt, and other symbols are same as before.

If at the head of the bolt an external axial tensile force T is applied, the tightening pressure σ_r on the surface of the rock decreases by an amount of $\Delta\sigma_r$. That is, the resultant pressure on the rock is now $\sigma_r - \Delta\sigma_r$. Simultaneously, the initially introduced prestressing tensile force T_{pr} in the bolt or anchor is increased by a yet unknown force ΔT (to be calculated), that is up to $T_{pr} + \Delta T$ (2).

The unknown force ΔT is calculated from force equilibrium after Birkenmeyer (2) as follows:

$$T = \Delta\sigma_r \cdot A_r + \Delta T \qquad (11\text{-}19)$$

or

$$\Delta T = T - \Delta\sigma_r \cdot A_r \qquad (11\text{-}20)$$

The necessary, yet missing, equation for determining ΔT is obtained from the elasticity condition that $\delta_r = \delta_s$. By Eqs. (11-17) and (11-18),

$$\delta_r = \delta_s = \Delta\sigma_r \cdot \frac{a}{E_r} = \frac{\Delta T}{A_s} \cdot \frac{L}{E_s} . \qquad (11\text{-}21)$$

This is to say that the deformation δ_s of the steel anchor rod because of ΔT must be equal to the rock deformation δ_r caused by $\Delta\sigma_r$. Hence, from Eq. (11-21), obtain $\Delta\sigma_r$ as:

$$\Delta\sigma_r = (\Delta T)\left(\frac{L}{a}\right) \cdot \left(\frac{1}{A_s}\right) \cdot \left(\frac{E_r}{E_s}\right) \qquad (11\text{-}22)$$

Substitution of Eq. (11-22) into the equilibrium equation (11-20), renders the sought value of ΔT:

$$\Delta T = \frac{T}{1 + \dfrac{L}{a} \cdot \dfrac{A_r}{A_s} \cdot \dfrac{E_r}{E_s}} \qquad \text{Q.E.D.!} \qquad (11\text{-}23)$$

$$= k \cdot T, \qquad (11\text{-}24)$$

where

$$k = 1 \bigg/ \left(1 + \frac{L}{a} \cdot \frac{A_r}{A_s} \cdot \frac{E_r}{E_s}\right) \qquad (11\text{-}25)$$

In practice, ΔT is very small, varying approximately from 0.001 to 0.01 of T.

Upon further increase of the externally applied axial tensile force T, a value of $T = N$ can be approached such that the value of the tightening compression stress σ_r on the rock surface, brought about by the prestressing force T_{pr}, becomes zero. When the tensile force T exceeds the value of N, the total force T must be taken up by the prestressed rock bolt or anchor. Upon this, the steel bolt elongates, and a gap opens between the anchor bearing plate and the surface of the rock face. This means that the entire applied tensile force $T \geq N$ must be carried by the now considerably thinner anchor steel.

The ultimate tensile force N can be calculated according to Birkenmeyer (2) from the following condition:

$$\sigma_r - \Delta\sigma_r = 0 \qquad (11\text{-}26)$$

Substitution of Eqs. (11-16), (11-22) and (11-23) into Eq. (11-26) renders

$$N = \frac{T_{pr}}{1-k} \qquad (11\text{-}27)$$

Also, before rupture of the bolt, the gap would open several centimeters wide. In other words, the rupture would be indicated by large displacements of the anchor head.

It should also be mentioned that because of plastic deformation and creep of the steel, the initially introduced prestressing force T_{pr} decreases in time. Usually, in practice the prestressing force is checked periodically.

For permanent rock bolting of rock faces, it is important that the performance of a rock bolt system should not deteriorate within a reasonable period of time, but that the system should insure a long term stability.

11-10. Pressure-Grouted Soil Anchors

The pressure-grouted soil anchor or tie-back is a special substructure anchoring element. It may be a steel rod, or a steel cable, or a multistrand of high-tension steel wires. Its purpose is to anchor, in one or several tiers, various temporary and permanent earth-retaining and foundation structures in cohesive and noncohesive soils to resist lateral, vertical, inclined, and hydrostatic uplift forces (3, 4, 5, 6, 7, 11, 12, 13, 16, 19, 22, 24, 26). A pressure-grouted soil anchor works in tension. The integrally performing wall, anchor, and soil form the so-called wall-anchor-soil system, frequently referred to as the tie-back system. Figure 11-17 shows a tie-back system for analyzing internal stability of free-end supported sheet piling wall. The type of soil anchor to use is determined by the nature of the construction site, the soil geotechnical properties, the load, the method of anchor installation, and other factors.

Generally, the tie-back must develop its anchorage within the stable soil far enough behind the Coulomb potential failure wedge that contributes to the active pressure on the earth-retaining wall (15, 18, 19).

ROCK REINFORCEMENT

The ultimate bearing capacity of the grouted soil anchor should be determined by testing them to failure or to a predetermined maximum load. Also, the anchor g-g should have a positive load transfer, over the embedded anchor length L, from the steel tension rod to the cement grout and then to the soil. Anchors should be protected against corrosion.

The installation of the anchor commences by drilling a hole into the soil of the vertical bank of the excavation. To prevent caving-in of the walls of the borehole, a metal casing is introduced. Then the steel anchor elements are inserted. Through the borehole, cement grout is injected under pressure around a certain length g-g of the rear-end part of the anchor shaft (Fig. 11-17). Upon grouting under pressure, an expanded, extruded body of cement grout in contact with the soil around the steel anchor shaft is formed. The grout forms a bond between the anchor steel shaft (rod) and the surrounding soil. After the grout has set, the head of the pressure-grouted soil anchor is connected in tension to the earth retaining structure to be anchored.

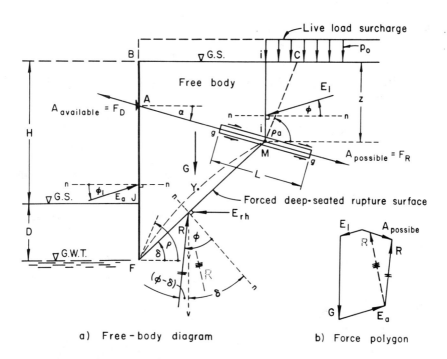

Fig. 11-17
The tie-back system.

The overall or external stability analysis of the groundbreak may be performed by using the usual analytical methods (plane or curvilinear rupture surfaces) as discussed in soil mechanics (16, 17, 19).

The stability in the deep-seated rupture surface $\overline{FM}$ serves mainly to establish the necessary length g-g of the grouted part of the anchor. Here one would start out with the assumption that the anchor yields with its surrounding soil and that therefore the wall inclines or yields toward the inside of the excavation and tends to slide down along a forced, deep-seated rupture surface $\overline{FM}$ (Fig. 11-17). Notice that, in this analysis, the length of the anchor proper is chosen approximately, and then, by the method of trial and adjustment, the available stability of the soil-anchor system is ascertained.

The driving force F_D or anchor pull of the anchor (= availabe force in the anchor, or $A_{available}$) is the one as determined from the analysis of the overall anchored sheet piling before the stability analysis of the deep-seated rupture test. The necessary resisting force F_R, also called the possible force, or $A_{possible}$, for equilibrium of the grouted soil anchor with the driving force is mobilized by the shear resistance between the contact surface of the grouted length g-g of the soil anchor and the adjacent soil (Fig. 11-17a), and is determined from the force polygon (Fig. 11-17b and Fig. 11-18) formed by the forces acting on the free-body (FBiMF) of the soil resting on the deep-seated rupture surface $\overline{FM}$ (or $\overline{FYM}$).

The internal stability analysis of the system involves evaluation of the driving and resisting forces, F_D and F_R respectively, acting on a designated free body (Fig. 11-17). For equilibrium condition of forces acting on the free body FBiMF (15, 18), one constructs a force polygon, Fig. 11-18 (19). From the force polygon, one scales off the magnitude of the maximum value of $A_{possible}$ ($= F_R$). Knowing the magnitude of the available anchor pull $A_{available}$ ($= F_D$), the degree of the internal stability of the system, i.e., the factor of safety η is calculated as

$$\eta = F_R/F_D = A_{possible}/A_{available} \qquad (11\text{-}28)$$

Expressed in terms of horizontal components,

$$\eta = F_{Rh}/F_{Dh} = A_{h\,possible}/A_{h\,available}. \qquad (11\text{-}29)$$

In Figs. (11-17) and (11-18), the forces acting on the free body FBiMF are:

G = self-weight of the free body

E_a = active earth pressure (shown as reaction to the free body at an angle of wall friction φ_1), the magnitude of which may be calculated as for flexible sheet piling walls (15, 18)

R = soil reaction from below as the combined effect from all the other forces acting on the free body (at an angle of internal friction of soil φ), against the deep-seated rupture surface $\overline{FM}$ (to be scaled off from the force polygon in Fig. 11-17b or Fig. 11-18)

E_1 = active earth pressure on the substitute (fictitious or imaginary) wall $i\text{-}i$ ($=i\text{-}M$), Fig. 11-17, acting at an angle of internal friction of soil φ (E_1 is to be calculated as Coulomb's active earth pressure on a massive wall)

a = angle of inclination of anchor with the horizontal

δ = angle of deep-seated rupture plane with the horizontal

= angle of Coulomb's rupture wedge with the horizontal

p_o = eventual uniformly distributed surcharge load (live load) on the ground surface

L = length of pressure grouted soil anchor.

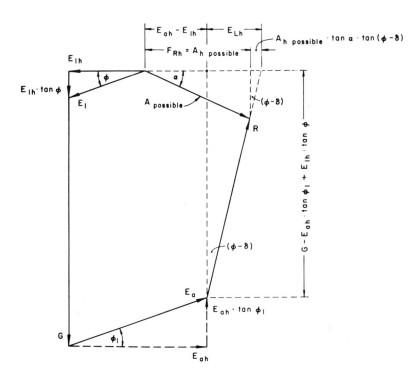

Fig. 11-18
Force polygon.

The magnitude of the force $F_{Rh} = A_{h\,possible}$ can also be computed from the force polygon (19) as

$$A_{h\,possible} = E_{ah} - E_{lh} + E_{Lh} - A_{h\,possible} \cdot \tan a \cdot \tan(\varphi - \delta) \qquad (11\text{-}30)$$

or

$$A_{h\,possible} = \frac{E_{ah} - E_{lh} + E_{Lh}}{1 + \tan a \cdot \tan(\varphi - \delta)} \qquad (11\text{-}31)$$

or

$$A_{h\,possible} = (E_{ah} - E_{lh} + E_{Lh}) \cdot C_{A_h} \qquad (11\text{-}32)$$

where

$$E_{ah} = E_a \cdot \cos\varphi_1 \qquad (11\text{-}33)$$

$$E_{lh} = E_l \cdot \cos\varphi \qquad (11\text{-}34)$$

$$C_{A_h} = 1/[1 + \tan a \cdot \tan(\varphi - \delta)] \qquad (11\text{-}35)$$

is a coefficient (the so-called anchor coefficient), and

$$E_{Lh} = [G - (E_{ah} \cdot \tan\varphi_1 - E_{lh} \cdot \tan\varphi)] \cdot \tan(\varphi - \delta) \qquad (11\text{-}36)$$

is an auxiliary horizontal force component or an auxiliary mathematical quantity in the force polygon as shown in Fig. 11-18.

The anchor coefficient C_{a_h} as a function of angle of inclination a of the anchor with the horizontal; angle of internal friction of soil φ, and angle of slope of the deep-seated rupture surface with the horizontal are tabulated in Table 11-1, and shown graphically in Fig. 11-19.

The internal stability calculations should be repeated several times, each time assuming a different length and different inclination of the anchor, until the position of the most dangerous deep-seated rupture surface has been found.

The minimum required factor of safety is $\eta_{min} \geq \eta_{allowable}$.

Some of the advantages of the tie-back soil anchors over conventional bracing systems are:

1. Each anchor can be prestressed. Prestressing of soil anchors reduces the movement of adjacent soil mass and results in no or insignificant settlement of streets and buildings located adjacent to the excavation.
2. The excavation pit can be kept free of all bracing and strutting so that there is freedom in the pit for excavation and construction activities un-

TABLE 11-1

Anchor Coefficient C_{A_h}

$(\varphi - \delta)°$	$\alpha = 0°$	$\alpha = 5°$	$\alpha = 10°$	$\alpha = 15°$	$\alpha = 20°$	$\alpha = 25°$	$\alpha = 30°$	$\alpha = 35°$
−20	1.000	1.032	1.068	1.108	1.152	1.204	1.266	1.342
−15	1.000	1.024	1.049	1.077	1.108	1.142	1.183	1.230
−10	1.000	1.015	1.032	1.049	1.068	1.089	1.113	1.140
−5	1.000	1.007	1.015	1.024	1.032	1.042	1.053	1.065
0	1.000	1.000	1.000	1.000	1.000	1.000	1.000	1.000
+5	1.000	0.992	0.984	0.977	0.969	0.960	0.951	0.942
+10	1.000	0.984	0.969	0.954	0.939	0.924	0.907	0.890
+15	1.000	0.977	0.954	0.933	0.911	0.888	0.866	0.842
+20	1.000	0.969	0.939	0.911	0.883	0.854	0.826	0.796
+25	1.000	0.960	0.924	0.888	0.854	0.821	0.787	0.753
+30	1.000	0.951	0.907	0.866	0.826	0.787	0.750	0.712
+35	1.000	0.942	0.890	0.842	0.796	0.753	0.712	0.671
+40	1.000	0.931	0.871	0.816	0.766	0.718	0.673	0.629

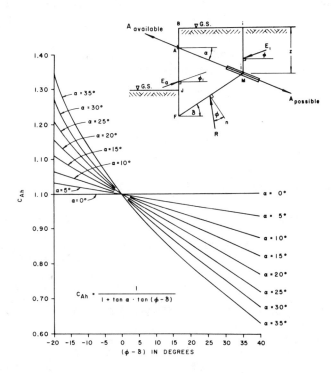

Fig. 11—19
Anchor coefficient C_{A_h} for calculating $A_{h\ possible}$.

cluttered and unobstructed by a forest of timber and steel. This reduces below-ground construction time and cost.

3. The construction and installation of the earth anchors can be accomplished without any noise pollution and without vibration harmful to adjacent structures.
4. Depending on the depth of excavation, soil anchors can be installed in one or several tiers.
5. Tie-backs can be installed in soil zones inaccessible from above because the anchors can be installed from the inside of the excavation.
6. Each of the selected anchors can be tested to failure or to a predetermined maximum load. Therefore their ultimate load-bearing capacity can be determined.
7. Defective anchors can be repaired.
8. Inadequate anchors can be replaced.

All in all, inclined pressure-grouted soil anchors may have economic advantages over conventionally braced or strutted retaining-wall systems, the former being competitive with the latter, which means savings in time and money.

One of the major disadvantages of soil anchors is the problem of corrosion of the steel rods and cables.

REFERENCES

1. Anonymous, 1974, The World Trade Center: A Building Project Like No Other (a pamphlet). New York, N.Y.: The Port Authority of New York and New Jersey, October, 1974.
2. BIRKENMEYER, M., 1953, "Vorgespannte Felsanker." *Schweizerische Bauzeitung*, November 21, 1953, No. 47, pp. 688—692.
3. CLOUGH, G. W., and Y. TSUI, 1974, "Performance of Tied-Back Walls in Clay." *Proc. ASCE, Journal of the Geotechnical Division*, December 1974, Vol. 100, No. GT 12, pp. 1259—1273.
4. CORDING, E. J., 1968, "Stability of Large Underground Openings at the Nevada Test Site," *ASCE Seminar on Rock Mechanics in Civil Engineering Practice*, held April 9—10 and May 7—8, 1968, in New York, N.Y., ASCE Metropolitan Section, Foundations and Soil Mechanics Group.
5. CORDING, E. J., 1976, "Observations and Design in the Construction of Rock Chambers in Urban Areas," *ASCE Rock Excavation Seminar*, held October 1976 in New York, N.Y., ASCE Metropolitan Section, Foundations and Soil Mechanics Group.
6. FARMER, I. W., 1968, *Engineering Properties of Rocks*. London, E. and F. N. Spon Ltd.

7. HANNA, T. H., and I. I. KURDI, 1975, "Studies of Anchored Flexible Retaining Walls in Sand." *Proc. ASCE, Journal of the Geotechnical Division*, August 1975, Vol. 100, No. GT 8, pp. 829—831.
8. HARTMAN, H., 1979, of the Executive Offices of the Norwegian State Railways in Oslo. Personal correspondence of August 7, 1979.
9. HEFLIN, L. H., 1968, "Engineering Geology — Its Role in the Development and Planning of the Washington Metro." *Journal of the Washington Academy of Sciences*, Vol. 58, 1968, pp. 165—170.
10. HELLBLOM, K., 1979, of the NYA ASFALT AB, Stockholm, Sweden. Personal correspondence, July 5, 1979.
11. HOBST, L. and J. ZAJÍC, 1977, *Anchoring in Rock*. Amsterdam: Elsevier Scientific Publishing Company.
12. HOEK, E., and J. W. BRAY, 1981, Revised 3rd edition, *Rock Slope Engineering*. London: The Institution of Mining and Metallurgy.
13. HOEK, E., and P. LONDE, 1974, "Surface Workings in Rock." *Advances in Rock Mechanics*. Proceedings of the International Society for Rock Mechanics, held at Denver, Co., in 1974. Washington, D.C.: National Academy of Sciences, Vol. 1A, pp. 612—654.
14. JUMIKIS, A. R., 1956, "Rupture Surface in Sand Under Oblique Loads," *Proc. ASCE, Soil Mechanics and Foundations Division*, Vol. 82, Paper 861, pp. SM-1 to SM-6.
15. JUMIKIS, A. R., 1964, "Fundamentals of Sheet Piling." *Mechanics of Soils — Fundamentals for Advanced Study*. Princeton, New Jersey: D. Van Nostrand Company, Inc., 1964, pp. 333—378.
16. JUMIKIS, A. R., 1965, *Stability Analyses of Soil-Foundation Systems*. New Brunswick, New Jersey: College of Engineering, Bureau of Engineering Research; Rutgers — The State University. Engineering Research Publication No. 44. (A design manual based on logarithmically spiralled rupture curves. Contains 13 mathematical-physical tables of relative values of logarithmic spiral elements).
17a. JUMIKIS, A. R., 1962, *Soil Mechanics*. Princeton, New Jersey: D. Van Nostrand Company, Inc., 1962.
17b. JUMIKIS, A. R., 1969, *Theoretical Soil Mechanics*. New York, N.Y.: Van Nostrand Reinhold Company, pp. 271—285.
18. JUMIKIS, A.R., 1971, "Sheet Piling." *Foundation Engineering*. Scranton, Pa.: INTEXT Educational Publishers, 1971, pp. 129—176.
19. JUMIKIS, A. R., 1978, "Concerning Pressure-Grouted Soil Anchors." *Transportation Research Record 690: Stabilization and Compaction*. Washington, D.C.: National Research Council — National Academy of Sciences, 1978, pp. 8—13.
20. KAPP, M. S., 1969, "Slurry-Trench Construction for Basement of World Trade Center." *Civil Engineering* (ASCE), April 1969, Vol. 39, No. 4, pp. 36—40.
21. LANG, T., 1962, "Theory and Practice of Rock Bolting," *Transactions American Institute of Mining and Metallurgical Engineers*, Mining Division, Vol. 223.
22. NELSEN, J. C., 1973, "Earth Tie-Backs Support Excavations 112-ft deep." *Journal of the Civil Engineering Division, ASCE*, November 1973, Vol. 43 C, E 11, pp. 40-43.
23. NYA ASFALT AB. *Design and Construction — Rock Cavern Oil Storage*. Stockholm, Sweden (Promotional brochure).
24. RANKE, A., and H. OSTERMEYER, 1968, "Beitrag zur Stabilitätsuntersuchung mehrfach verankerter Baugrundumschließungen." *Die Bautechnik*, 1968, No. 10, pp. 341—350.

25. Schleicher, F., 1926, "Zur Theorie des Baugrundes." *Der Bauingenieur,* 1926, No. 48, pp. 931—935, and No. 49, pp. 949—952.
26. Schousboe, J., 1975, "Some Applications of Prestressing in Foundation Construction." Proc. ASCE, Journal of the Construction Division, June 1975, Vol. 101, No. CO 2, pp. 403—413.
27. *Williams Form and Rock Bolt Engineering 1979—80.* Grand Rapids, Michigan: Williams Form Engineering Corp., 1978, Catalog No. 7980.

OTHER RELATED REFERENCES

Herzog, M., 1980, "Bemessung des Tunnelbaus auf den Quelldruck des Gebirges." *Schweizer Baublatt,* Febr. 5, 1980, No. 11, pp. 3—4.

Herzog, M., 1980, "Wirklichkeitsnahe Bemessung der Anker von Gebirgstragringen." *Die Bautechnik,* 1980, No. 5, pp. 162—164.

Jäger, B., 1974, "Die Felssicherungen am Drachenfels." *Mitteilungen der Landesstelle für Naturschutz und Landschaftspflege in Nordrhein-Westfalen,* September 1974, Vol. 3, No. 2, pp. 38—41.

Londe, P., 1973, "The Role of Rock Mechanics in the Reconnaissance of Rock Foundations." *Quarterly Engineering Geology,* 1973, Vol. 6, pp. 57—74.

Londe, P., 1973, "Water Seepage in Rock Slopes." *Quarterly Engineering Geology,* 1973, Vol. 6, pp. 75—92.

Londe, P., 1973, "Analysis of the Stability of Rock Slopes." *Quarterly Engineering Geology,* 1973, Vol. 6, pp. 93—127.

Ortlepp, W. D., 1969, "An Empirical Determination of the Effectiveness of Rockbolt Support Under Impulse Loading." Oslo: ISLPUO.

Pender, E. P., A. D. Hosking, and R. H. Mattner, 1963, *Grouted Rock Bolts for Permanent Support of Major Underground Works.* Institute of Engineers, Australia, July—August, 1963, Vol. 35.

Piteau, D. R., and Associates Limited, Vancouver, B.C., 1979, *Rock Slope Engineering.* Washington, D.C.: Federal Highway Administration, Office of Development, Implementation Division, January, 1979, Reference Manual. FHWA — TS — 79— 208.
Part A. Engineering Geology Considerations and Basic Approach to Rock Slope Stability Analysis for Highways; Part B. Methods of Obtaining Geologic Structural, Strength and Related Engineering Geology Data; Part C. Approach and Techniques in Geologic Structural Analysis; Part D. Slope Stability Analysis Method; Part E. Rock Slope Stabilization, Protection, and Warning-Instrumentation Measures and Related Construction Considerations; Part F. Blasting for Rock Slopes and Related Excavation Considerations; Part G. (Field Manual). Description of Detail Line Engineering Geology Mapping Method by D. R. Piteau and D. C. Martin, D. R. Piteau and Associates Limited, West Vancouver, B.C., July 1977; Part H. ("Appendix"). Chapter 9 of Landslides: Analysis and Control (TRB Special Report' 176).

Prestressed Concrete Institute, 1974, *"Tentative Recommendations for Prestressed Rock and Soil Anchors."* Chicago, September, 1974.

Proceedings of a Conference on Site Exploration in Rock for Underground Design and Construction, held March 29—31, 1978 at Alexandria, Virginia. Springfield, Virginia: The National Technical Information Service. Sponsored by the U.S. Department of Transportation and the ASCE, National Capital Section, Washington, D.C. Final Report, July 1979 (Report No. FHWA — TS — 79 — 221) (98 pages).

REDLINGER, J. F. and E. L. DODSON, 1966, "Rock Anchor Design." *Proceedings of the (1st) Congress of the International Society of Rock Mechanics,* held September 25 to October 1, 1966 at Lisbon, Portugal. Published by the Laboratório Nacional de Engenharia Civil, Lisbon, Portugal, 1966, Vol. 2, pp. 171—174.

SCHMIDT, A. E., 1956, "Rock Anchors Hold TV Tower on Mt. Wilson." *Civil Engineering,* January, 1956, Vol. 26, No. 1, pp. 56—58.

SELTZ-PETRASH, A., 1979, "Washington Metro: A People's Eye View." *Civil Engineering — ASCE,* June, 1979, Vol. 49, No. 6, pp. 59—63.

SMITH, R., 1959, "At Oahe: Contractor Licks Shale by Building Backwards." *Engineering News Record,* July 23, 1959, Vol. 163, No. 4, pp. 42—43.

TAIT, R. G. and H. T. TAYLOR, 1975, "Rigid and Flexible Bracing Systems on Adjacent Sites." *Proc. ASCE, Journal of the Construction Division,* June 1975, Vol. 101, No. CO 2, pp. 365—376.

WARE, K. R., 1973, 1974, 1975, "Tieback Wall Construction — Results and Control." *Proc. ASCE, Journal of the Soil Mechanics and Foundations Division:* Dec. 1973, Vol. 99, No. SM 12, pp. 1135—1152. Oct. 1974, Vol. 99, No. GT 10, p. 1167. May 1975, Vol. 99, No. GT 5, p. 495.

WEBER, E., 1966, "Injektionsanker, System Stump Bohr AG für Verankerungen im Lockergestein und Fels." *Schweizerische Bauzeitung,* Feb. 10, 1966, No. 6, pp. 120—124.

WINDOLF, G., 1976, "Sweden's Underground Millions." *Tunnels and Tunneling.* Sept.—Oct., 1976, Vol. 8, No. 6, pp. 24—26.

CHAPTER 12

ROCK SLOPES

12-1. Definitions

In highway, foundation, and hydraulic structures engineering in rock, there is a need for assessing the degree of stability of man-made as well as natural rock slopes.

Highway construction contributes to rock slides during excavation and after completion.

A slope is a vertical or inclined boundary surface between air and rock, or the body of an earthwork such as a dam, cut, or fill.

In geotechnics, the topic "stability of slopes" is dealt with from two engineering viewpoints, namely:

1. The design of man-made slopes of cuts and fills in advance of new rock or earthwork construction, with prescribed safety requirements against failure, and
2. The study of stability of existing slopes of rocks and earthworks, slopes which are potentially unstable, slopes which have failed, or slopes which have to be redesigned.

The concept of the "stability" of a slope is an indeterminate one, because no slopes made in or of rock or soil can be regarded as fully guaranteed for their stability during their service over a period of many years. Climatic, hydrologic, and tectonic conditions, and man's activities in the immediate and/or adjacent area of the structure, underground opening, or earthwork may bring about, years later, changes affecting the stability of man-made and natural slopes. In particular, one should not overlook the possibility of the rock and soil becoming saturated by water with time.

The term "failure of slope" refers to any slope instability that affects its performance, and man's operation in a natural, geologic environment.

In conjunction with man's work in rock, instability of rock slopes usually results from an unfavorable, critical attitude and the location and frequency of the various discontinuities of the rock formations in question. Conventionally, the concept of failure refers to the stress condition in rock, when the externally applied stress exceeds the strength of the rock. If this happens, the rock material breaks down, and the material is finished.

12-2. Factors Contributing to Slope Failure

Theoretically, by means of the geostatic pressure formula $\sigma = \gamma \cdot H$, an ideal, sound, unfissured rock, such as a granite, for example, whose uniaxial compressive strength is $\sigma_u \approx 2200$ kg$_f$/cm² = 2.2×10^4 t$_f$/m² = 215.75 MN/m² and whose unit weight $\gamma = 2.6$ t$_f$/m³ = 25.5 kN/m³, would stand vertically at a height of $H = \sigma_u/\gamma = (2.2 \times 10^4)/2.6$ t$_f$/m³ = 215.750/25.5 $\approx$ 8460 m. However, in nature there is no ideal rock whose vertical slope can stand this high.

Observations in the field have shown that many gentle slopes of much less height have failed. This means that the critical height of a rock slope is governed not merely by the shear strength of the rock alone, but also by the various rock defects, such as jointing, cracks, fissures, and other possible weaknesses. Also, the physical characteristics of a rock mass, the mechanism involved in rock failure, the effect of water on rock — all these factors affect the engineering structures made in rock.

Continuous fissures and jointing systems separate the rock mass partly or completely into blocks. The shear strength along these planes of weakness is considerably low. Therefore they offer low resistance to displacement of the rock blocks, and consequently low resistance to failure. This is even more true if the orientation of the rock stratification system is inclined outward of the rock mass. Besides, through the jointing system water usually circulates. These conditions merely indicate that without special engineering measures, seldom are dry, leakproof, underground openings and slopes in rock encountered. Gravity, weathering, and erosion brought about by a humid climate are factors contributing eventually to the instability of rock slopes. Exposed to weathering effects directly, vertical and steep slopes of rock weather relatively quickly. Also, water flowing out from rock slopes influence, in time, their stability adversely. To this, frost action adds its share to the instability of rock slopes.

Slope design in rock and the stability of slopes of walls of rock excavations depend upon the following main factors:

1. The type of rock of which or in which the slope is made;
2. The structure, stratification, and attitude of the rock formations (angle of dip of rock strata, for example);
3. Presence of a potential failure surface in rock (frequency of geological discontinuities) and the steepness of its angle of dip toward the excavation or valley;
4. The geometry of the cross-section of the slope (height, slope angle, berms, for example);
5. The degree of shattering of unsound rock in the slope;
6. The presence of zones of breccia, if any;
7. Unit weight of the rock slope material (gravity is one of the principal causes of all rock slides);
8. Magnitude of externally applied loads on the rock slope structure;
9. Distribution of weight and loads;
10. In-situ strength properties of slope rock (yielding of rock; creep; breaking, crushing, and rock-bursting of the rock mass brought about by high-intensity loads and stress concentrations in rock and by excavation-induced rock deformations);
11. Loosening up and separation of rock strata and sliding;
12. Loosening of rock and subsequent heaving because of stress relaxation in rock brought about by removal of rock constraint;
13. Loosening of rock strata from blasting;
14. The initial or primary stress regimen (stress field) in a rock mass before excavation;
15. Distribution of secondary, induced stresses in rock upon and after excavation;
16. Position of groundwater table;
17. Amount of moisture content (degree of saturation) in the rock (water is one of the most aggressive factors contributing adversely to the instability of slopes);
18. Vibrations and seismic forces;
19. Various environmental conditions and processes sculpturing the face of slopes (rock-weathering, frost, and chemical action of porewater on rock minerals).

All of the above-listed factors (and therefore also rock properties) may vary from place to place, from point to point or from one time to another.

Rock-slope failures may be classified according to the mode in which the applied stress exceeds the strength of the rock. Examples of slope failures are:
1. Rock fall,
2. Block flow,

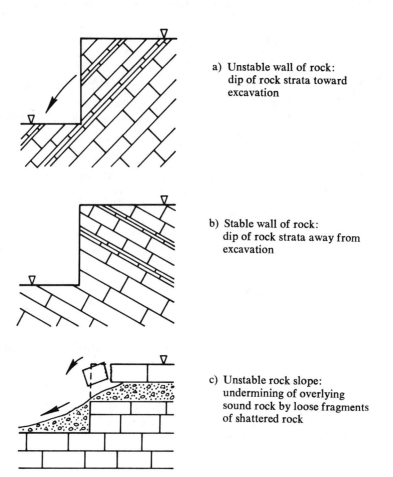

a) Unstable wall of rock: dip of rock strata toward excavation

b) Stable wall of rock: dip of rock strata away from excavation

c) Unstable rock slope: undermining of overlying sound rock by loose fragments of shattered rock

Fig. 12-1
Unstable and stable excavation walls in rock

3. Slide along a pre-existing, dipped plane of discontinuity (bedding planes, faults, and joints, for example), and
4. Rotational slides.

Rock falls from cliffs and talus slopes have been studied in the field, and theoretically by rock trajectory evaluation. Detailed criteria for design of ditch sections along highways and of rock fences for protection against rock falls may be found in Reference 31. The criteria include fallout areas for energy dissipation, steep off-shoulder slopes to cope with angular momentum generated after impact, and rock fences acting as flexible buttresses.

Figures 12-1a, b, and c illustrate some unstable and stable rock slopes as a function of attitude of rock formations.

12-3. Stability of Rock Slopes

Almost all rock-slope failures occur along pre-existing, predetermined, natural rock discontinuities or planes of weakness, such as a fault or a shear zone, for example.

In rock-slope stability problems, the actual failure or sliding surface depends upon the spatial orientation, frequency, and distribution of the discontinuities, and the involved shear and interlock resistance to shear along them. Upon shearing, the geologic rock formation becomes unlocked, resulting in sliding (8, 9, 12, 13, 38, 39).

Figure 12-2 shows a pre-existing geological discontinuity oriented at an angle of dip α with the horizontal. If the weight W of the rock mass superimposed upon the inclined discontinuity A-A_1 is large enough, and the

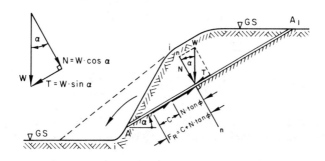

Fig. 12-2
Slope stability problem involving sliding over a geological discontinuity

resistance to shear F_R (or against sliding down of the rock mass) in this discontinuity is not sufficiently large enough, the superimposed rock mass above the discontinuity may become detached from the rock below the discontinuity and slide down over the discontinuity into the cut. Thus the man-made slope of the cut becomes ruined.

Basically, stability calculations of a rock slope over a pre-existing discontinuity such as a sliding surface is a problem of statics. For example, in Fig. 12-2, one determines the driving force F_D per unit length of run of slope

$$F_D = T = W \cdot \sin \alpha \qquad (12\text{-}1)$$

causing shearing in the rock discontinuity $A\text{-}A_1$. Then, one computes the resisting force F_R to shear:

$$\begin{aligned} F_R &= N \cdot \tan \varphi + C \\ &= W \cdot \cos \alpha \cdot \tan \varphi + C \end{aligned} \qquad (12\text{-}2)$$

where T = tangential (shear) force in the plane of discontinuity,
W = weight of the sliding rock mass over the shear plane,
α = angle of dip of slope,
φ = angle of friction,
$C = (c)\,(\overline{A - A_1})\,(1.0)$ = total cohesive (resisting) force along the shearing surface $(\overline{A - A_1})\,(1.0)$, and
$c = C/(\overline{A - A_1})\,(1.0)$ = cohesion per unit area.

The safety or stability of the rock slope in this example is expressed by means of the so-called factor of safety η against sliding (viz., shear) as a ratio of the resisting force F_R to the driving force F_D as

$$\eta = \frac{F_R}{F_D} = \frac{W \cdot \cos \alpha \cdot \tan \varphi + C}{W \cdot \sin \alpha} \qquad (12\text{-}3)$$

When $C = 0$, then Eq. (12-3) simplifies to

$$\eta = \frac{W \cdot \cos \alpha \cdot \tan \varphi}{W \cdot \sin \alpha} = \frac{\tan \varphi}{\tan \alpha} \qquad (12\text{-}4)$$

In such a case — for noncohesive material — the factor of safety is independent of the height of the slope and the form of the wedge.

If the total pore-water pressure U on the sliding surface is known, then

$$\eta = \frac{(W \cdot \cos \alpha - U) \cdot \tan \varphi}{W \cdot \sin \alpha} \qquad (12\text{-}5)$$

If the factor of safety turns out to be unacceptable, the slope must be redesigned adopting a different angle of slope, or reinforced by means of rock anchors, or buttressed by means of a retaining wall. Provision for drainage facilities may decrease pore-water pressure in rock, as well as uplift forces.

12-4. Reinforcement of Rock Slopes

Rock-bolting can be used on natural or artificial slopes to prevent blocks of rock from falling away from the main mass when isolated by joints and faults. If the rock cut as shown in Fig. 12-2 is unstable ($\eta < 1$), then the rock slope can be stabilized by reinforcing it by means of a rock-anchoring system. The purpose of anchors is to transmit the stabilizing force into the rock slope comprehending the rock joints, viz., discontinuities. By means of

a) Rock-anchored vertical slope

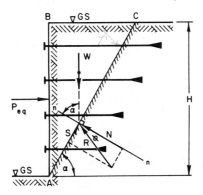

b) Force triangle-solving for P_{eq}

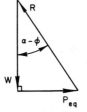

Fig. 12-3
Rock-anchor system

the installed rock anchorage, the joints close more or less. Thus the friction between the surfaces of separation in the anchorage zone is increased.

Now, for the purpose of illustrating determination of the necessary anchor force, refer to Figs. 12-3a and b. Assume an unstable, vertical slope. To maintain force equilibrium of this slope, a horizontal rock-anchor system is introduced. These anchors supply the necessary force P_{eq} on the rock for equilibrium. The magnitude of this necessary force P_{eq} for equilibrium can be conveniently solved graphically by means of the force triangle, as shown in Fig. 12-3b.

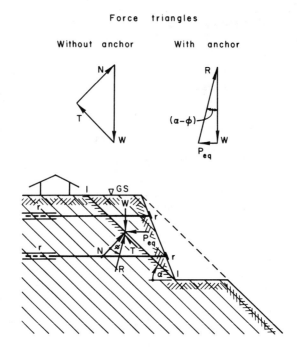

Fig. 12-4
Anchoring of a rock slope
$1-1$: Potential failure line
$r-r$: rock anchor
    W: weight of ruptured rock wedge above plane $1-1$
    φ: angle of friction along plane $1-1$
    N: normal force
    T: tangential shear resistance
    R: reaction
    P_{eq}: anchor force

The actual total rock-anchor force P is obtained by multiplying the equilibrium force P_{eq} by a prescribed factor of safety η:

$$P = \eta \cdot P_{eq} \qquad (12\text{-}6)$$

For a certain number of chosen anchors, the force P determines their diameter as well as their spacing in the face of the slope. The length of the anchors is that which is necessary to cross the discontinuity plus the needed fixing (keying) or bonding length of the anchor in the rock. Thus the rock anchors would prevent massive falls of loose rock fragments from the steeply cut, reinforced rock face.

High slopes may be constructed by parts, separating the upper and lower slopes by a berm (Fig. 12-4).

Unstable rock slopes may be kept stable, also, by means of retaining walls. Figure 12-5 shows a rock slope stabilized by means of a retaining wall and several rock anchors.

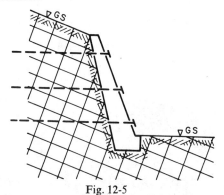

Fig. 12-5
Anchoring of a rock slope with several anchors, and stabilization by means of a retaining wall

If the retaining wall is founded on a firm rock, a single anchor may be used, depending upon the height of the wall. By means of anchoring, a considerable reduction in thickness of the retaining wall may be achieved. The rock slopes may also be reinforced by means of anchors directly, without constructing any retaining walls.

Some applications of rock slope stabilization by means of anchors are illustrated in Figures 12-5a, 12-5b, and 12-5c. Figure 12-5a shows an anchored rock slope retaining wall along a road. Figures 12-b and 12-5c show a method of rock slope stabilization by means of anchors along a road.

Sometimes fissured rock may also be stabilized by means of cement grouting.

Fig. 12-5a
Anchored rock slope retaining wall along a road.
(Photo by Andris A. Jumikis)

Fig. 12-5b
Method of rock slope stabilization by means of anchors.
(Photo by Andris A. Jumikis)

Fig. 12-5c
Method of rock slope stabilization by means of anchors.
(Photo by ANDRIS A. JUMIKIS)

Depending upon the geologic nature and jointing pattern of the rock, in evaluating the stability of rock slopes each individual rock stabilization project requires an individual approach, analysis, and solution (31, 38, 39).

12-5. Effect of Precipitation on Stability of Rock Slopes

The permeability of rock to water depends upon the presence of open and continuous fissures and seams within the rock mass. Part of the precipitation water intrudes through the surface cracks into the rock. Depending upon the intensity of precipitation, this water may bring about a high or low water table.

Generally, the difference in water table in the ground during dry and wet seasons may vary by several meters. However, in rock, the position of the water table is indeterminate because of the various kinds, frequency, and distribution of seams, cracks, and other kinds of voids in the rock.

During intensive rainfall, all of the rainwater cannot be absorbed by the rock. Most of the water runs down along the sloping ground surface.

Also, during the maximum spring thaw of snow, most of the meltwater usually runs down, because the cracks near the ground surface of the slope are still frozen. Also, the meltwater of the ice from within the fissured rock cannot escape. Frequently, these conditions, too, impair adversely the stability of slopes in fissured rock, bringing about rock bursts and rock slides.

12-6. Rock-Slope Stability Based on Rock Strength

Nowadays, the most frequently used theory for evaluating, theoretically, the strength and stability of a homogeneous isotropic material is the one put forward by MOHR (29). According to MOHR's theory, rupture of a material takes place along surfaces where the tangential, viz., shear, stresses attain a definite value. At limit equilibrium, by COULOMB's shear strength equation, the magnitude of the shear stress τ is a function of the effective normal stress, namely:

$$\tau = f(\sigma_{n_{eff}}) = \sigma_{n_{eff}} \cdot \tan \varphi + c \qquad (12\text{-}7)$$

where $\sigma_{n_{eff}}$ = effective normal stress on the rupture plane (Fig. 12-6),

$\tan \varphi$ = coefficient of friction, and

c = test parameter known as the cohesion.

In practice, the limiting stability or equilibrium condition of a homogeneous rock element can be conveniently expressed with a satisfactory degree of precision by means of principal stresses σ_1 and σ_3 in the form of COULOMB's shear strength formula, Eq. (12-7).

a) Rock element

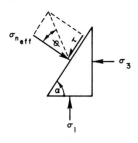

b) MOHR's stress circle

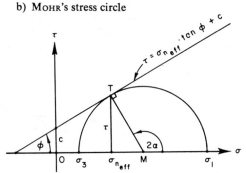

Fig. 12-6
COULOMB-MOHR strength condition

From MOHR's stress circle, Fig. 12-6, obtain:

$$\tau = \frac{\sigma_1 - \sigma_3}{2} \cdot \cos \alpha = \left(\frac{\sigma_1 + \sigma_3}{2} - \frac{\sigma_1 - \sigma_3}{2} \cdot \sin \varphi \right) \cdot \tan \varphi + c \qquad (12\text{-}8)$$

After rearrangement, the major principal stress σ_1 is written as

$$\sigma_1 = \sigma_3 \left(\frac{1 + \sin \varphi}{1 - \sin \varphi} \right) + \frac{2 \cdot c \cdot \cos \varphi}{1 - \sin \varphi} \qquad (12\text{-}9)$$

or

$$\sigma_1 = \sigma_3 \cdot K_\mathrm{p} + 2 \cdot c \cdot \sqrt{K_\mathrm{p}} \qquad (12\text{-}10)$$

or
$$\sigma_1 = \sigma_3 \cdot \tan^2\left(45° + \frac{\varphi}{2}\right) + 2 \cdot c \cdot \tan\left(45° + \frac{\varphi}{2}\right) \quad (12\text{-}11)$$
where
$$(1 + \sin\varphi)/(1 - \sin\varphi) = K_p \quad (12\text{-}12)$$
and
$$\cos\varphi/(1 - \sin\varphi) = \sqrt{K_p} \quad (12\text{-}13)$$

K_p is the COULOMB's passive earth pressure (or earth resistance) coefficient.

The degree of stability of a homogeneous rock slope at any point of the slope is usually expressed by means of a factor of safety η if the principal stresses and rock strength parameters φ and c are known. The factor of safety η is then expressed as the ratio of the rock strength,

$$\sigma_R = \sigma_1 = \sigma_3 \cdot K_p + 2 \cdot c \sqrt{K_p} \quad (12\text{-}14)$$

where σ_R = resisting stress at failure to the acting major stress, and

$$\sigma_D = \sigma_{1\,\text{acting}} \quad (12\text{-}15)$$

where σ_D = driving stress:

$$\eta = \frac{\sigma_R}{\sigma_D} = \frac{\sigma_3 \cdot K_p + 2 \cdot c \cdot \sqrt{K_p}}{\sigma_{1\,\text{acting}}} \quad (12\text{-}16)$$

the assumption herein being that the rock strength parameters φ and c do not depend upon the stress condition of the rock massif, nor upon the depth of the point under consideration in the rock below the ground surface.

When $\sigma_3 = 0$, then

$$\tau_{max} = \sigma_1/2 \quad (12\text{-}17)$$

and

$$\sigma_1 = 2 \cdot \tau_{max} \quad (12\text{-}18)$$

Substitution of these values into the η-equation, Eq. (12-16), renders the factor of safety η against rupture of a rock slope in its lower part as

$$\eta = \frac{c \cdot \sqrt{K_p}}{\tau_{max}} = \frac{c \cdot \tan\left(45° + \frac{\varphi}{2}\right)}{\tau_{max}} \quad (12\text{-}19)$$

Thus, for an unconfined condition ($\sigma_3 = 0$), the factor of safety η of a rock slope is a function of the maximum shear stress τ_{max}.

By means of the η-equation, it is possible to calculate factors of safety for a number of assumed points in the slope, i.e., $\eta = f(a_a)$, where a_a is the angle between the horizontal and a line drawn from the toe-point of the slope to the assumed point in the slope. Connecting points of the same factor of safety by means of a curve renders a field of isocurves, viz., zones of relative strength of the slope of the rock.

12-7. Closing Remarks about Stability of Rock Slopes

Evaluation of stability of rock slopes may be made if the nature, magnitude, and distribution of natural, internal stress conditions in the slope-forming rock massif are known. Unfortunately, in rock engineering, up to now, the problem of determining natural stress conditions in rock has been worked out very little methodologically. The solution to this problem requires a great deal of theoretical as well as experimental research. One should rcognize that the factor "stressed condition" in rock is one of the

Fig. 12-7
Concentration of stresses in the lower part of a slope
(Redrawn from Hoek and Gralewska, 1969)

486   ROCK MECHANICS

determining factors in the behavior of rock slopes, and thus in the degree of stability of slopes.

Research has shown that, basically, there occur large, internal stress concentrations in the lower part of the slope in the neighborhood around and below the toe of the slope as shown qualitatively in Fig. 12-7, probably because of the great weight of the rock near the slope toe, and because of residual tectonic stresses in the rock. The large stress concentrations in this part of the slope may be the source of rock bursts and other kinds of rock failure.

Open-pit mining engineering practice has shown that rock in the lower part of the slope of a quarry fails in shear brought about by tangential or shear stresses. Therefore, in studying stressed conditions of rock slopes, major attention should be devoted to determining the magnitude of the maximum shear stresses in the lower part of the slope of the rock.

12-8. Stability Analyses of Soil Slopes

Stability analyses of soil slopes may be performed based on the most disadvantageous position of a plane rupture surface (3, 19, 28).

Stability analyses of slopes made in soils and in highly weathered, disintegrated, loosely fragmented, or very jointed rock masses or rock fills

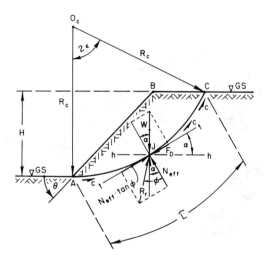

Fig. 12-8
Circularly curved rupture surface in a (φ-c)-soil.

ROCK SLOPES

can also be performed based on curved rupture surfaces (2, 9, 14, 19, 30, 32, 33). Many weathered shales have engineering characteristics like those of soils. Therefore stability analyses in such shales may be performed as in soils.

Figure 12-8 shows a rotational slope failure of a slope of a (φ-c)-soil over a circularly curved rupture surface.

The factor of safety η against the circular rotational mode of rupture of soil or weathered rock material is calculated based on the position of the most dangerous (=critical) rupture surface (19) as

$$\eta = \frac{M_R}{M_D} = \frac{F_R \cdot r_c}{F_D \cdot r_c} \tag{12-20}$$

$$= \frac{\tau \cdot A \cdot r_c}{W \cdot a} = \frac{\tau \cdot (2\epsilon) \cdot r_c^2}{W \cdot c} \tag{12-20a}$$

$$= \frac{(N_{\text{eff}} \cdot \tan\varphi + C) \cdot r_c}{F_D \cdot r_c} \tag{12-20b}$$

$$= \frac{N_{\text{eff}} \cdot \tan\varphi + C}{F_D} \tag{12-21}$$

$$= \frac{W \cdot \cos\alpha \cdot \tan\varphi + C}{W \cdot \sin\alpha} \tag{12-21a}$$

$$= \cot\alpha \cdot \tan\varphi + \frac{C}{W \cdot \sin\alpha} \tag{12-22}$$

$$= \frac{\tan\varphi}{\tan\alpha} + \frac{C}{W \cdot \sin\alpha} \tag{12-22a}$$

That circle which gives the least factor of safety among the circles analyzed is the critical rupture surface.

In these equations,

M_R = resisting moment about the center O_c of the critical circle
M_D = driving moment about point O_c

$F_R = \tau \cdot A$ = resisting force against sliding
$F_D = W \cdot \sin\alpha$ = driving force promoting sliding
r_c = radius of critical circle
2ϵ = central angle at point O_c
$\tau = \sigma_{n_{eff}} \cdot \tan\varphi + c =$
 = shear strength of slope material
$\sigma_{n_{eff}}$ = effective normal stress on the rupture surface
φ = angle of internal friction of soil
$\tan\varphi$ = coefficient of friction
c = cohesion per unit area of rupture surface
$C = c \cdot A$ = total amount of cohesion in the rupture surface
$A = (\widehat{A_c}) \cdot (1.0) = (\widehat{L}) \cdot (1.0) = (2\epsilon) \cdot (r_c) \cdot (1.0)$ = shear area
$\widehat{L} = (2\epsilon) \cdot r_c$ = length of the rupture surface arc $\widehat{AJA_c}$
W = weight of the rupture wedge ABA_cJA
a = moment arm of weight W with respect to the critical center O_c of the critical circular surface $\widehat{AA_c}$
$N_{eff} \cdot \tan\varphi + C = \tau \cdot A = F_R$ = total resisting shear force against rupture of soil
$N_{eff} = N - U$ = effective normal force
$N = W \cdot \cos\alpha$ = normal component of weight
α = angle between the horizontal and tangent t-t at point J on the rupture curve (Fig. 12-8)
U = uplift pressure (if any)
θ = angle of inclination of face of slope to the horizontal
h = vertical height of slope.

If $\eta = 1.0$, the earth slope system is at equilibrium, i.e., rupture is just, just impending

If $\eta > 1.0$, the earth slope may be considered as stable as concerns pure static calculations

If $\eta < 1.0$, the system should be regarded as unstable

If $\varphi = 0$, then, by Eq. (12-22a),

$$\eta = C/W \cdot \sin\alpha \qquad (12\text{-}23)$$

If $c = 0$, then $\eta = \tan\varphi/\tan\alpha$ \qquad (12-24)

If the factor of safety turns out to be unacceptable, the slope must be redesigned adopting a different angle θ of slope, or benching the slope, or buttressed by means of a retaining wall, or any other engineering means.

It has also been observed that the rupture surfaces in φ and $(\varphi - c)$-soils tend to follow an arc of a logarithmic spiral of the general equation of

$$r = r_i \cdot e^{\pm \omega \, \tan\varphi} \tag{12-25}$$

[See Fig. 12-9, and consult Sections 4-16 and 8-7, and Fig. 11-3a], (Refs. 5, 15-19).

Here
- r = variable radius vector of spiral
- r_i = initial radius vector of spiral
- ω = amplitude (angle between r_i and r)
- φ = angle of internal friction of sand.

In pure c-soils, $\varphi = 0$, the spiral degenerates into a circle ($r = r_i$ = const), and the c-soil rupture takes place along a circularly curved cylindrical rupture surface.

In his analyses of clay slopes, COLLIN (1) used cycloides as the form of rupture surface.

Seepage Force

If the slope is subjected to seepage, the factor-of-safety equation should also include the seepage force D, see Fig. 12-10 (19):

$$D = \gamma_w \cdot i \cdot A_{\mathrm{AMKNA}} \cdot (1.0) \tag{12-26}$$

where
- γ_w = unit weight of water
- i = hydraulic gradient
- A = wet area AMKNA = cross-sectional area of that part of the slope material which is below the uppermost seepage line $\widehat{AMK}$.

The hydrodynamic force D is a body force. It is applied at the centroid (C.G.) of the wet part of the soil mass underneath the uppermost seepage line, and is directed along the tangent drawn to that flow line which passes through the centroid.

For the case where the seepage force D is present, the factor of safety η for the critical circle is calculated as

$$\eta = \frac{M_R}{M_D} = \frac{(N_{\mathrm{eff}} \cdot \tan\varphi + C) \cdot r_c}{\left(F_D + D \cdot \dfrac{r}{r_c}\right) \cdot r_c} \tag{12-27}$$

Fig. 12-9
Logarithmically spiraled rupture surface curves in sand. Author's studies (15-19).
a) Rupture of a slope
b) Groundbreak

a) Seepage force D

b) Partly submerged slice

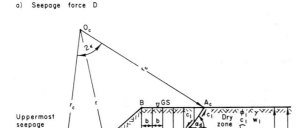
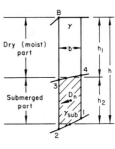

Fig. 12-10
Method of slices

where

r = moment arm of seepage force D with respect to the critical point of rotation O_c, and all other symbols are the same as before.

Because part of the weight of soil is submerged under water, and part of the soil above the uppermost flow line is "dry" (Fig. 12-10), the stability analysis of such a soil system is conveniently performed by the method of slices (19). In this method, the total weight W of the slices is calculated as

$$W = \sum_{1}^{x} (w) = (b)(1.0) \left[\sum_{1}^{d} (\gamma \cdot h_1) + \sum_{1}^{s} (\gamma_{sub} \cdot h_2) \right], \quad (12\text{-}28)$$

where

x = total number of slices

d = number of dry parts of slices

s = number of submerged parts of slices

w = total weight of one slice (includes the submerged part (w_2), and the dry part (w_1), of that slice)

b = width of slice

γ_w = unit weight of water

$\gamma_{sub} = \gamma - \gamma_w = (1-n)(G-1) \cdot \gamma_w$ = unit weight of the submerged material

n = porosity
G = specific gravity of soil particles
h_1 = height of slice above the uppermost flow line
h_2 = height of submerged part of slice
$h = h_1 + h_2$ = total height of a slice.

The resisting moment M_R is then generally indicated as

$$M_R = \left\{ \left[\sum_1^d (w_i \cdot \cos\alpha_d) \right]_{dry} \cdot \tan\varphi_1 + \right.$$

$$+ \left[\sum_1^s (w_1 + w_2) \cdot \cos\alpha_s \right]_{wet} \cdot \tan\varphi_2 +$$

$$\left. + \sum_1^d [c_1 \cdot L_1 \cdot (1.0)] + \sum_1^s [c_2 \cdot L_2 \cdot (1.0)] \right\} \cdot r_c \qquad (12\text{-}29)$$

The driving moment M_D is:

$$M_D = \left[\sum_1^d (w_1 \cdot \sin\alpha_d) + \sum_1^s (w_1 + w_2) \cdot \sin\alpha_s \right] \cdot r_c + \sum_1^s (D) \cdot r$$

$$(12\text{-}30)$$

The factor of safety is

$$\eta = M_R / M_D \qquad (12\text{-}31)$$

In Eqs. (12-29) and (12-30),

α_d = variable angle of slope of the tangent drawn at the midpoint of the base of the dry slices

α_s = same for the submerged slices

$\tan\varphi_1$ = coefficient of friction in the dry part $\widehat{KA_c}$ of the circular rupture surface

$\tan\varphi_2$ = coefficient of friction along the submerged part $\widehat{AJK}$ of the circular rupture surface

c_1 = cohesion in the dry part $\widehat{KA_c}$ of the rupture surface

c_2 = cohesion in the submerged part $\widehat{AJK}$ of the rupture surface

$L_1 = \widehat{KA_c}$ = length of the dry part of the rupture surface curve

$L_2 = \widehat{AJK}$ = length of the submerged part of the rupture surface curve

r_c = length of radius of the critical circle, viz., moment arm of forces acting in the circular rupture surface with respect to point of rotation O_c, and

r = moment arm of the total seepage force D with respect to the point of rotation O_c.

12-9. Stability Analyses of Rock Slopes

Rock slope slides along curvilinear rupture surfaces, such as those which occur in soil or in loose, weathered rock, are almost nonexistent in solid rock. The formation of a curvilinear, or even plane rupture surface in hard, solid rock is very difficult because the solid, segmented rock blocks or slabs would have to be sheared through.

Relative to rock, the sliding of sloped, stratified, broken loose rock would take place over existing inclined planes of separation, viz., planes of discontinuity between beddings of rock strata. Slidings are particularly facilitated if the gouge in the discontinuities is of low shear strength as compared to that of the rock material. However, for a given compressive strength of rock, this condition is valid only up to a certain limiting height of the slope (refer to Section 12-2).

In practice, rock slope stability analyses are performed relative to resistance to sliding of rock blocks and/or slabs on geologically predetermined plane sliding surfaces inclined at various angles of inclination α to the horizontal, angles of inclination θ of the face of a rock slope and the prevailing water regimen in the rock mass (11, 20, 27).

The magnitude of the allowable factor of safety to use in stability analyses of rock slopes depends upon the technological importance of the structure built on or in rock and the physical and mechanical properties of the rock.

12-10. Sliding of a Rock Block on Geologically Predetermined Planar Discontinuity

Figure 12-11 shows a rock block ABJKA separated from the rock massif by a vertical, water-filled tension crack $\overline{JK} = z$ in the upper face of the slope. The block rests on an inclined, geologically predetermined planar discontinuity $\overline{AK}$ filled with a permeable gouge. The block, whose weight is W, is therefore subjected 1) to a horizontal hydrostatic water pressure $H = (1/2) \cdot \gamma_w \cdot z^2$ from the water in the tension crack, and 2) to uplift pressure U. The sliding of the rock block is resisted by the shear strength of the gouge material. The shearing of the gouge and the sliding of the block depends also upon the roughness of the sliding planes (see Section 4).

The total shear resistance T of the gouge in the inclined discontinuity $\overline{AK} = L$ is given as

$$T = N_{\text{eff}} \cdot \tan\varphi + C \qquad (12\text{-}32)$$

$$= (N - U) \cdot \tan\varphi + C \qquad (12\text{-}33)$$

Herein

$N_{\text{eff}} = N - U$ = effective normal pressure component (12-34)
$N = W \cos\alpha$ = normal component of the weight W of the block to the sliding surface
$W = (1/2) \cdot \gamma \cdot [(b + B) \cdot h - (b \cdot h) - (X \cdot z)]$

$$= (1/2) \cdot \gamma \left[\frac{1 - (z/h)^2}{\tan\alpha} - \frac{1}{\tan\theta} \right] \qquad (12\text{-}35)$$

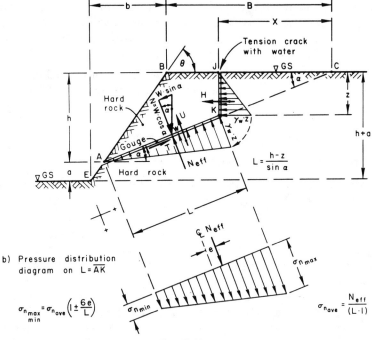

Fig. 12-11
Rock block on an inclined planar discontinuity filled with a permeable gouge

= weight of the sliding block
α = angle of inclination of the sliding surface to the horizontal

$$U = (1/2) \cdot \gamma_w \cdot z \cdot (h - z)/\sin\alpha \qquad (12\text{-}36)$$

= uplift (neutral) pressure
φ = angle of friction
$\tan\varphi$ = coefficient of friction

$$C = c \cdot A = c \cdot L \cdot (1.0)$$

= total cohesive resistance of the gouge in the discontinuity
c = cohesion of gouge per unit area; cohesion of gouge is a function of its moisture content. The more moisture, the more plastic the gouge is; that is, the resistance to shear of the gouge decreases with increase in its moisture content
A = total shear area of gouge under the sliding block of rock
L = length of discontinuity
γ = unit weight of sliding rock material
b, B, X, h = slope dimensions as shown in Fig. 12-11
z = depth of tension crack
θ = angle of face of slope to the horizontal
γ_w = unit weight of water

The system's resultant resisting force F_R is:

$$\begin{aligned} F_R &= T - (H \cdot \sin\alpha) \cdot \tan\varphi = \\ &= N_{\text{eff}} \cdot \tan\varphi + C) - H \cdot \sin\alpha \cdot \tan\varphi \\ &= [(N - U) - (H \cdot \sin\alpha)] \cdot \tan\varphi + C \\ &= (W \cdot \cos\alpha - U - H \cdot \sin\alpha) \cdot \tan\varphi + C \qquad (12\text{-}37) \end{aligned}$$

The system's resultant driving force F_D is:

$$F_D = W \cdot \sin\alpha + H \cdot \cos\alpha \qquad (12\text{-}38)$$

Factor of safety η against sliding:

$$\eta = \frac{(W \cdot \cos\alpha - H \cdot \sin\alpha - U) \cdot \tan\varphi + C}{W \cdot \sin\alpha + H \cos\alpha} \qquad (12\text{-}39)$$

If $\eta = 1.0$, the rock-slope system is at equilibrium (or sliding is just, just impending)
If $\eta < 1.0$, the system is unstable
If $\eta > 1.0$, the system is stable

For temporary rock slopes that are required to remain stable only for a short period of time, a factor of safety of $\eta = 1.3$ may be generally adequate. For rock slopes of structures of important technological significance, intended to last for a long period of time, or for slopes made adjacent to existing important installations, the allowable factor of safety against rupture and sliding is usually taken as $\eta \geq 1.5$.

If the calculated factor of safety η turns out to be unacceptable, the slope face angle θ to the horizontal should be decreased, or the slope should be benched, or the rock block should be anchored to the mass of the rock below the discontinuity. A purposely designed drainage facility for the wet rock-slope system may reduce the hydrostatic pressure H on the block, as well as the uplift force U.

The factor-of-safety equation (12-39) can be specialized for various conditions, such as, for example:

for $H = 0$, $U = 0$ (dry, or completely drained slope)
for $H \neq 0$, $U = 0$ (water in tension crack only)
for $\varphi \neq 0$ and $c = 0$ (dry or wet condition)
for $\varphi = 0$ and $c \neq 0$ (dry or wet condition)

The η-equation (12-39) also indicates that the value of the factor of safety η can be increased in several ways:

1. by effective drainage of the slope. This would eliminate the hydrostatic pressure H and the uplift pressure U, that is, letting $H \to 0$ and $U \to 0$;
2. by grouting of the jointed rock; grouting increases the strength of the rock mass, and excludes water from joints and cracks;
3. by increasing the system's resisting forces against sliding by tying or anchoring the sliding rock block or slab to the rock mass below the discontinuity. The rock anchoring — or reinforcement — may be accomplished either by means of tensioned anchor rods or by tensioned cables (see Section 11).
4. The increase of the value of the factor of safety may also be achieved by reducing the weight W of the rock block or slab (flattening, benching, decreasing the height of the slope). Observe that a decrease in weight W brings about a decrease in the normal force N, and thus a decrease in the system's resisting force F_R. On the other hand, Eq. (12-39) shows that a decrease in weight W simultaneously decreases the system's driving force F_D. Hence, the effect of decreasing W on the factor safety η should be studied very carefully.
5. The reduction of externally applied loads (from structures, foundations, where applicable) may also result in a higher factor of safety.

Generally, rock slopes should be designed for a load that will not exceed the bearing capacity of the rock.

Relative to the use of cohesion in earth and rock slope stability calculations in earth pressure computations, and in performing stability analyses on retaining walls, it is apropos here to say that in practice the cohesion is often omitted (19). This is so done because upon the increase in soil, viz., gouge, moisture content, cohesion may decrease considerably, thus endangering the earthworks and rock-slopes. Cohesion in rock slope design may be used only when the discontinuity planes, viz., sliding surfaces, are bonded, and when a definite shear stress is needed to bring about failure of the bonding cement at zero effective normal stress.

It should also be mentioned that the resistance to sliding down on a gouge depends not only upon the nature of the gouge and its consistency, but also upon the degree of separation of the joints, the form of the fault and the roughness of the walls of the discontinuities. Thus, upon sliding, not only must the cohesion c of the gouge be considered (if any), but the entire resistance of the gouge-fault system should be taken into account — a very complex problem, indeed.

Because rock slopes in nature are of various forms of geometry and stratification, it is practically impossible to give any overall stability computation recipe or a method of procedure for all these cases, nor is it possible to comprehend and treat analytically all of the numerous parameters and variables imposed by nature. Therefore, stability analyses should really be performed individually for each kind of slope and stratification system of the rock on hand.

The factor of safety η against sliding for the rock-block system as shown in Fig. 12-11 can also be determined graphically from the force polygons in Figs. 12-12a, b, and c.

a) If the force polygon closes, Fig. 12-12a, then the rock-block force system is at equilibrium. Here the necessary cohesive force C_{nec} is equal to the available cohesive force C, i.e., $C_{nec} = C$, and the factor of safety against sliding is

$$\eta = [N_{eff} \cdot \tan\varphi + C]/F_D = 1.00$$

b) If the available cohesive force C is less than the necessary one for equilibrium, i.e., if $C < C_{nec}$, Fig. 12-12b, the force polygon does not close, and the system is unstable:

$$\eta = [N_{eff} \cdot \tan\varphi + C]/F_D < 1.00$$

c) If the available cohesive force C is greater than the necessary one for equilibrium, i.e., if $C > C_{nec}$, Fig. 12-12c, then

$\eta = [N_{eff} \cdot \tan\varphi + C]/F_D > 1.00$, and the system is regarded as stable.

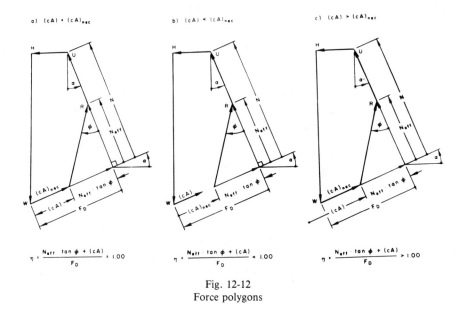

Fig. 12-12
Force polygons

12-11. Anchoring of Rock Slopes

Unstable rock slopes may be stabilized by means of rock bolts and rock anchors fixed into the rock mass below the discontinuity.

The solution to the anchoring problem depends upon the following factors:

1. the most favorable position (angle of inclination ω) of the anchor
2. the necessary capacity of the anchor P_{min}, and
3. the effective level of tensioning of the anchor.

For a dry rock slope (Fig. 12-13) the angle of inclination ω can be determined analytically by the method of extremes.

Expressing from the η-equation (12-39) the anchor force P for $\eta = 1.0$, forming its first derivative $dP/d\omega$ with respect to angle ω, and setting this derivative equal to zero ($dP/d\omega = 0$), obtain the angle $\omega = 90° - \varphi$ at which the anchor force P is at a minimum. Substituting the so found angle ω into the general P-equation, the magnitude of P_{min}, viz., the capacity of the anchor, is computed. Whether the P-function has a minimum, this must be ascertained by forming $d^2P/d\omega^2$. If $d^2P/d\omega^2 > 0$, the P-function has a minimum (P_{min}).

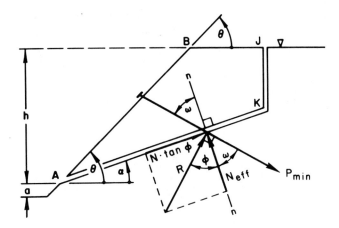

Fig. 12-13
Angle of inclination ω of anchor P_{min}

The orientation of the least anchor force P_{min} is perpendicular to the force vector R resulting from the normal force N_{eff} and the friction force $N_{eff} \cdot \tan\varphi$ (Fig. 12-13).

The symbol P (or P_{min}) is here understood to be the sum of *all forces in all anchors* necessary for rock block or slab reinforcement at force equilibrium ($\eta = 1.0$).

At large angles of inclination ω of the anchor, the necessary resisting force against sliding becomes relatively very large; the anchor bolt is stressed in bending, and the anchor bolt bears against the rock wall of the anchor drill hole subjecting the rock to compression. Most rocks are sensitive to such kind of loading.

According to rock engineering practice, all rock anchors are post-tensioned. Post-tensioning is done gradually to the full load capacity of the anchors as soon as possible after their installation in order to prevent from the very beginning even the least movement of the rock. Without tensioning, the anchors will deploy their tensile force only when the rock blocks or slabs have already begun to slide. In such a case, only the mobilized sliding friction on the sliding surface will be effective. The sliding friction is considerably less than the static friction at rest.

By means of measuring the tension in the steel of the anchor, and the deformations of the compressed rock (see Section 11), it can be ascertained whether the reinforcement is or is not satisfactory.

For a sliding rock block system including H, U and the anchor force P, the system's overall factor of safety η calculates as

$$\eta = \frac{F_R}{F_D} = \frac{[W \cdot \cos\alpha - H \cdot \sin\alpha - U + P \cdot \cos\omega] \cdot \tan\varphi + C}{W \cdot \sin\alpha + H \cdot \cos\alpha - P \cdot \sin\omega} \quad (12\text{-}40)$$

If in Eq. (12-40) $C = 0$; $H = 0$; $U = 0$, and $P = 0$, then

$$\eta = \frac{\tan\varphi}{\tan\alpha} \quad (12\text{-}41)$$

For equilibrium ($\eta = 1.0$),

$$\alpha = \varphi \quad (12\text{-}42)$$

The expression $\alpha = \varphi$ is the criterion for failure impending, and the rock block has to be anchored.

Relative to calculation of the system's overall factor of safety by Eq. (12-40), there are some theoretical aspects involved. For a discussion of those, refer to HOEK and BRAY (Reference 12, Appendix 3, pp. 396—398). Because in nature the anchor tension P, the frictional resistance due to φ, the cohesive strength c, and the various water pressures cannot be assumed to be all fully mobilized simultaneously, P. LONDE suggests (see HOEK and BRAY, Ref. 12) that

"instead of using a single factor of safety for defining the stability of the slope, different factors of safety should be used, depending upon the degree of confidence which the designer has in the particular parameters being considered. High factors of safety can be applied to ill-defined parameters (such as water pressures and cohesive strengths), while low factors of safety can be used for those quantities (such as the weight of a wedge) which are known with a greater degree of precision".

For a typical problem, LONDE suggests:

η_c = 1.5 for cohesive strengths (c)
η_f = 1.2 for frictional strengths (φ)
η_U = 2.0 for water pressures
η_W = for weights and forces."

With these ideas in mind, the stability equation for the sliding block on an inclined planar discontinuity is written as

$$(W \cdot \cos\alpha - \eta_U \cdot H \cdot \sin\alpha - \eta_U \cdot U + P \cdot \sin\omega)\frac{\tan\varphi}{\eta_f} + \frac{C}{\eta_c}$$

$$= W \cdot \sin\alpha + \eta_U \cdot H \cdot \cos\alpha - P \cdot \cos\omega \quad (12\text{-}43)$$

Solution of Eq. (12-43) for P renders the anchor tension necessary for anchoring the rock block to the rock mass to prevent the block from sliding down the incline.

For three-dimensional analyses of rock slides, refer to References 10, 12, 35—37.

12-12. Earthquake Effects

General Notes

The sense of security man gets from the feel of the earth's solid crust is ill-founded. It should be remembered that terra firma is nothing else but a changeable crust, about 120 km thin, and the uneasy innards of the globe which this crust covers never stop rumbling and stirring. In other words, earthquakes are possible in almost any region, but the range of probability varies enormously. When the probability exceeds a certain critical value, precautionary measures in design should be taken.

One should also be aware that much more knowledge is needed for a better understanding of the performance of man-made structures such as dams, hydraulic structures, hydropower plants, tunnels and other kinds of underground openings, electric power transmission towers and other structures subject to earthquake action. Slopes of cuts and fills, waterfront structures, bridge piers, water mains, oil pipelines and earth retaining structures should also be designed and constructed to cope with seismic forces and earthquake oscillations.

Seismic Forces

In regions of seismic activity the stability calculations of slopes of earthworks should also include seismic forces, because they reduce the margin of safety, or may even bring about the collapse of a structure. Under certain conditions, both horizontal and vertical seismic forces must be considered. The magnitude of a horizontal seismic force F_s is calculated as

$$F_s = m \cdot a = \left(\frac{W}{g}\right) \cdot a = \left(\frac{W}{g}\right) \cdot \alpha_e g = \alpha_e \cdot W \qquad (12\text{-}44)$$

where m = mass of structural body being studied
$\qquad a$ = seismic acceleration, such as the acceleration of the earthquake wave
$\qquad W$ = weight of the structural body whose mass is m
$\qquad g$ = 981 cm/s = accelerations of gravity, and
$\qquad \alpha_e$ = seismic (earthquake) factor

In the United States, the value of the horizontal seismic factor for α_e is taken as 0.75, or $a = (0.75)$ g for rock foundations and $a = (0.10)$ g on sand and other soil foundations (6). If a dam is built near a live or active fault, or if it is founded on loose soil material, higher values of horizontal acceleration are in order subject to aseismic building code regulations. It is interesting to note that structures in Japan which were designed for $a = (0.10)$g have survived the most severe earthquakes. The Tokyo Bridge Building Code requires a vertical acceleration of $a = \left(\dfrac{1}{6}\right) \cdot g$ coupled with a horizontal acceleration of $a = \left(\dfrac{1}{3}\right) \cdot g$. In California the horizontal acceleration a is calculated with a seismic factor of $\alpha_e = 0.25$, i.e., $a = (0.25)$ g, (7). For the design of the San Francisco-Oakland Bridge, engineers reckoned on a horizontal acceleration of the soil of $a = (0.10) \cdot g$ with a period of oscillation of 1.5 seconds, corresponding to an oscillation of 56 mm.

Because of seismic vibration, the shear strength of the soil and joints of rock under foundations, around tunnels and in slopes of cuts and fills has frequently been exhausted. Also, too steep slopes, natural and/or man made, may easily slide down because of vibration on roads induced by all kinds of rolling stock.

In earthquake-prone regions, the area of a contemplated engineering structure should be subjected to very careful geologic investigation, particularly in respect to fault zones. With reference to a possible vibration of a structure brought about by earthquakes, or other kinds of tremors and shocks, in no instance should a structure be constructed across live faults or in the immediate proximity of such faults.

Although in many regions of the world vertical earthquake acceleration in stability analyses of structures is usually neglected, and regardless of the casual figures for seismic effects as cited above, it is imperative that in regions of seismic activity the design of any structure should meet all of the seismic requirements prescribed by the corresponding building codes.

Earthquake Effect on Earth Retaining Walls

Based on its studies, the Tennessee Valley Authority (34) reports that the weight W of COULOMB's rupture wedge of soil and the seismic force F_s combine into a resultant force W_s (Fig. 12-14).

The length of the active earth pressure vector E_a in the force polygons indicates clearly that the seismic force $F_s = m \cdot a = \left(\dfrac{W}{g}\right) \cdot a$ increases the magnitude of the active earth pressure E_a considerably. Here m = mass; a = $\alpha \cdot g$ = earthquake acceleration; α_e = coefficient of earthquake acceleration; $\alpha_e W$ = equivalent static force; g = 981 cm/s is the acceleration of gravity; ϱ = angle of rupture; φ_1 = angle of wall friction; F_v = vertical seismic force, and a_v = vertical acceleration.

In the interest of public safety, and to avoid damage to structures by resonance, special building codes for aseismic structural design are put into effect. Such codes regulate the height of the buildings, seismic factors to use, kind of static systems of structures, fire protection, and other safety measures.

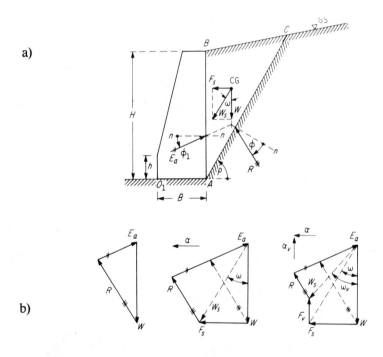

Fig. 12-14
Effect of seismic force on the magnitude of active earth pressure.
a) Earth retaining wall
b) Force polygons

The seismic force F_s is applied at the center of gravity (C.G.) of the sliding-impending block of rock (Fig. 12-15). This seismic force increases the driving force.

The factor of safety η for an unreinforced rock-block system is written as

$$\eta = \frac{F_R}{F_D} = \frac{(W \cdot \cos\alpha - F_s \cdot \sin\alpha - H \cdot \sin\alpha - U) \cdot \tan\varphi + C}{W \cdot \sin\alpha + F_s \cdot \cos\alpha + H \cdot \cos\alpha} \quad (12\text{-}45)$$

If $\eta \leq 1.0$, the slope must be redesigned, or anchors should be used.

Before stability calculations with respect to seismic influences from blasting, it is first necessary to locate the dangerous planes of separation, viz., discontinuities of the rock in the area concerned. Then the shear strength τ between the planes of separation and the magnitude of seismic acceleration should be determined.

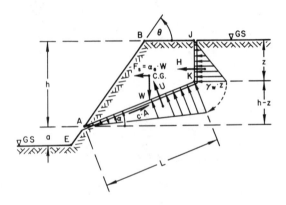

Fig. 12-15
Position of horizontal seismic force F_s.

12-13. Stabilization of Rock Slopes by Means of "Reticulated Root Piling."

Rock and soil slope stabilization against rupture can also be accomplished by means of the so-called "Reticulated Root Piles" (Italian: pali radici). A reticulated root pile system is a three-dimensional lattice structure — a cluster of a closely spaced network of mini-piles, called root piles. The root piles correspond to the steel reinforcement.

The term "Root Piles" associated with its widely used method of rock reinforcement and stabilization is a registered trademark, and the method itself has been patented by the Naples-based Italian firm FONDEDILE, S.p.A. (14, 21-26).

The basic element of the reticulated root pile structure is the root pile — an example of the "mini" or "micro" pile. It is a small diameter (8 cm and larger) cast-in-place reinforced concrete pile.

In the cluster, root piles are installed in the ground at an angle one against each other, and are grouted into boreholes under a high pressure.

According to the opinion of Dr. F. Lizzi, Technical Manager of FONDEDILE (25), the use of rock anchors for landslide and rockslide control should be considered with caution for many reasons, because of:

1. "The potential for introducing concentrated stresses in a mass which often has already been seriously disturbed by blasting or natural fractures. These stresses could worsen rather than improve the situation."
2. "The excessive length of the ties to reach stable anchoring strata (if any exists)."
3. "The risk of the anchor becoming completely useless as a consequence of relaxation and/or corrosion."
4. "The necessity of heavy connecting structures on the surface in order to distribute the stresses imposed by the anchors; in many cases these structures are environmentally unacceptable."

The so installed cluster of root piles draws into active work practically the entire thickness of the rock mass to be stabilized.

At their tops, the root piles are connected by relatively thin connecting concrete cap-beams.

The cluster of the reticulated root piles takes full advantage of the in-situ existing soil, increasing its natural compressive strength, and more important, such a pile system improves the shear strength of the soil. Because of its high skin friction, the root pile forms a tight bond with the soil into which it is cast. The soil-reticulated root pile system performs as an integral unit. It is assumed here that the stresses transferred from a structure above act on the entire lattice structure and are completely distributed to the root piles and to the soil.

One important advantage of stabilizing the ground by reinforcing it by means of root-piles is that there is no need for any prior excavation. Reticulated root pile structures can be installed in any type of soil (dense, loose, permeable, impermeable) or rock. It is preferable to use a large number of small-diameter root-piles rather than a small number of large-diameter piles. Field and model test experiments with root piles are reported by Lizzi in References 24 to 26.

Reticulated root piles can be used as gravity retaining walls; for encompassing and stabilizing excavation walls; for soil and rock slope

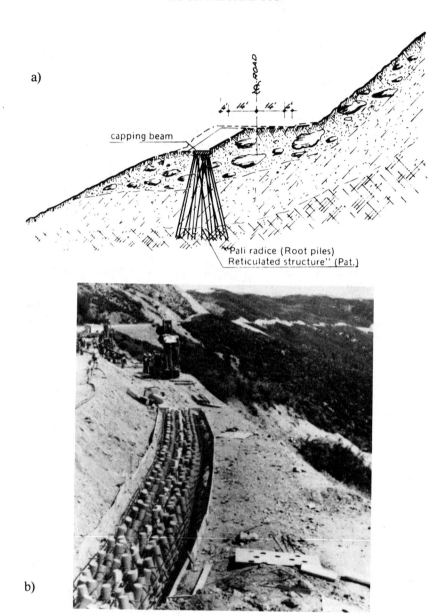

Fig. 12-16
Stabilization of landslide on the Mendocino National Forest Highway 7 in California
(Refs. 24 and 25)
a) Typical cross section
b) Top view of reticulated pile structure.
(Photo courtesy of Dr. F. LIZZI, FONDEDILE: Naples, Italy)

protection against sliding; for underpinning of foundations of heavy structures; bridge piers; monuments; for stabilizing landslides (Refs. 21—26); for reinforcing of steep, high overhanging cliffs; tunnels, and for other possible applications.

For example, in 1975—1976 a reticulated root pile structure was constructed in Jackson, Mississippi to support and stabilize the abutment and pier foundation of a bridge on the Interstate Route 55 (24, 25). In 1977, a reticulated root pile structure was completed in the Mendocino National Forest in Glenn County, California to stabilize a landslide area along the Forest Highway 7 (Fig. 12-16a and b, Refs. 24 and 25).

A very extensive application of the reticulated root piling method has been made in Brazil to prevent the occurring of several dangerous landslides along the new road Santo-São Paulo (Rodovia dos Imigrantes) (LIZZI, 24, 25), see Figs. 12-17a and b.

The various applications of reticulated root piles in tunnel engineering are described in References 4 and 21.

12-14. Methods of Remedy Against Rock Slides

To cope with rock slides in fresh cuts, and to correct rock slides, the following methods of remedy are applied.:

1. installation of appropriate drainage systems; drainage of the hillside;
2. injection and grouting of cracks and joints in fractured rock, and guniting and shotcreting of rock slope surfaces. Grout not only adds cohesive bond, but also excludes water;
3. flushing clean of all joints and discontinuities of gouge and clay films to increase contact friction on rock surfaces;
4. increase of frictional resistance in the sliding surface by electrochemical stabilization of soil and/or gouge;
5. unloading of the head of the rock slope by removing some rock material which tends to cause failure;
6. reducing the slope angle;
7. providing for some external support to hold back the toe of the slope by means of a slope-toe counterweight; this will help to halt the down-movement of rock;
8. wire mesh;
9. gabions;
10. freezing of the ground (as a temporary measure);

and a combination thereof.

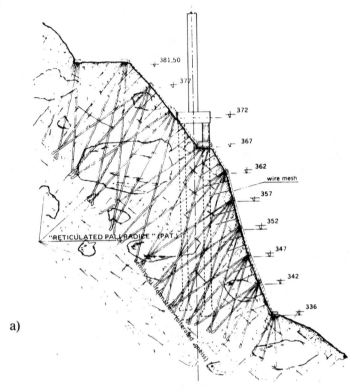

Fig. 12-17
Rodovia dos Imigrantes.
a) Stabilized rock slope with root piles
b) Root piles installed in altered gneiss (Refs. 24 and 25).
(Photo courtesy of Dr. F. Lizzi, FONDEDILE: Naples, Italy)

The method to choose for rock slide control depends upon the understanding of *what* sliding is; *how much* material is involved; *where* and *how* the material is distributed, and *why* the material slides down.

Rockfall Control by Means of Wire Mesh

To prevent the loose rock fragments, boulders, and rockfall debris from plunging down onto a road or railroad, a wire mesh can be draped over the face of a steep rock slope. Figure 12-18 shows the installation of a wire mesh over a high, steep rock slope along the Flåm Railroad in Western Norway.

Fig. 12-18
A wire mesh is being draped over a face of a high, steep rock slope to prevent rockfall along a critical section of the Flåm Railroad in Western Norway.
(Photo by ARNE HOFSBETH. Photo courtesy of Mr. H. HARTMARK of the Main Administration, Norwegian State Railroads, Oslo)

Rock Slope Treatment by Gabions and Wire Mesh

One of the rock slide preventions and rock slope face protection methods is by means of gabions combined with drapes of wire mesh.

A gabion is a long, hollow, strong wire basket or a large cage of wickerwork or strap iron filled with stones. Gabions are used as revetment in highway, railway, hydraulic structures and mining engineering. Gabions are flexible and permeable.

Figure 12-19 shows a gabion and wire mesh treatment along Interstate Route 70 east of Glenwood Springs, Colorado.

Fig. 12-19
Gabion and wire mesh treatment of a rock slope along Interstate Route 70 east of Glenwood Springs, Colorado.
(Photo courtesy of Colorado Department of Highways)

REFERENCES

1. COLLIN, A., 1846, *Landslides in Clays*. Translated from the French into English by W. R. Schriever, 1956. Toronto: University of Toronto Press.
2. FELLENIUS, W., 1936, "Calculation of Stability of Earth Dams," *Transactions, 2nd Congress on Large Dams*, held 1936 in Washington, D.C. Washington, D.C.: U.S. Government Printing Office, 1938, Vol. 4, pp. 445—462.
3. FELLENIUS, W., 1947, *Erdstatische Berechnungen mit Reibung und Kohäsion (Adhäsion) und unter Annahme Kreiszylindrischer Gleitflächen* (3rd edition). Berlin: Wilhelm Ernst und Sohn, pp. 7-29.
4. Fondedile (no date), *Construction of Tunnels (for Roads, Rails, etc.) in the Underground of Built Up Areas* (23 pages).
5. FRÖHLICH, O. K., 1953, "The Factor of Safety with Respect to Sliding of Soil Along the Arc of a Logarithmic Spiral." *Proceedings of the 3rd International Conference for Soil Mechanics and Foundation Engineering*, held in 1953 at Zürich, Switzerland. Vol. 2, 1953, pp. 230—233.
6. HARRIS, C. M., and C. E. CREDE (editors), 1961, *Shock and Vibrations Handbook*. New York, N.Y.: McGraw-Hill Book Company. Three volumes, 1961.
7. HECK, N. H., 1936, *Earthquakes*. Princeton, New Jersey: Princeton University Press, p. 202.
8. Highway Research Board, 1958, E. B. Eckel, Editor, *Landslides and Engineering Practice*. Washington, D.C.: Highway Research Board, Special Report 29. Publication 544.
9. Highway Research Board, 1963, *Stability of Rock Slopes*. Washington, D.C.: *Highway Research Record No. 17*, National Research Council (52 pages).
10. HOEK, E., 1973, "Methods for the Rapid Assessment of the Stability of Three-Dimensional Rock Slopes." *Quarterly Journal Engineering Geology*. Vol. 6, No. 3, 1973.
11. HOEK, E., 1976, "Rock Slopes." *Rock Engineering for Foundations and Slopes*. New York, N.Y.: American Society of Civil Engineers, Vol. 2, 1977, pp. 157—171.
12. HOEK, E., and J. W. BRAY, 1977, *Rock Slope Engineering* (2nd ed.). London: The Institution of Mining and Metallurgy.
13. JAEGER, J. C., 1971, "Friction of Rocks and Stability of Rock Slopes." Eleventh Rankine Lecture, *Géotechnique* (London), Vol. 21, No. 2, pp. 97—134.
14. JANBU, N., 1957, "Earth Pressures and Bearing Capacity Calculations by Generalized Procedures of Slices." Proceedings of the 4th International Conference on Soil Mechanics and Foundation Engineering, held in Paris, 1957. Paris: Dunod. Vol. 2, 1957, pp. 207—212.
15. JUMIKIS, A. R., 1956, "Rupture Surface in Sand Under Oblique Loads." *Proceedings ASCE, Journal of the Soil Mechanics and Foundations Division*, Vol. 82, No. SM1, January, 1956, Paper 861, pp. SM1 to SM26.
16. JUMIKIS A. R., 1961, "The Shape of Rupture Surface in Dry Sand." *Proceedings, 5th International Conference on Soil Mechanics and Foundation Engineering*, held July 7—22, 1961 in Paris. Paris: Dunod, 1961, Vol. 1, paper 3A/23, pp. 693—698.
17. JUMIKIS, A. R., 1965, *Stability Analyses of Soil-Foundation Systems* (Based on Logarithmically Spiralled Rupture Curves). New Brunswick, New Jersey: College of Engineering, Rutgers — The State University, Engineering Research Publication No. 44, 1965. Contains 13 tables for the construction of logarithmically spiralled rupture surface curves (55 pages).

18. JUMIKIS, A. R., 1969, *Theoretical Soil Mechanics*. New York, N.Y.: Van Nostrand Reinhold Company, pp. 271—285.
19. JUMIKIS, A. R., 1982, *Soil Mechanics*. Melbourne, Florida: R. E. Krieger Publishing Co., Inc.
20. LADD, G. E., 1935, *"Landslides, Subsidences and Rock Falls for the Railroad Engineer."* Proceedings of the American Railway Engineering Association, Vol. 36, 1935, p. 1091—1162.
21. LIZZI, F., 1971, *Special Patented Systems of Underpinning and, More Generally, Subsoil Strengthening by means of "Pali Radice" (Root Piles with Special Reference to Problems Arising from the Construction of Subways in Built-up Areas)*. Naples: Fondedile S.p.A. (A lecture delivered at the University of Illinois, Champaign; Massachusetts Institute of Technology, Cambridge, Mass.; U.S. Bureau of Reclamation, Denver, Colorado, and Club of the Civil Engineers of Vancouver, B.C., Canada, April—May 1971).
22. LIZZI, F., 1973, *Less "Pieux Racines FONDEDILE" et les "Réseaux de Pieux-Racines."* Rueil Malmaison, France: Fondedile, France. Conference held May 4, 1973, at l'Ecole Centrale des Arts et Manufactures at Chatenay-Malabri.
23. LIZZI, F., 1976, *Static Restoration of Historic Monuments, Using Fondedile Techniques*. Naples: Fondedile, May 31, 1976 (6 pages).
24. LIZZI, F., 1977, "Practical Engineering in Structurally Complex Formations (The "In-Situ Reinforced Earth.")". *Proceedings of the International Symposium on The Geotechnics of Structurally Complex Formations*, held July 1977 at Capri. Associazione Geotechnica Italiana, Vol. 1, pp. 327—333.
25. LIZZI, F., 1978, "Reticulated Root Piles — to Correct Landslides." Preprint 3370 of paper for the ASCE Annual Convention and Exposition, held October 16—20, 1978 at Chicago (25 pages).
26. LIZZI, F., and G. CARNEVALE, 1979, "Reticulated Root Pile Structures for In-Situ Soil Strengthening. Theoretical Aspects and Model Tests." *Proceedings of the International Conference on Soil Reinforcement, held March 20—22, 1979, in Paris, pp. 317—324*.
27. LONDE, P., 1973, "The Role of Rock Mechanics in the Reconnaissance of Rock Foundations, Water Seepage in Rock Slopes and in the Analysis of the Stability of Rock Slopes." *Quarterly Journal of Engineering Geology*, Vol. 5, 1973, pp. 57—127.
28. MENCL, V., 1966, "Mechanics of Landslides with Non-Circular Slip Surfaces with Special Reference to the Vaiont Slide." *Géotechnique* (London), Vol. 16, No. 4, 1966, pp. 329—337.
29. MOHR, O., 1900, "Welche Umstände bedingen die Elastizitätsgrenze und den Bruch eines Materials?" in *Zeitschrift des Vereins Deutscher Ingenieure,* Vol. 44, p. 1524.
30. PETTERSON, K. E., 1955, "The Early History of Circular Sliding Surfaces." *Géotechnique* (London): The Institution of Civil Engineers. Vol. 5, 1955, pp. 275—296.
31. RITCHIE, A. M., 1963, "Evaluation of Rockfall and its Control." *Highway Research Record* No. 17, Highway Research Board, May, 1963, pp. 13—28.
32. Statens Järnvägars Geotekniska Kommision, 1922. *Slutbetänkande*. Stockholm: 1922, p. 57.
33. TERZAGHI, K., 1950, "Mechanics of Landslides." *Application of Geology to Engineering Practice* (Berkey Volume). Geological Society of America, pp. 83—123.
34. U.S. Tennessee Valley Authority, 1951, *The Kentucky Project*. Knoxville, Tennessee. Technical Report No. 13, 1951.

35. WITTKE, W., 1964, "Ein rechnerischer Weg zur Ermittlung der Standsicherheit von Böschungen in Fels mit durchgehenden, ebenen Absonderungsflächen." Wien und New York, N.Y.: Springer-Verlag, *Grundfragen auf dem Gebiete der Geomechanik — Principles in the Field of Geomechanics*. Rock Mechanics and Engineering Geology — Supplementum I; 14th Symposium of the Austrian Regional Group (i.f.) of the International Society for Rock Mechanics held Sept. 27-28, 1963, at Salzburg; pp. 103—129.
36. WITTKE, W., 1965, "Verfahren zur Berechnung der Standsicherheit belasteter und unbelasteter Felsböschungen." Wien und New York, N.Y.: Springer-Verlag. Supplementum II, 15th Symposium of the Austrian Regional Group (i.f.) of the International Society for Rock Mechanics held September 24—25, 1964 at Salzburg; pp. 52.
37. WITTKE, W. et al., 1972, "Three Dimensional Calculation of the Stability of Caverns, Tunnels, Slopes and Foundations in Anisotropic Jointed Rock by Means of the Finite Element Method," in *Deutsche Beiträge zur Geotechnik*. Essen: Deutsche Gesellschaft für Erd- und Grundbau, Vol. 1, No. 1.
38. ZARUBA, Q., and V. MENCL, 1969, *Landslides and their Control*. Amsterdam: Elsevier.
39. ZISCHINSKY, U., 1966, "On the Deformation of High Slopes." *Proceedings* (First) *Congress of the International Society for Rock Mechanics*, held 1966 at Lisbon, Portugal. Vol. 2, 1966, p. 179.

OTHER RELATED REFERENCES

ASCE — *Rock Engineering for Foundations and Slopes*. New York, N.Y.: Vol. 1, 1976; Vol. 2, 1977.

CEDERGREN, H. R., 1967, *Seepage, Drainage and Flow Nets*. New York, N.Y.: John Wiley and Sons.

GOODMAN, R. E., 1976, *Methods of Geological Engineering in Discontinuous Rocks*. St. Paul, Minnesota: West Publishing Co., 1976 (476 pages).

HANDIN, J., and STEARNS, D. W., 1964, "Sliding Friction of Rock," *Transactions of the American Geophysical Union*. Vol. 45, 1964, p. 103.

HEUZE, F. E., and GOODMAN, R. E., 1971, "Three-Dimensional Approach for the Design of Cuts in Jointed Rock." Proceedings, 13th Symposium on Rock Mechanics, Urbana, Illinois, 1971.

HOCKING, G., 1976, "A Method for Distinguishing Between Single and Double Plane Sliding of Tetrahedral Wedges." *Int'l. Journal for Rock Mechanics and Mining Sciences*. Vol. 13, 1976, pp. 225—226.

HOEK, E., and A. GRALEWSKA, 1969, *A Rock Mechanics Information Service*, London: Imperial College Rock Mechanics Report No. 9, May, 1969.

HOEK, E., and E. T. BROWN, (Circulation Draft, 1978), *Underground Excavation Engineering*. To be published by the Institute of Mining and Metallurgy, London.

JOHN, K. W., 1968, "Graphical Stability Analysis of Slopes in Jointed Rock," *Proceedings ASCE*, Soil Mechanics and Foundations Divisions, Vol. 94, No. SM2, pp. 497—526.

KOMARNITSKII, N.I., 1968, *Zones and Planes of Weakness in Rocks and Slope Stability*. New York, N.Y.: Consultants Bureau.

LADANYI, B., and G. ARCHAMBAULT, 1970, "Simulation of Shear Behavior of a Jointed Rock Mass," *Proceedings 11th Symposium on Rock Mechanics*, New York pp. 105—125.

LEGGETT, R. F., 1962, *Geology and Engineering*. New York, N.Y.: McGraw-Hill Book Co.

LONDE, P., VIGIER, G., and R. VORMERINGER, 1969, "Stability of Rock Slopes, a Three-Dimensional Study." *Proceedings ASCE*, Vol. 95, No. SM 1, Paper 6363, January, 1969, pp. 235—262.

NEWMARK, N. M. and E. ROSENBLUETH, 1971, *Fundamentals of Earthquake Engineering*. Englewood Cliffs, New Jersey: Prentice-Hall, Inc.

PETZNY, H., 1967, "Über die Stabilität von Felshängen," in *Felsbau in Theorie und Praxis — Rock Engineering in Theory and Practice*, L. MÜLLER and C. FAIRHURST *(editors)*. Wien: Springer-Verlag, pp. 37—57.

SCHROTER, G. A., and R. O. MAURSETH, 1960, "Hillside Stability." *Civil Engineering*, June, 1960.

SHARPE, C. F. S., 1938, *Landslides and Related Phenomena*. New York, N.Y.: Columbia University Press.

Statens Järnvägars Geotekniska Kommission, 1922, *Slutbetänkande 1914—1922. Stockholm, 1922.*

WOOD, A. M., and N. E. V. VINER-BRADY, 1955, "Folkstone Warren Landslip Investigation, 1948—1950", "Folkstone Warren Landslip Remedial Measures, 1948—1950". London: Institute of Civil Engineers. *Proceedings,* Vol. 4.

APPENDICES

APPENDIX 1

GREEK ALPHABET

A	α	Alpha	N	ν	Nu
B	β	Beta	Ξ	ξ	Xi
Γ	γ	Gamma	O	o	Omicron
Δ	δ	Delta	Π	π	Pi
E	ε	Epsilon	P	ϱ	Rho
Z	ζ	Zeta	Σ	$\sigma\varsigma$	Sigma
H	η	Eta	T	τ	Tau
Θ	ϑ	Theta	Y	υ	Upsilon
I	ι	Iota	Φ	φ	Phi
K	$\varkappa$	Kappa	X	χ	Chi
Λ	λ	Lambda	Ψ	ψ	Psi
M	μ	Mu	Ω	ω	Omega

APPENDIX 2

KEY TO SIGNS AND NOTATIONS

Symbol

A	Area; circumferential shear area; constants; rock element
A_o	Initial, unnecked, cross-sectional area of a rock test specimen
ASCE	American Society of Civil Engineers
ASTM	American Society for Testing and Materials
a	A coefficient; length; a ratio: $a = r_o/r_i$; cross-sectional area of anchorage; *are* (1 *are* = 1 a = 100 m²)
B	A constant
b	A coefficient; width of gauge section; length
C	Coefficient; compression; total amount of cohesion
Cal	Large or kilo-calorie = 1000 gram-calories
C-fl	Point of counter-flexure
₵	Center line
c	Cohesion; a coefficient; test parameter; distance
cal	Small calorie
c_{eff}	Cohesion on effective stress basis
c_m	Mass heat capacity
c_{mi}	Mass heat capacity of ice
c_{mr}	Mass heat capacity of rock and dry soil mineral particles
c_{mw}	Mass heat capacity of water
const	A constant
c_v	Volumetric heat capacity
c_{vf}	Volumetric heat capacity of frozen soil or rock

c_{vu}	Volumetric heat capacity of unfrozen soil or rock
D	Diameter; width of opening; width of footing
d	Diameter
dr	Differential radius
ds	Differential arc
dT	Differential temperature
du	Differential displacement
dV	Differential volume
$d_{\omega=0}$	$= r - r_i =$ thickness of tensile zone above the crest
Δd	Diametral deformation
Δdr	Change in differential radius
$d\sigma_r$	Differential radial stress
$d\omega$	Differential angle
E	Modulus of elasticity
E_{ave}	Average modulus of elasticity
E_c	Compression modulus or modulus of deformation; modulus of elasticity of a confined rock
E_i	Initial tangent modulus at zero load
E_{rock}	Modulus of elasticity of rock
E_s	Secant modulus of elasticity for a particular point
E_t	Tangent modulus of elasticity on the stress-strain curve at a particular point
E_{t50}	50 %-tangent modulus
e	Void ratio; base of the system of natural logarithms
e-e	An envelope line
e_{el}	$= \varepsilon_{el}/\varepsilon_{total} =$ ratio of elastic strain to total strain
e_{pl}	$= \varepsilon_{ir}/\varepsilon_{total} =$ ratio of plasticity strain to total strain
F	Force; frictional force; resistance to shear force
F_D	Driving force
F_R	Resistance to shear; resisting force
f	A function of ...; coefficient of static friction: $f = \tan \varphi$
f'	Coefficient of dynamic friction
G	Specific gravity
G_b	Bulk or apparent specific gravity
G.S.	Ground surface
GW	Groundwater
GWT	Groundwater table
g	Acceleration of gravity; gram

H	Symbol for rating of hardness of minerals; horizontal force
HRB	Highway Research Board
h	Hour; pressure head; height; thickness
Δh	Deformation; compression; change in height
I	Moment of inertia (axial)
I_T	Toughness index of rock
ISRM	International Society of Rock Mechanics
i	Hydraulic gradient; thermal gradient; pressure gradient; angle of slope of an inclined shear plane; imbrication
i_{cr}	Critical hydraulic gradient
K	Coefficient of thermal conductivity
K_a	COULOMB's active earth pressure coefficient
K_p	COULOMB's passive earth pressure coefficient
k	Kilo; permeability to water; kip (1 k = 1000 lb)
kg	Mass kilogram
kg_f	Force kilogram
kg/m³	Mass kilogram per cubic meter
kg_f/cm^2	Force kilogram per square centimeter
kN/m²	Kilo newtons per square meter
kN/m³	Kilo newtons per cubic meter
L	Length, distance; latent heat of fusion = 80 cal/g
L_t	New length upon thermal expansion
ΔL	Change in length; linear deformation; displacement
dL	Differential length
l	Liter
l/s	Liters per second
M	Mega; bending moment
M_R	Modulus ratio
MN/m²	Mega newtons per square meter
MN/m³	Mega newtons per cubic meter
m	Meter; POISSON's number: $m\ 1/\mu$; displacement coefficient; mass
max	Maximum
min	Minimum
m_r	POISSON's number of rock
N	Newton; normal load
N/m	Newtons per square meter

n	Porosity; relative volume of voids; number of specimens; number of bolts; function of support rigidity and strength of rock mass
n_a	Relative volume of air
n_m	Relative volume of water
n_s	Relative volume of solids
n_w	Relative volume of water
n'	Function of rock mass quality and the kind of support method
P	Load; force; ultimate punching load; total rock anchor force
P_N	Normal pressure
P. L.	Proportional limit; plastic limit
P_{eq}	Force on rock for equilibrium; anchor force
p	Pressure
p_h	Horizontal stress
p_i	Initial stress $= c \cdot \cot \varphi$; inside pressure; bolt pressure on rock
p_{max}	Maximum pressure
p_{min}	Minimum pressure
p_o	Outside pressure
psi	Pounds per square inch
p_v	Vertical stress
Q	Discharge; amount of heat flow; heat energy
Q_L	Total latent heat of fusion
R	Radius; resultant force; reaction
r	Radius; polar coordinate
r_i	Initial radius; inside radius
r_o	Outside radius
Δr	Change in radius
S	Degree of saturation; total shear resistance
S_f	Softening factor
S_t	Dry/wet factor
s	Second; shear strength
s_{ave}	Average punching shear strength
s_1, s_2, s_3, s_4	Shear strength
T	Temperature; horizontal shear force; tangential force; torque; tension

APPENDIX

$\overline{T}$	Effective shear force
T_f	Freezing temperature
T_i	Initial temperature
T_o	Initial temperature
T_s	Surface temperature
ΔT	Temperature increment; temperature decrement; temperature difference; rise in temperature; change in temperature
T_1	Final temperature
t	Symbol for a metric force ton; time; thickness of gauge section; thickness of specimen
t_f	Metric force ton
t_f/m^3	Metric force ton per cubic meter
$\tan \varphi$	Coefficient of friction
$t\text{-}t$	A tangent line
U	Neutral force; uplift pressure; hydrostatic pressure; groundwater pressure
u	Neutral stress; pore water pressure; displacement; radial displacement; deformation
u_i	Inside displacement
u_o	Outside displacement
V	Volume; total volume
V_a	Volume of air or gas
V_s	Volume of solids
V_t	Total volume; new volume after thermal expansion
V_v	Volume of voids
V_w	Volume of water
ΔV	Volume element
v	Velocity; index for vertical
v_{cr}	Critical velocity
W	Weight; total weight; weight of loosened rock; weight of ruptured rock wedge
W_{air}	Weight in air
W_d	Dry weight
W_s	Weight of solids
W_{sat}	Weight of saturated soil (rock)
$W_{sat\ in\ air}$	Weight of saturated soil (rock) in the air
$W_{sat\ in\ water}$	Weight of saturated soil (rock) in water
W_w	Weight of water
w	Natural moisture content; displacement; deformation
w_o	Surface displacement of rock

w_{sat}	Saturation moisture content
X	Force-field component
x	An unknown; length; a coordinate
Y	Yield point; force-field component
y	Deflection; length; an unknown; a coordinate
Z	Section modulus = I/c; force-field component
z	An unknown; depth; a coordinate
α	Alpha. Angle; a constant; a ratio: $\alpha = r_0/r$; coefficient of thermal diffusivity or temperature conductivity; angle of rupture; angle of dip; angle of slope
α_a	Angle
α_t	Coefficient of linear thermal expansion
β	Beta. Coefficient of volumetric thermal expansion; a constant; angle
γ	Gamma. Unit weight
γ_d	Dry unit weight
γ_s	Unit weight of solids
γ_{sat}	Saturated unit weight
γ_{sub}	Submerged unit weight
γ_w	Unit weight of water
Δ	Delta. Difference; increment; decrement
ΔH, ΔL	Change in length; deformation
ΔT	Change in temperature
Δd	Change in diameter; diametrical deformation
Δdr	Change in elementary radius dr
Δh	Absolute deformation
Δp	Pressure increment
Δr	Absolute radial deformation
Δr_d	Thickness of zone of blasting disturbance
Δt	Time increment
Δu	Absolute deformation

Δx	
Δy	Coordinate increments
Δz	
$\Delta \varepsilon$	Change in strain
$\Delta \sigma$	Change in stress; standard deviation
δ	Delta. Angle; deformation
ε	Epsilon. Strain; relative deformation; angle
ε_{el}	Elastic strain
ε_h	Horizontal strain
ε_{ir}	Irreversible, permanent strain
ε_{pl}	$= \varepsilon_{ir} =$ permanent strain
ε_t	Tangential strain component
ε_{total}	Total strain
ε_v	Vertical strain
ε_x	Lateral strain
$\varepsilon_y, \varepsilon_z$	Longitudinal strains
ε_1	Longitudinal strain; bending tensile strain
ε_2	Bending compression strain
ε_3	Lateral or transverse strain
ζ	Zeta. A coefficient; zeta potential
η	Eta. Factor of safety. Absolute or dynamic viscosity [g/cm·s]
θ	Theta. Angle.
λ	Lambda. A coefficient; ratio
λ_o	A constant
μ	Mu. Poisson's ratio: $\mu = 1/m$; coefficient of friction
ν	Nu. Kinematic viscosity: $\nu = \eta/\varrho$ [cm²/s]
ξ	Xi. A coefficient
ϱ	Rho. Radius; angle; density
Σ	Sigma. Summation of ...
ΣF_r	The sum of projected forces in radial direction
σ	Sigma. Stress; normal stress; compressive stress
$\sigma_A, \sigma_B, \sigma_D$	Stresses at point A, B, and D, respectively
σ_D	Driving stress
$\sigma_{P.L.}$	Stress at plastic limit
σ_R	Resisting stress at failure
σ_Y	Yield stress
σ_{ave}	Average stress; average strength of rock specimen
σ_c	Compressive stress
σ_{cu}	Ultimate compressive strength of an equivalent cubical rock specimen whose ratio $b/h = 1.0$

σ_{eff}	Effective stress
σ_h	Horizontal stress
σ_{max}	Maximum stress
σ_{min}	Minimum stress
σ_n	Normal stress
$\bar{\sigma}_n$	Effective normal stress
$\sigma_{n_{eff}}$	Effective normal stress
$\sigma_{n1}, \sigma_{n2}, \sigma_{n3}, \sigma_{n4},$	Normal stresses
σ_o	Contact pressure on rock (soil)
σ_{pr}	Precompression stress
σ_r	Radial stress
σ_{re1}	Radial elastic primary stress
σ_{re2}	Radial elastic secondary stress
$\sigma_{r\,el}$	Radial stress in the elastic zone
$\sigma_{r\,max}$	Maximum radial stress
σ_{ro}	Radial stress for the outer contour
σ_{rp}	Radial stress in the plastic zone
σ_t	Tensile stress; tangential stress
σ_{te2}	Secondary tangential elastic stress
$\sigma_{t\,el}$	Tangential stress in the elastic zone
$\sigma_{t\,max}$	Maximum normal tangential stress
$\sigma_{t\,min}$	Minimum normal tangential stress
σ_{to}	Tangential stress for the outer contour
σ_{tp}	Tangential stress in the plastic zone
$\sigma_{t_{ult}}$	Uniaxial ultimate tensile strength
σ_u	Unconfined compressive strength
σ_{ult}	Ultimate stress
σ_{unc}	Unconfined compressive strength
σ_v	Vertical stress; primary stress
$\sigma_x, \sigma_y, \sigma_z$	Stress components
σ_1	Major principal stress
$\sigma_1 - \sigma_3$	Deviator stress
σ_2	Intermediate principal stress
σ_3	Minor principal stress
$\left.\begin{array}{c}\sigma_{1_{eff}}\\ \sigma_{3_{eff}}\end{array}\right\}$	Effective principal stresses
τ	Tau. Shear stress; shear strength;
$\bar{\tau}$	Effective shear stress
τ_e	Total elastic shear stress in elastic-plastic boundary

τ_{e1}	Primary elastic shear stress
τ_{e2}	Secondary elastic shear stress
τ_{el}	Shear stress in elastic zone
τ_{max}	Maximum shear stress
τ_p	Shear stress in plastic zone
τ_{rt}, τ_{tr}	Shear stresses in polar coordinate system
$\tau_{xy}, \tau_{yz}, \tau_{zx}$	Shear stresses
Φ	Phi. AIRY's stress function
φ	Angle of friction; test parameter
φ_{eff}	Angle of friction on effective stress basis
φ_r	Angle of residual shear resistance
φ_u	Test parameter
φ_{unc}	Angle of friction obtained from unconfined compression test
ω	Omega. Angle; amplitude; a polar coordinate

APPENDIX 3

GLOSSARY OF TERMS

Abutment The place of abutting. The part of a structure that directly receives thrust or pressure (as of a supporting zone of an arch, vault, end-pier of a bridge abutted against soil).

Adit A nearly horizontal passage from the surface in a mine.

Aeolian (Eolian) Pertaining to wind. Designates rocks and soils whose constituents have been carried and laid down by atmospheric currents.

Age
1. Any great period of time in the history of the earth of unspecified duration.
2. A time when a particular event occurred, such as the Ice Age.
3. Refers to the position of anything in the geologic time scale.

Amphibole The generic name of bisilicate minerals whose chief rockmaking member is hornblende. General formula: $A_{2\text{-}8}B_5(\text{Si}, \text{Al}_4) \cdot O_{11}\ (OH)$. A is mainly Mg, Fe″, Ca, and Na. B is mainly Mg, Fe″, Al, and Fe‴. The amphiboles are common rock-forming minerals.

Amphibole group The amphibole group has hornblende as its most familiar member.

Amphibolite	A faintly foliated metamorphic rock developed during the regional metamorphism of simatic rocks. Composed mainly of hornblende and plagioclase feldspars.
Andesite	An extrusive rock, equivalent of diorite, and shows mostly feldspar grains.
Anhydride of	A compound derived from another (as an acid), by removal of the elements of water.
Anhydrite	A mineral $CaSO_4$ consisting of the anhydrous calcium sulfate that is usually massive and white or slightly colored. Differs from gypsum in lacking water.
Anisotropy	A condition of a material having different properties in different directions. For example: the state of geologic strata of transmitting sound waves with different velocities in the vertical and in the horizontal directions. Situation of a material having different moduli of deformation in different directions.
Anthracite	Compact, dense, hard black coal with a hardness of $H = 2$ to 2.5, and $G = 1.4$ to 1.8. Burns with a short flame and great heat.
Anticline	An arch-like fold (up-fold) in rocks, with the beds dipping in opposite directions from the crest.
Apron	A floor or lining of concrete, timber, or other resistant material at the toe of a dam, bottom of a spillway, or chute, to prevent erosion from falling water or turbulent flow. Also, a wall of cement grout in a rock foundation on the upstream of a massive dam for purpose of reducing seepage underneath the dam and to reduce uplift pressure on the base of the massive dam.
Arch	The configuration of the upper part of a tunnel section above the springline. The crown. The curved roof of an underground opening (tunnel).
Are	Area (1 are = 1 a = $100 \, m^2$)
Arkose	A variety of sandstone containing more than 25 or 30 % of feldspar and usually derived from the disintegration of granite or other acid rocks of granular texture.
Artesian water	Groundwater that is under sufficient pressure to rise above the level at which it is encountered by a well, but which does not necessarily rise to or above the surface of the ground.

Ash	In geology, the finest rock material from volcanic explosions.
Atmosphere	1. The whole mass of air surrounding the earth.
2. The air of a locality.
3. Surrounding influence or environment.
4. A unit of pressure equal to the pressure of the air at sea level or approximately $14.7\,lb/in.^2 = 1.0335\,kg_f/cm^2 = 10.1352\,N/cm^2$.
5. 1 technical atmosphere (at) $= 1\,kg_f/cm^2 = 736\,torr$
6. 1 physical atmosphere (atm) $= 1.0335\,kg_f/cm^2 = 760\,torr$
Attapulgite	A light-green absorbent fuller's earth. A basic hydrous silicate of magnesium and aluminum. $(OH)_2 \cdot H_2\,(Mg,\,Al_4O_3)\,Si_3H_4O_{10}$.
Attitude	Position with respect to the direction of the compass and to the horizontal.
Basalt	An extrusive rock. Generally, any fine-grained, dark-colored igneous rock.
Base exchange	Ion exchange. The process by which certain cations (positively charged atoms or groups of atoms, such as H^+, Na^+, NH_4^+) replace other cations. The term is used in two different ways: (1) to describe the action of zeolites and other water softeners by which calcium and other undesirable bases in the water replace sodium in the water softener compound; (2) the exchange of cations on or within parts of the crystal lattice of certain clay minerals, such as montmorillonite.
The nature of the exchangeable base in clays of this type very greatly influences their physical properties, i.e., clays having sodium as the exchangeable base generally have much higher plasticity indexes than those in which calcium is the exchangeable base.
The clay particle with its cations may be regarded as a kind of salt in which the colloidal clay particle is the anion. Certain cations may replace others making the clay more flocculent. The cation replacement is known as the "base exchange."
Basement rock	1. A name commonly applied to metamorphic or igneous rocks underlying the sedimentary sequence.
2. Metamorphic and igneous Precambrian rock.

Bearing capacity   Load per unit area which can be safely supported by the ground (rock and/or soil).

Bed   A layer of rock or soil.

Bedding   The property of a rock mass of being composed of beds.

Bedding plane   Plane dividing sedimentary rocks of the same or different lithology. The division planes which separate the individual layers, beds, or strata marking the boundary between a bed and the bed above and/or below it.

Bedding surface   A conspicuous surface within a sedimentary rock mass, representing an original surface of deposition.

Bedrock
1. Any layer of rock underlying soils.
2. Geologically, the term denotes material underlying drift deposits.
3. It is the solid, undisturbed rock in place either at the ground surface or beneath surficial deposits of gravel, sand, or clay.

Bending strength   An alternative term for flexural strength.

Bentonite   A clay formed from the decomposition of volcanic ash and is predominantly composed of high contents of the clay mineral montmorillonite, usually characterized by high swelling upon wetting.

Biotite   "Black mica," ranging in color from dark brown to green. It is a common rock-forming silicate mineral of the mica group: K (Mg, Fe''), (Al, Fe''') Si_3O_{10} $(OH)_2$.

Bluff   A high, steep bank. A cliff.

Bond strength   Stress required to rupture a bond in a material, or two kinds of material cemented one to another (a concrete block cemented on rock).

Borings
1. Subsurface investigation by drilling down into the ground to the desired depth, removing the material penetrated so that it can be examined at the surface, recording the elevation at which changes in material are found, obtaining samples from the various strata, and preparing a log or profile of the boring data.
2. Boreholes drilled into rock or soil.

Break   Separation into parts with suddenness or violence. A burst. Fragmentation of solid rock as a result of blast.

Breakdown of structure	1) Occuring often quite gradually under an increasing load. 2) Displacements of crystal grains accompanied by a partial destruction of the cohesion, occuring under an increasing load. This results in a gradual loosening or tearing apart of the structure of the material under increasing stress.
Breaking load	Load which causes fracture in compression, tension, flexure or torsion tests.
Breccia	Sedimentary rock consisting largely of angular fragments embedded in a fine-grained matrix.
Brittle	A material is said to be in a brittle state or brittle under conditions in which its stability to resist load decreases with increasing deformation.
Brittle fracture	Failure or rupture of a material with little or no plastic flow of deformation of a rock crystal lattice. Usually this type of failure is associated with impact loads. However, many materials at low temperatures also show brittle fracture failures under static loads.
Brittleness	A property of material that ruptures or fractures with little or no plastic flow.
Burst	To break suddenly into pieces from impact or from pressure within.
Calcareous	Containing calcium carbonate.
Calcite	A mineral, calcium carbonate ($CaCO_3$). The principal constituent of limestone.
Calcium carbonate	A solid, $CaCO_3$, occuring in nature (calcite, for example).
Calyx	1. A coring barrel, provided with a special device for collecting coarse rock cuttings which are falling back from the upward flowing drilling fluid. 2. A very large diameter borehole drilled by the shot-core drilling method. 3. The actual method of shot-core drilling, called the "calyx" boring.
Canyon	A steep-walled valley or gorge in a plateau or mountainous area.
Carbonate	A compound containing the radical CO_3-.
Cavern	A subterranean hollow; an underground cavity. Most frequent in limestones and dolomites.
Chlorite	1. A green stone 2. Any of a group of monoclinic usually green minerals associated with and resembling the micas.

Clastic rock	Composed principally of detritus transported mechanically into its place of deposition (sandstones and shales as distinct from limestones and anhydrites).
Clastic (adj.)	Composed of broken fragments of minerals or rocks.
Clay	In soil mechanics usage, soil consisting of inorganic material, the particle size of which have diameters smaller than 0.002 mm (by the International Soil Classification System).
Clay minerals	One group of finely crystalline, hydrous silicates with a two- or three-layer crystal structure; the major components of clay minerals; the most common minerals belong to the kaolinite, montmorillonite, attapulgite, and illite groups.
Cohesion	The property of rock (soil) particles to bind together.
Colloid	1. A suspension of very fine particles in a liquid phase (usually water) too small for resolution with an ordinary light microscope. 2. A discrete mineral particle less than 0.002 microns in diameter. 3. A finely divided dispersion of one material, the dispersed phase, in another, the dispersion medium.
Competent ground	Ground that does not require support when an underground opening (tunnel) is excavated through it.
Competent rock	A rock which is capable of sustaining an underground opening or a steep slope at the earth's surface without artificial support. Also, a rock which is sufficiently strong to transmit forces under given conditions. Rocks which are sufficiently plastic to deform without fracturing are incompetent.
Conformity	The relations of adjacent beds not separated by sedimentary discontinuity.
Conglomerate	Sedimentary rock consisting largely of rounded pebbles cemented together by another mineral substance.
Consolidation	Geologically, any or all of the processes whereby loose, soft, or liquid earth materials become firm and coherent. In soil mechanics, the adjustment of a saturated soil

APPENDIX 535

	in response to increased static load. The process involves the squeezing of water from the pores of the soil and a decrease in its void ratio in time.
Continuity	Something absolutely continuous and homogeneous of which no distinction of content can be affirmed except by reference to something else (as duration). An uninterrupted, ordered sequence. An identity of substance-uniting discrete parts.
Continuum	Continuum is a mathematical abstraction applied to a large collection of material particles.
Coulee	Commonly, in the northern plains of the western United States, any gully, gorge, dry wash, or intermittent stream valley of considerable size. (Can. French coulée) — flow of lava. Also, a thick sheet or solidified stream of lava.
Crataceous period	(See Geologic Time Table 2-1).
Creep	Time-dependent deformation or strain. Deformation that occurs over a period of time when a material is subjected to constant stress at constant temperature. Slow deformation that results from long application of stress. An imperceptibly slow, more or less continuous, downward and outward movement of slope-forming rock or soil.
Creep limit	Alternative term for creep strength.
Creep strength	Maximum stress required to bring about a specified amount of creep in a specified time.
Critical stress	Maximum and minimum compressive stresses on the boundary of an opening.
Crust (of the earth)	That part of the earth lying above the Mohorovičić disconformity.
Crystal	A solid with orderly atomic arrangement.
Crystal form	The geometric form taken by a mineral, giving an external expression to the orderly internal arrangement of atoms.
c-**soil**	Cohesive soil.
Cutoff	A wall or diaphragm of concrete or steel, or grout wall, or a slurry trench for the purpose of reduction of seepage under the dam and rock foundation material, and to reduce uplift pressure on the base of a dam and other structural foundation.
Decomposition	The breaking down of minerals by themselves or in rocks through chemical processes, usually related to weathering.

Density	In the physical sense, density ϱ of a substance is defined as its mass m per unit volume V of a body: ϱ = mass/volume = m/V (kg/m^3). However, in civil engineering usage, the term density is tacitly assumed to mean the unit weight of a material.
Detritus	Material produced by the weathering and disintegration of rocks that has been moved from its site of origin. A deposit of such a material.
Devonian Age	See Geologic Time Table, Table 2-1.
Diabase	See "Trap rock."
Diastrophism	All movements of the solid part of the earth, resulting in its displacement (faulting) or deformation (folding).
Dihedral	A figure formed by two intersecting planes.
Dike	1. A tabular-shaped (thin and flat) body of igneous rock that has been injected while molten into a fissure. 2. A bank, dam, wall or mound built around a low-lying area to control or confine water, or to prevent flooding.
Dilatancy	Refers to a relative increase in volume. Packed sand expands when sheared.
Dilation	Deformation, that is, change in volume but not in shape.
Dilatometer	An instrument for measuring thermal expansion.
Diorite	Plutonic rock composed essentially of sodic plagioclase and hornblende, biotite, or pyroxene. Small amounts of quartz may be present.
Dip	1. The vertical angle that a stratum (or fault plane, or any planar feature) makes with the horizontal. The dip is at right angles to the strike. 2. The slope of a bed of rock relative to the horizontal. 3. A pronounced depression in the land surface.
Disconformity	Unconformity between parallel strata. Unconformity in which the beds below and above unconformable contact are parallel.
Discrete	Composed of distinct parts; discontinuous.
Disperse	Finely divided.
Dolerite	Any coarse basalt.
Drainage	Removal of surface or groundwater from a given area either by gravity or by pumping.
Drainage well	A vertical shaft constructed in masonry dams to

APPENDIX 537

intercept seepage before it appears at the downstream face of the dam.

Drift
1. A horizontal underground passage.
2. Rock material of any sort deposited in one place after having been moved from another.

Ductility
Extent to which a material can sustain plastic deformation without breaking or rupture. Elongation and reduction of area are common indices of ductility.

Dynamic or absolute viscosity, η
The tangential force per unit area of two parallel planes at unit distance apart when the space between them is filled with a fluid and one plane moves with a unit velocity in its own plane relative to the other.

Elastic
Capable of sustaining stress without permanent deformation.

Elasticity
The property or quality of being elastic.

Elastic limit
The least stress without undergoing permanent deformation. Also called the yield point.

Electrolyte
A substance which becomes an electric conductor when dissolved in water. Electrolytes include acids, bases, and salts.

Emulsion
The phenomenon of holding finely divided particles of a liquid in suspension with the body of another liquid.

Energy
Ability to do work.

Engineering geology
The application of geologic data and principles to engineering problems dealing with naturally occurring rock and soil for the purpose of assuming that geological factors are recognized and adequately interpreted in engineering practice.

Eolian
See aeolian

Epoch
A division of geologic time less than a period and greater than an age. A geologic time unit. Subdivision of a period. An instant of time or a date selected as a point of reference.

Era
1. The largest unit of geologic time. Eras are subdivided into periods (see Geologic Time Table, Table 2-1).
2. One of the major divisions of geologic time, including one or more periods. The eras usually recognized are the Archeozoic, Proterozoic, Paleozoic, Mezozoic and Cenozoic.

Erosion	Processes whereby earthy or rock material is loosened or dissolved and removed from any part of the earth's surface. It includes weathering, solution, and transportation.
Erosion agents	Running water, waves, moving ice, and wind.
Escarpment or scarp	1. A steep cliff or ridge that is formed by sudden earth movements — usually vertical but sometimes also horizontal — along fault lines. 2. Any line of cliffs, or abrupt slope breaking the continuity of a land surface. 3. A fault-line scarp is one formed along the line of a fault.
Extrusive (extrusion)	This term is applied to igneous rocks which emerge at the surface. Those igneous rocks were derived from magmas or magmatic materials poured out or ejected at the earth's surface, as distinct from the intrusive or plutonic igneous rocks that have been injected into older rocks at depth without reaching the surface. (Effusive rocks; volcanic rocks).
Face	The solid surface of the unbroken part of the rock at the advancing end of the working place in a tunnel or adit.
Failure	Failing to perform an expected action. In a general sense, failure includes both fracture and flow.
Fall line	1. A line characterized by numerous waterfalls, as the edge of a plateau in passing which the streams make a sudden descent. 2. A line drawn through the falls on successive rivers is called the fall line.
Fatigue	Permanent structural change that occurs in a material subjected to fluctuating stress and strain. In general, fatigue failure can occur with stress levels below the elasticity limit.
Fault	A fracture in the rock of any dimension.
Fault trace	The line of intersection of a fault plane with the surface.
Fault zone	A wide shatter belt rather than a fault plane.
Feldspars	An aluminum silicate mineral group in which the principal cations are K^+, Na^+, Ca^+, Ba^+, and Sr^+. Potash feldspar: $K[AlSi_3O_8]$; Soda feldspar: $Na[AlSi_3O_8]$; Lime feldspar: $Ca[Al_2Si_2O_8]$.
Fill	Deposit of soil, rock, or other materials placed by man.

Fissure	An extensive crack, break, or fracture in the rock.
Flow	Unrestricted plastic deformation.
Fly ash	Fine solid particles of noncombustible ash carried out by the draft into the air by burning pulverized coal.
Fold	A bend in a rock of any form.
Force	Any cause of action that alters or tends to alter a body's state of rest or of uniform motion in a straight line.
Force field	A region of space or a medium where at every point at any time there acts a definite, coordinated force — the so-called force field.
Force majeure	Superior or irresistible force.
Formation	A group of rocks with recognizable and traceable boundaries, sufficiently alike lithologically to be mapped as a unit.
Foundation	1. The lower part of a structure that transmits the structural and applied loads to the rock or soil. 2. The artificially laid base upon which a structure stands, or on which any erection is built up.
Fracture	A general term for discontinuities. A surface of discontinuity formed by rock failure under stress. Rock cracked or broken into fragments along planes other than joints or bedding. Breaks in rocks caused by intense faulting and folding. Fracture includes faults, joints, cracks, and other breaks. The term fracture is used in the sense of brittle fracture to imply a complete loss of cohesion across a surface. Failure by parting of the material. Distinct surfaces of separation in the material.
Friction	Forces offering resistance to relative motion between surfaces in contact.
Fuller's earth	A natural, fine-grained earthy material, such as clay, with high absorbent power. Consists mainly of hydrated aluminum silicates. Used as absorbent in refining and decolorizing oils, as a catalyst, such as bleaching agent.
Gabbro	A plutonic rock consisting of calcic plagioclase. Apatite and magnetite or limonite are common accessories.
Gabion	A long wicker basket containing stones to serve the purpose of protecting engineering works and hydraulic structures against erosion, and used in slope stabilization work.

Gallery A subsidiary passage in a cave at a higher level than the main passage.
A horizontal or nearly horizontal underground passage either natural or artificial.

Geiger counter An instrument used to detect gamma rays given off by radioactive substances.

Gel
1. A jelly-like material formed by coagulation of a colloidal dispersion.
2. A semisolid colloidal solution.
3. A form of matter in a colloidal state that does not dissolve, but nevertheless remains suspended in a solvent from which it fails to precipitate without the intervention of heat or of an electrolyte.

Geologic time scale See Table 2-1.

Geology The study of the earth. It is the science which treats the origin, history, and structure of the earth, as recorded in the rocks; together with the forces and processes now operating to modify rocks.

Geotechnics
(géotechnique) The discipline of foundation engineering. In its modern form this term is synonymous with "Soil Mechanics" or "Earthworks" (such as roads, tunnels, foundations, dams) in any way associated with engineering in rock and soil.

Geotectonic Pertaining to the form, arrangement, and structure of the rock masses composing the earth's crust.

Geothermic
(geothermal) Pertaining to the heat of the earth's interior.

Geothermic degree The average depth within the earth's crust corresponding to an increase of one degree in temperature.

Geothermal gradient The change in temperature of the earth with depth, expressed either in degrees per unit depth or in units of depth per degree.

Geyser Literally, a roarer; intermittent hot springs or fountains. Columns of water are thrown out at intervals with great force.

Gneiss A coarsely banded metamorphic rock.

Gouge Finely abraded material occurring between the walls of a fault, the result of grinding movement, and foreign infilling materials such as calcite deposited between the structural discontinuities.
Filling material such as silt, clay, rock flour and

APPENDIX

other kind of geological debris in joints, cracks, fissures, faults, and other discontinuities in rock.

Granite — A coarse-grained, igneous rock with abundant quartz and feldspar.

Gravity — The force exerted by the earth and by its rotation on unit mass or the acceleration imparted to a freely falling body in the absence of frictional forces.

Ground — The material through which a tunnel is driven, whether it is a solid rock, unconsolidated rock, or running mud.

Ground arch — The rock located immediately above a tunnel which transfers the overburden load onto rock located on both sides of the tunnel.

Groundbreak — Rupture of soil: the failure of soil in shear, viz., loss of soil bearing capacity.

Groundwater — That part of the subsurface water that has collected to form an underground "reservoir" as determined by the porosity and the degree of fissuring of rocks and by the underlying impervious strata. The waterlogged strata below the groundwater surface is the zone of saturation. The groundwater surface itself is the groundwater table.
1. Phreatic water. That part of the subsurface water which is in the zone of saturation.
2. Water which fills continuously and competely the voids of the earth's crust and is subjected to gravity and hydrostatic pressure only.
3. The water contained in the ground below the water table.
4. Gravitational water. Water that moves downwards under the action of gravity, from soil surface to the watertable.

Grout — Thin mortar: a fluid mixture of cement and water, or a mixture of cement, sand, and water.

Grout curtain — A row of vertically drilled holes filled with grout under pressure to form the cutoff wall under a dam, or to form a barrier around an excavation through which water cannot seep or flow.

Grouting — The act or process of applying grout or of injecting grout into grout holes or crevices of a rock.

Gunite — A sprayed concrete, mortar, projected with a pneumatic air gun in thin layers on rock surface to seal off light seepage.

Hard rock	Rock which requires drilling and blasting for its economical removal.
Heterogeneous	A characteristic of a medium or a field of force that signifies that the medium has properties that vary with position within. Having unlike qualities. Opposed to homogeneous.
Homogeneous	Of the same nature, consisting of similar parts or elements of a like character. Having the same properties at all points. Uniform and continuous in all parts. A uniform composition throughout, as opposed to heterogeneous.
Hooke's law	Strain is proportional to stress.
Hoop stress	Circumferential stress in a material of cylindrical form subjected to internal or external pressure.
Hornblende	A mineral approximately $Ca_2Na(Mg, Fe)_4(Al, Fe, Ti)_3Si_6O_{22}(O_1OH)_2$ which is the common dark variety of aluminous amphibole.
Humus	Dark-colored, organic, well-decomposed soil material consisting of the residues of plant and animal materials together with synthesized cell substances of organisms and various inorganic elements.
Hydrostatic uplift pressure	Water pressure in an upward direction against the bottom of a structure, as a dam, a road slab, or a basement floor.
Igneous rocks	Strictly speaking, rocks which have been molten at some time in their history.
Impact	Collision of two masses.
Impact strength	The ability of a material to withstand shock loading. The work done in fracturing a test specimen in a specified manner under shock loading.
Impermeable	Impervious. Having a texture that does not permit water to move through it perceptibly under the head differences ordinarily found in subsurface water.
Impulse	The product of the average value of a force and the time during which it acts being a quantity equal to the change in momentum produced by the force.
Incompetent rock	Rock incapable of standing in underground openings or steep slopes at the surface without support.
Injection	In rock engineering, the process of introducing, under pressure, a liquid or plastic material (grout) into fissures, cracks or voids in a rock formation.
In situ	In its natural position or place.
Intact rock	A material which can be sampled and tested in the

APPENDIX 543

	laboratory, and which is free of large-scale structural features such as joints, bedding planes, shear zones, and other kinds of rock defects.
Intensity	In seismics, the degree of shaking at a specified place.
Internal friction	Forces within a rock or soil which offer resistance to externally applied shear forces.
Internal (residual) stress	Stress which exists in virgin rock before excavation (natural stressing, or rock prestressing).
Intrusive (intrusion)	Applied to igneous rocks which have been emplaced below the surface. Igneous rocks that invade older rocks.
Intrusive rock	Rock that consolidated from magma beneath the surface of the earth.
Ion	An electrically charged atom, molecule, or group of atoms or molecules. Anion: an ion that carries a negative charge. Cation: an ion that having lost one or more electrons has a net positive charge. Electron: the elementary particle of mass 9×10^{-28} grams and unit electrical charge (4.80×10^{-10} e.s.u.). e.s.u. = Abbreviation for electrostatic units. A centimeter-gram-second system of electric units in which the unit of charge is that charge which exerts a force of 1 dyne on another unit of charge when separated from it by a distance of 1 cm in vacuum.
Ion exchange	See base exchange.
Isotropic	Having the same properties in all directions (said of a medium with respect to elasticity, conduction of heat or electricity, or radiation of heat or light).
Joint	A fracture or parting plane along which there has been little if any movement parallel with the walls.
Joint system	A group of two or more intersecting sets of joints.
Kaolin	A white or slightly discolored clay rock formed by the decomposition of material high in feldspar.
Kaolinite	The mineral characteristic of kaolin, and having the general formula $Al_2O_3 \cdot 2SiO_2 \cdot 2H_2O$. As the mineral particles are extremely small there are no reliable field methods whereby kaolinite can be distinguished from other clay minerals.
Karst topography	A type of landform developed in a region of easily soluble limestone bedrock. The limestone rocks are so honeycombed by natural tunnels and openings

dissolved out by groundwater that much of the drainage is underground. Streamless valleys are common, and valleys containing streams often end abruptly where the latter plunge into underground tunnels and caverns, sometimes to reappear as great springs elsewhere. The type locality is the Karst, a limestone plateau on the eastern coast of the Adriatic Sea.

Kinematic viscosity ν The ratio of dynamic viscosity η to density ϱ: $\nu = \eta/\varrho$.

Landslide The perceptibly downward and outward movement of slope-forming materials such as rock, soil, artificial fill, or a combination of these materials. Also, earth and rock that become loosened from a hillside by water or snow or an earthquake, and slide or fall down the slope.

Lath A long, thin mineral crystal.

Lava Molten volcanic rock extruded at the surface.

Law of superposition The law that underlying strata must be older than overlying strata. Upon this law all geological chronology is based.

Leak The escape of water through a crack, joint, hole, crevice, fault.

Leakage The uncontrolled loss of water from artificial structures due to hydrostatic pressure.

Limestone A sedimentary rock composed largely of calcium carbonate ($CaCO_3$).

Limonite A general term for broken hydrous iron oxide.

Liquefaction The change in the phase of a substance (say a mass of a dry sand) to the fluid state (say dry sand as a loose, moving, pouring body). A saturated non-cohesive soil mass can be in a fluid state if no intergranular pressure exists within it. Such a fluid state is commonly called liquefaction.

Lithology The physical character of rock. The composition and texture of rock.

Loam
1. A detrital deposit containing nearly equal proportions of sand, silt, and clay, these terms referring to the grain sizes of the particles.
2. Clayey earth.

Log The detailed record of the rocks passed through in drilling. When accurate, logs constitute a valuable source of information.

APPENDIX

Macadam	1. Macadamized roadway or pavement, esp. with a bituminous binder. 2. The broken stone used in macadamizing.
Macadamize	To construct or finish (a road) by compacting into a solid mass a layer of small broken stone on a convex well-drained roadbed.
Magma	Naturally occurring hot, molten, mobile rock material (liquid silicate melt) generated within the earth and capable of intrusion and extrusion, from which igneous rocks have been derived by cooling and crystallization.
Marble	A metamorphosed, recrystallized limestone.
Marl	1. A calcareous clay, or intimate mixture of clay and particles of calcite or dolomite, usually fragments of shells. 2. Marl in America is chiefly applied to incoherent sand, but abroad compact, impure limestone is called marl.
Mass movement	Unit movement of a portion of land surface as in creep, landslide, or slip.
Massive rock	Said of a rock if the strength of the bond across partings or joints is comparable to the rock strength.
Mass-wasting	The slow downslope movement of rock debris. A general term for a variety of processes by which large masses of earth materials are moved by gravity either slowly or quickly from one place to another.
Meerschaum	See sepiolite. A tough, compact, hydrous magnesium silicate.
Metamorphic rocks	Rocks which have been altered considerably by heat, pressure, chemical action, and other factors.
Modulus of bending	Rate of maximum fiber stress to maximum strain within the elastic limit of a stress-strain diagram obtained in flexure test.
Modulus of compression	E_c: obtained from repeated loading-unloading tests.
Modulus of deformation	See modulus of compression.
Modulus of elasticity	Rate of change of strain as a function of stress = Young's modulus of elasticity.
Modulus ratio	E_{t50}/σ_{ult}.
Modulus of rigidity	Rate of change of strain as a function of stress in a specimen subjected to shear or torsion loading.

Moho, or Mohorovičić disconformity or discontinuity	1. A point ranging from about three miles beneath the ocean basin floor to about 25 miles beneath the continental surface at which seismological studies indicate a transition in earth materials from those of the earth's crust to those of subjacent mantle. 2. A level of major change in the interior of the earth. It is found just beneath the earth's crust at depths ranging from 8 to 32 km.
Mohr's circle	A graphical construction showing the state of stress at a point in a material.
Mohr's failure	Failure of a material by fracture or permanent deformation when the shear stress on a real or potential fracture or slip plane has increased to a value that depends on the stress acting normal to the plane, or when the maximum principal stress has attained a certain limiting value.
Mohs hardness scale	A relative scale of the hardness of minerals arbitrarily reading form 1 to 10.
Momentum	The product of the mass and the velocity of a particle.
Monomer	The simple unpolymerized form of a chemical compound.
Monomeric	Refers to simple rather than compound molecular structure.
Moraine	Drift, deposited, accumulated chiefly by direct glacial action.
Mudstone	1. A hardened mud, clay. 2. Hard argillaceous rock, including more or less laminated shale or slate. 3. An indurated non-laminated sediment composed of clay minerals and other constituents of the mud grade.
Muskovite	White mica (potash mica).
Mylonite	A fine-grained, laminated rock formed by extreme microbrecciation and milling (grinding) of rocks during movement on fault surfaces. Mylonitic seams may be very thin and difficult to find.
Neutral pressure u	See pore-water pressure.
Ooze, to	1. To pass or flow slowly through or as if through small openings or interstices. 2. To move slowly or imperceptibly.
Ordovician	In geology, a period. It is the second period of the

	Paleozoic era, and the rocks formed during that time. The ordovician period is preceded by the Cambrian period and succeeded by the Silurian period. Ordovician rocks are well exposed in the eastern part of the United States and consist mainly of limestones and shales.
Orogeny	The process of forming mountains, particularly by folding and thrusting. A major disturbance or mountain-building movement in the earth's crust.
Orthoclase	A potash feldspar $K_2O \cdot Al_2O \cdot 6SiO_2$. Contains 16.9% potash, K_2O. See feldspars.
Outcrop	1. That part of a stratum which appears at the surface. 2. Actual exposure of bedrock.
Overburden	The mantle of earth overlying rock formations.
Packer	In rock drilling, a device lowered in the liner pipes, which expand automatically, or can be used to expand by manipulation from the surface at the correct time, to bring about a watertight joint against the sides of the borehole or the casing, thus entirely exluding water from higher horizons.
Perched (or false) water table	Caused by a layer of impervious material preventing the water from percolating to its natural level.
Period	A chronological division. A historical period. A division of geologic time longer than an epoch and included in an era (see geologic time table, Table 2-1).
Permafrost	Permanently, or perennially, or ever frozen ground.
Permanent set	Extent to which a material is permanently deformed by a specified load.
Permeability	Capacity of rock or soil for transmitting a fluid under a hydraulic gradient. The flow per unit area under unit hydraulic gradient. Darcy's law of permeability for saturated, noncohesive soils: $v = ki$, where v = velocity of flow; k = coefficient of permeability, and i = hydraulic gradient.
Petrology	The study of rocks.
pH	The negative logarithm of the hydrogen ion activity. For example, pH7 indicates an H^+ concentration (activity) of 10^{-7} mole/liter.
Phenocrysts	Phenocrysts are chiefly the common igneous-rock making minerals such as augite, biotite, feldspar, hornblende, olivine, and quartz.

Phreatic water	Groundwater. That part of the subsurface water which is in the zone of saturation.
Pillar	In-situ rock support between multiple openings.
Plagioclase	A feldspar, having calcium or sodium in its composition.
Plastic	Yielding or flowing under steady load. The term "plastic" is frequently used to describe processes involving yield and ductility. The term is used here in the sense defined in the mathematical theory of plasticity.
Plastic deformation	Deformation that remains after the load causing it is removed. It is the permanent part of the deformation beyond the elastic limit of a material. It is also called plastic strain and plastic flow. Irrecoverable deformation.
Plasticity	The property of a material that enables it to undergo permanent deformation without appreciable volume change or elastic rebound and without rupture.
Plutonic rock	An igneous rock that has formed beneath the surface of the earth at great depth by consolidation from magma.
Poisson's ratio	Ratio of lateral strain to axial strain in an axially loaded specimen.
Pore water pressure	Pressure of water in the pores (voids) of a saturated medium of soil. Also known as neutral pressure u.
Porosity	The ratio of the volume of voids in a rock or soil to its total volume.
Pozzolan	1. A pulverulant siliceous or siliceous and aluminous substance that reacts chemically with slaked lime at ordinary temperature and in the presence of moisture to form cement. 2. A loosely compacted siliceous rock of volcanic origin, or tuff. 3. A mortar used in ancient Roman buildings, of lime mixed with the dust of volcanic tuff found near Puteoli = Pozzuoli.
Precambrian Age	Pertaining to or designating all rocks formed prior to the Cambrian period (see Table 2-1).
Profile	1. A vertical columnar section through the rock or soil of the underlying strata below the ground surface along any fixed line, or a tabular statement of the succession of the various kinds of material and their thicknesses.

APPENDIX 549

3. The successions of horizons down to the parent material.
4. A graphical representation of elevation plotted against distance.
5. A drawing showing a vertical section of the ground exposing its various zones of rock or soil and their inclusions.

Proportional limit   The greatest stress at which a material is capable of sustaining load without deviation from proportionality of stress and strain (Hooke's law).

Pyroxene   Any of a group of igneous-rock forming silicate minerals that contain calcium, sodium, magnesium, iron, or aluminum.

Q.E.D.   Quod erat demonstrandum (which was to be proved or shown).

Quarry   An open or surface working, usually for the extraction of building stone, such as granite, limestone, marble, sandstone, slate, for example.

Quartzite   A hard, granulose, metamorphic rock composed essentially of quartz. Sandstone cemented by silica.

Quasi-elastic   Quasi: as if, as though it were. Seemingly almost elastic.

Quaternary   The latest and current period of geologic time, the time of extensive glaciation in North America. See Table 2-1.

Relaxation   Rate of reduction of stress in a material due to creep. An alternative term is stress relaxation.

Residual soil   Soil formed in place by the disintegration and decomposition of rocks and the consequent weathering of the mineral materials.

Residual stress   (Internal stress). Stress, which exists in virgin rock before excavation. Known also by the term natural stressing or rock prestressing.

Resilience   Measure of recoverable elastic energy in a deformed material. It is the amount of energy released when a load is removed from a specimen.

Rheology   Science dealing with the flow or deformation of matter. The study of time-dependent strain in both solids and liquids. Sometimes rheology describes the study of all types of deformation — elastic, plastic, and viscous, particularly the plastic flow of solids and flow of non-Newtonian liquids.

Rift   1. An obscure foliation, either vertical (or nearly

so) or horizontal, along which a rock splits more readily than in any other direction.
2. See fault trace.

Rift valley A relatively long and narrow trough-like valley formed by the sinking of a strip of the earth's crust between two approximately parallel and opposed normal faults or zones of faulting.

Rigidity Lack of flexibility. The property of stiffness or of not yielding.

Rise of an arch The maximum height that the undersurface of an arch rises.

Rock
1. In the popular sense and also in the engineering sense, the term rock refers to any hard, solid matter derived from the earth. In a strictly geological sense, a rock is any naturally formed aggregate or mass of mineral matter, whether or not coherent, constituting an essential and appreciable part of the earth's crust. A few rocks are made up of a single mineral, as, for example, a very pure limestone. Two or more minerals are usually mixed together by geological processes to form a rock.
2. To the engineer, the term rock signifies firm and coherent or consolidated substances that cannot normally be excavated by manual methods alone.

Rock burst A sudden and often violent failure of masses of rock in quarries, tunnels, mines, and other underground openings, and steep slopes of open cuts.

Rock engineering The practical, technical use of engineering application of rock mechanics in the design of engineering structures such as various underground openings: power plants, storage spaces, shelters, fortifications, shafts, adits, mines, tunnels, support for structural foundations.

Rockfall The relatively free falling of a newly detached segments of bedrock of any size from a cliff, steep slope, cave, or arch.

Rock mass A much used term with an obvious meaning. It tends to imply something larger, less individualized, less distinctly marked off from its surroundings.

Rock, massive See Massive rock.

Rock mechanics The study of the theoretical and applied behavior of rocks.

	It is that branch of mechanics concerned with the response of rock to the force fields of its environment.
Rock slide	The downward and usually rapid movement of newly detached segments of the bedrock sliding on bedding planes, joint, or fault surfaces or any other plane of separation.
Roof	Top or ceiling of an underground excavation.
Rotational slide	See slip.
Rupture	Failure of a material with development of fractures or shears. Deformation characterized by loss of cohesion.
Sandstone	Sedimentary rock composed of cemented sand grains, usually quartz.
Scarp	1. A line of cliffs produced by faulting or erosion. 2. A low, steep slope along the margin of a plateau or terrace, or along a beach caused by wave action. 3. To cut down vertically or to a steep slope.
Schist	A medium or coarse-grained metamorphic rock with subparallel orientation of the micaceous minerals that dominate its composition.
Scour	Erosion, especially by moving water.
Sedimentary rocks	Rocks formed by the accumulation of sediment derived from the breakdown of earlier rocks, by chemical precipitation, or by organic activity.
Seepage	1. The slow movement of water through a porous medium. 2. The slow movement of water through small openings and spaces in the surface of unsaturated soil into or out of a body of surface or subsurface water.
Self-supporting system	A system in which the structural stresses are carried on the walls, pillars, and other unexcavated parts of the opening rather than on lining, packs, struts, braces, and chucks placed within the opening.
Semi-elastic	Incompletely elastic.
Seismic	Refers to characteristics of, or brought about by, earthquakes or earth vibration.
Seismic exploration	A subsurface rock and soil exploration method without excavation utilizing the variation in the rate of propagation of shock waves in geological strata. The seismic velocity indicates the kind of soil and rock traversed by the shock waves.

Seismograph	An instrument for detecting vibrations in the earth.
Sepiolite	Compact, with smooth feel, and fine earthy texture, or clay-like. Microscopically is shown to be a mixture of fine fibrous material and an amorphous substance of apparently the same composition. The mixture forms the variety Meerschaum.
Shaft	A vertical excavation of limited area and comparatively great depth.
Shale	A naturally densified, laminated sedimentary rock consisting predominantly of silt and clay. Shale splits easily parallel to the bedding.
Shearing	An action of applied shear force causing or tending to cause two contiguous parts of a material to slide relative to each other in a direction parallel to their plane of contact resulting in a displacement of adjacent elements along a flat or curved surface.
Shear plane	A plane within a rock along which shear stress has resulted in actual fracture and slip.
Shear strength	A measure of the resistance of a material to shear stress.
Shear stress	A stress applied parallel to the surface or area. Maximum shear stress that can be sustained by a material before rupture. Applied stress that brings about rock and soil particles to slide on one another.
Shear zone	A zone in which shearing has occurred on a large scale, so that the rock is crushed and brecciated.
Shock	A sudden and violent blow, or a nonperiodic excitation, in the form of a pulse, a transient vibration; mechanical shock.
Shotcoring	A combination of rotary abrasion and wear of rock by means of steel shot fed into the wash-water of the casing. Some of the shot embed themselves into the soft steel rock-coring barrel; some get outside, some under, and some shot get inside of the rotating coring barrel.
Shotcrete	A pneumatically applied concrete mortar projected on uneven rock surface. Shotcrete provides an effective protection of rock against weathering.
Silica	The chemical composition of oxygen and silicon, which are the two most common elements in the earth's crust. Silicon dioxide: SiO_2. In the natural crystalline form, occurs chiefly as the mineral quartz.

APPENDIX 553

Silicate	A chemical compound of silica with one or more metallic oxides. Hence silicate minerals.
Siliceous	Containing, or consisting of silica.
Silicon	A tetravalent, nonmetalic element (Si) that occurs combined as the most abundant element next to oxygen in the earth's crust, and is used especially in alloys.
Silicones	Resinous materials derived from organosiloxane polymers.
Silt	The term silt applies to unconsolidated material finer than sand and coarser than clay. In the nontechnical sense, silt is the muddy, fine sediment carried and laid down by rivers or by the ocean in bays and harbors.
Siltstone	A consolidated silt.
Silurian age	The third in order of age of the geologic periods comprised in the Paleozoic era, in the nomenclature in general use. Also, the system of strata deposited during that period. In the United States, Silurian rocks consist largely of marine sediments with abundant limestone. See Table 2-1.
Sima	The basic outer shell of the earth. It is comparatively heavy basic igneous rock characterized by silica and manganese. Hence **simatic rocks.**
sine qua non	(Without which not). An absolutely indispensable or essential thing.
Sink	1. Any slight depression in the land surface, especially one having no outlet; one of the hollows in limestone regions (limestone sink) often communicating with a cavern or subterranean passage so that water running into it is lost. 2. To bore (drill) or put down a borehole.
Sinkhole	See sink.
Slate	A metamorphosed clay rock with a pronounced cleavage along which it readily splits.
Slip	A relative displacement of points on opposite sides of a fault, measured on the surface of the fault. A minor fault. A slide.
Slip lines	Orthogonal curves, whose directions at any point bisect the angles between the principal axes at that point.
Slurry	A suspension of pulverized solid in a liquid, such as cement, clay, clay-cement grouts, for example.

Soil	All unconsolidated earth material of whatever origin that overlies bedrock that has been in any way altered or weathered.
Soil engineering	The application of the principles of soil mechanics in the investigation and analysis of the engineering properties of earth materials for purpose of design and construction of engineering structures.
Soil mechanics	1. A discipline of engineering science dealing with the study and utilization of physical and mechanical properties, behavior, and performance of soils as a construction material. 2. The application of the laws of mechanics and hydraulics to engineering problems dealing with sediments and other unconsolidated accumulations of solid particles produced by the mechanical and chemical disintegration of rocks, regardless of whether or not they contain an admixture or organic constituents.
Sol	A colloidal dispersion of a solid in a liquid.
Solifluction	The slow downhill flowage or creep of soil and other loose material that is saturated with water. It is especially active in sub-arctic regions and some high mountains during snow melting periods.
Spring line	Place on the side of a tunnel where the tunnel starts curving into the arch.
Stiffness	The ratio of change of force to the corresponding change in deflection of an elastic element ($F \cdot L^{-1}$).
Strain	Change in length per unit length (relative deformation).
Strain hardening	Increase in strength caused by plastic deformation in cyclic loading.
Stratum	A layer or bed of rock or soil.
Strength	The maximum stress that a material can withstand during a normal short-time experiment without failing by rupture or continuous deformation. Rupture strength or breaking strength refers to the stress at the time of rupture.
Stress	Force per unit area.
Stress field—biaxial	One where the medium is subjected to a compressive or tensile stress in two directions (dimensions).
Stess field—hydrostatic	One where the medium is subjected to equal stresses in three mutually perpendicular directions.
Stress field—	One where the medium is subjected to compressive

APPENDIX 555

unidirectional (unconfined)	or tensile stresses in one direction.
Stress relaxation	Decrease in stress in a material subjected to prolonged constant strain at a constant temperature. Stress relaxation behavior is determined in a creep test.
Stress-strain diagram	Graph of stress as a function of strain.
Stress trajectories (or isostatics)	An orthogonal system of curves whose directions at any point are the directions of the principal axes. They, therefore, intersect a free boundary at right angles.
Strike	Strike describes the geometry of attitude of inclined beds. When an inclined layer of unfolded or folded rock intersects a horizontal surface (land or water), the compass direction of the line formed by its intersection is called the strike. The dip of a fold is the acute angle between the horizontal and the axial plane, measured perpendicularly to the strike.
Structural bond	A bond that joins basic, load-bearing parts of an assembly. The load may be static or dynamic.
Structure	The overall character of a rock observable on a hand specimen or outcrop scale.
Substance	Material of which a thing is made. That which constitutes anything what it is.
Subsurface structure	Any excavated or natural subsurface opening, or system of openings that is virtually self-supporting.
Surrounding	The region outside an object, outside a system.
Suspension	(See slurry). The floating of finely dispersed solid particles in a liquid in which it does not dissolve.
Suspension grout	A grout in which fine-particled solid material is suspended in water.
Syncline	A trough-like fold (down-fold) in the rocks, with the beds dipping inward on either side.
System	A particular collection of matter which is being studied. An entity subject to study. An ordered grouping of certain elements and/or facts in a field of knowledge according to certain principles.
Talus	Fragments which are broken off by the action of the weather from the face of a steep rock, as they accumulate at its foot forming a sloping heap, called talus.
Tectonic	Of, pertaining to, or designating the rock structure and external forms resulting from the deformation

	of the earth's crust by deep-seated crustal and sub-crustal forces in the earth.
Tectonic creep	Fault creep of tectonic origin; also called slippage.
Tectonics	The study of the broader structural features of the earth and their causes. The phenomena associated with rock deformation and rock structures generally.
Tectonism	Crustal instability.
Tensile strength	Ultimate strength of a material subjected to tensile loading.
Terrace	Any level-topped surface, with a steep escarpment, whether it be solid rock or loose material. It is step-like in character.
Thermal stress	Internal stress, caused in part by uneven heating.
Time scale	An arbitrary system of organizing geologic events or subdividing geologic time usually presented in the form of a chart, see Table 2-1.
Torsional deformation	Angular displacement of a specimen caused by a specified torque in torsion test.
Torsion test	Method for determining behavior of materials subjected to twisting loads (torque).
Toughness	Extent to which a material absorbs energy without fracture.
Trap	Old name for lava flow.
Traprock	Basaltic lava fields. Any of various dark-colored, fine-grained igneous rocks such as basalts used in building of roads.
Triassic Age	The first period of the Mesozoic era. Table 2-1.
Triaxial compressive strength	A measure of the strength of a material under a direct stress when the material is confined by a fluid under pressure, or by an enclosing jacket of metal or other material.
Tuff	A rock formed of the finer kinds of volcanic detritus. The compacted volcanic fragments are generally smaller than 4 mm in diameter.
Tunnel	A passage in rock (soil) open at both ends.
Ultimate strength	The greatest engineering stress developed in material before rupture.
Unconfined compressive strength	A measure of the strength of a material under a directed, uniaxial stress.
Underground opening	Natural cavities or man-made excavations under the surface of the earth.
Vadose water	Subsurface water suspended in the unsaturated zone of aeration of soil between ground surface and

groundwater. It is the water in the earth's crust above the permanent groundwater table.

Vesicle
1. A small, circular, enclosed space.
2. A small cavity in an aphanatic or glassy igneous rock, formed by the expansion of a bubble of gas or steam during the solidification of the rock.

Vesicular Characteristic of, or characterized by, pertaining to, or containing vesicles.

Viscoelastic A material or a condition where strain resulting from stress is partly elastic and partly viscous.

Viscosity
1. That property by which a fluid offers a resistance to shear stress. It is independent of pressure.
2. The property of liquids and solids that causes them to resist instantaneous change of shape and to produce strain that is dependent on time and the magnitude of stress.

Volcanic Pertaining to volcanoes or any rocks associated with volcanic activity at or below the surface.

Water A chemical compound consisting of two volumes of hydrogen and one volume oxygen (H_2O).

Water, chemically pure Chemically pure water, H_2O, is the chemical compound consisting of hydrogen and oxygen in the ratio of about eight parts of oxygen to one part of hydrogen by weight.

Water, connate Water that has got into a rock formation by being entrapped in the interstices of the rock material (either sedimentary or extrusive igneous) at the time the material was deposited. It may be derived from either ocean water or land water.

Water glass (Liquid glass). A glassy or stony substance consisting of silicates of sodium or potassium, or both, soluble in water forming a viscous liquid. Silicic acid.

Water, interstitial Water that exists in the interstices or voids in a rock, soil, or other kinds of porous medium.

Water, juvenile Water from the interior of the earth which is new or has never been a part of the general system of groundwater circulation. See magmatic water.

Water, magmatic Water that exists in a magma or molten rock, believed to be one of the constituents of a magmatic solution.

Water, meteoric Water that is in or derived from the atmosphere. The term has been used in various ways, sometimes

	to include all subsurface water of external origin and sometimes to include only that derived by absorption, excluding especially the connate ocean water.
Water, phreatic	See groundwater.
Water, vadose	Seepage water occurring below the surface and above the groundwater table. Contrasted with phreatic, which refers to the groundwater below the groundwater table.
Weathering	The physical disintegration and chemical decomposition of earth materials at or near the earth's surface by the elements.
Yield	The onset of plastic deformation. That stress in a material at which plastic deformation occurs.
Yield point	Stress at which strain increases without accompanying increase in stress. Also, see Elastic limit.
Yield strength	Stress at which strain increases without accompanying increase in stress. It is an indication of maximum stress that can be developed in a material without causing plastic deformation. It is the stress at which a material exhibits a specified permanent deformation, and is a practical approximation of elastic limit.
Young's modulus	Alternative term for the modulus of elasticity in tension or compression. It is a constant for each elastic material that equals the ratio of stress to strain ($E = \sigma/\epsilon$).
Zeolite	A generic term for a group of minerals occurring in cracks and cavities of igneous rocks, especially the more basic lavas. Zeolites are hydrous silicates of aluminum with either sodium or calcium, or both.
Zeta potential ζ (cross potential)	Electric potential across the double layer or interface between the solid and liquid phases of a soil or rock systems.
φ-soil	Frictional (noncohesive or cohesionless) soil.
(φ-c)-soil	Frictional-cohesive soil.
c-soil	Pure cohesive soil.

REFERENCES

American Geological Institute, 1974, *Dictionary of Geological Terms*. Garden City, N.Y.: Anchor Press/Doubleday.

CHALLINOR, J., 1978, *A Dictionary of Geology* (5th ed.). Cardiff: University of Wales Press.

COATES, D. F., 1970, "Glossary" Appendix B), in *Rock Mechanics Principles*. Department of Energy, Mines, and Resources, Ottawa, Canada. Mines Branch Monograph 874 (revised 1970).

"Glossary of Terms and Definitions in Soil Mechanics." *Proceedings ASCE,* No. SM 4, 1958.

GRAHAM, A. C., Editor, 1966, *The Basic Dictionary of Science*. New York, N.Y.: The McMillan Company.

HUNT, V. D., 1979, *Energy Dictionary*. New York, N.Y.: Van Nostrand Reinhold Company.

JUMIKIS, A. R., 1977, *Glossary of Terms in Thermal Soil Mechanics*. New Brunswick, New Jersey: College of Engineering, Bureau of Engineering Research. Rutgers, The State University of New Jersey. Engineering Research Publication No. 57.

KRYNINE, D. P., and W. R. JUDD, 1957, *Principles of Engineering Geology and Geotechnics*. New York, N.Y.: McGraw-Hill Book Company, Inc.

LAPEDES, D. N., Editor, 1978, *McGraw-Hill Dictionary of Scientific and Technical Terms*, 2nd ed. New York, N.Y.: McGraw-Hill Book Company.

McGraw-Hill Encyclopedia of Science and Technology, 1977, New York, N.Y.: McGraw-Hill Book Company (15 volumes).

POUGH, F. H., 1976, *A Field Guide to Rocks and Minerals* (4th ed.). Boston: Houghton Mifflin Company.

PRINZ, M., G. HARLOW, and J. PETERS, Editors, 1978, *Simon and Schuster's Guide to Rocks and Minerals*. New York, N.Y.: Simon and Schuster.

RICE, C. M., 1955, *Dictionary of Geological Terms*. Princeton, New Jersey: C. M. Rice.

RUNCORN, S. K., General Editor, 1967, *International Dictionary of Geophysics*. Oxford-New York, N.Y.: Pergamon Press (2 volumes).

STOKES, W. L., and D. J. VARNES, 1955, *Glossary of Selected Geologic Terms*. Denver, Colorado: Colorado Scientific Society Proceedings Volume 16.

The Encyclopedia Americana, 1968 (International edition), New York, N.Y.: Americana Corporation.

The New Encyclopaedia Britannica, 1977 (15th edition). Chicago-London: Encyclopaedia Britannica, Inc.

UVAROV, E. B., and D. R. CHAPMAN, 1971, *A Dictionary of Science*. Baltimore, Maryland: Penguin Books, Inc.

Van Nostrand's Scientific Encyclopedia, 5th ed., 1976, D. M. Considine, ed. New York, N.Y.: Van Nostrand Reinhold Company.

VOLLMER, E., 1967, *Encyclopedia of Hydraulics, Soil and Foundation Engineering*. Amsterdam, Holland: Elsevier Publishing Company.

Webster's Seventh New Collegiate Dictionary, 1966. Springfield, Massachusetts: G. and C. Merriam Company, Publishers.

APPENDIX 4

ROCK DEFECTS

Term	Description
1	2
Bedding joint	Joint parallel to bedding.
Bedding plane	In sedimentary or stratified rocks, the division planes which separate the individual layers, beds, or strata.
Crack	A break without parting. To crack: to break so that fissures appear on the surface.
Fault	The displacement of rocks along a zone of fracture.
Fissure	An extensive crack, break, or fracture in the rocks.
Fold	A flexure or bend in strata or any planar structure when rocks are in a plastic condition.
Fracture	The manner of breaking and appearance of a mineral when broken. Breaks in rocks due to intense folding of faulting.
Joint	Fracture in rock, generally more or less vertical or transverse to bedding, along which no appreciable movement has occurred.

Joint set	A group of more or less parallel joints.
Parting	A small joint in coal or rock. A thin depositional layer separating thick deposits, as shale in a coal seam; also a joint or fissure.
Seam	A quarryman's term for a joint, cleft, or fissure in rocks. In mining and geology, the term designates a thin layer of stratum, usually of coal, clay or other material.

APPENDIX 5

CONVERSION FACTORS FOR UNITS OF MEASUREMENT

A. Use of SI Units in Geotechnical Engeneering

B. Conversion Factors

A. USE OF SI UNITS IN GEOTECHNICAL ENGINEERING

acceleration, a	m/s²
acceleration of gravity, g	9.80665 m/s²
velocity, v	m/s
coefficient of permeability, k	cm/s
length, L; depth L; d	m; cm; mm
displacement; settlement, s; deflection, ΔL	cm
distance, L	km; m; cm
area, A	cm²; m²
volume, V	cm³; m³
section modulus, Z	cm³; m³
moment of inertia, I, J	cm⁴; m⁴
forces of all kinds	N

force per unit length, P; surcharge, p	N/m
modulus of elasticity, E	N/m², or in Pa
allowable pressure on soil and rock, σ_{all}	N/m², or in Pa
contact pressure on soil, σ_o	N/m², or in Pa
stress in soil, σ; surcharge, p, q	N/m², or in Pa
shear stresses, τ, s; cohesion c	N/m², or in Pa
mantle (skin) friction; adhesion f	N/m², or in Pa
subgrade reaction of soil, C	N/cm³
unit weight of soil, γ	N/m³
total earth pressures and resistances, E_a, E_o, E_p	N/m or N

B. CONVERSION FACTORS

LENGTH

1 mm	= 0.03937 in. 1 cm = 10 mm = 0.3937 in.
1 m	= 100 cm = 1000 mm = 39.37 in. = 3.28083 ft.
1 in.	= 2.54 cm 1 ft = 0.304801 m 1 yd = 0.91443 m

AREA

1 cm²	= 0.15500 in.² = 0.001076387 ft² = 0.0001 m² = 100 m²
1 m²	= 10.76387 ft² = 10^{-4} ha = 10^4 cm²
1 ha	= 2.4710624 acres = 10,000 m² = 100 ares
1 in.²	= 6.45164 cm² = 6.45164 × 10^{-4} m²
1 ft²	= 0.0929 m² = 2.296 × 10^{-5} acre = 144 in.²
1 acre	= 43,560 ft² = 40,468420 ares = 0.4046842 ha
1 are	= 100 m²

VOLUME

1 cm³	= 10^{-6} m³ = 0.06102 in.³ = 3.531 × 10^{-5} ft³
1 m³	= 10^6 cm³ = 6.102 × 10^4 in.³ = 35.31445 ft³
1 in.³	= 5.787 × 10^{-4} ft³ = 16.38716 cm³ = 1.6387 × 10^{-5} m³
1 ft³	= 1728 in.³ = 2.832 × 10^4 cm³ = 0.02832 m³

APPENDIX

CAPACITY

1 liter	$= 1000 \text{ cm}^3 = 0.001 \text{ m}^3 = 61.02 \text{ in.}^3 = 3.531 \times 10^{-2} \text{ ft}^3$
	$= 0.26417$ U.S. gallons
1 U.S. gallon	$= 231 \text{ in.}^3 = 0.1337 \text{ ft}^3 = 3785 \text{ cm}^3 = 3.78543$ liters
	$= 0.00379 \text{ m}^3 = 0.83267$ Imperial gallons
1 cm^3	$= 0.001$ liters $= 2.642 \times 10^{-4}$ gallons
1 m^3	$= 1000$ liters $= 264.17047$ gallons $= 35.31 \text{ ft}^3$
1 ft^3	$= 7.481$ gallons $= 28.32$ liters
1 Imperial gallon	$= 277.4191 \text{ in.}^3 = 0.16054 \text{ ft}^3$
	$= 4545.9 \text{ cm}^3 = 4.5459$ liters
	$= 0.0045459 \text{ m}^3 = 1.20095$ U.S. gallons

FORCE

Force unit: newton (N).   Formula: (kg) (m/s^2)

Conversion factors for force apply only under standard acceleration of 9.80665 m/s^2 due to gravity.

Notations:

N	—	newton
Pa	—	pascal
g$_f$	—	force gram
kg$_f$	—	force kilogram
t$_f$	—	metric force ton
ton	—	short ton = 2000 lb
k	—	kip = 1000 lb

1 N	=	$(1.00000)(10^5)$	dynes
	=	$(1.01971)(10^2)$	g$_f$
	=	$(1.01971)(10^{-1})$	kg$_f$
	=	$(1.01971)(10^{-4})$	t$_f$
	=	$(2.24809)(10^{-1})$	lb
	=	$(2.24809)(10^{-4})$	k
	=	$(1.12405)(10^{-4})$	ton
1 kN	=	$(1.00000)(10^3)$	N
	=	$(1.00000)(10^{-3})$	MN
	=	$(1.00000)(10^8)$	dynes
	=	$(1.01971)(10^5)$	g$_f$
	=	$(1.01971)(10^2)$	kg$_f$
	=	$(1.01971)(10^{-1})$	t$_f$
	=	$(2.24809)(10^2)$	lb
	=	$(2.24809)(10^{-1})$	k
	=	$(1.12405)(10^{-1})$	ton

1 MN	=	$(1.00000)(10^6)$	N
	=	$(1.00000)(10^3)$	kN
	=	$(1.01971)(10^8)$	g_f
	=	$(1.01971)(10^5)$	kg_f
	=	$(1.01971)(10^2)$	t_f
	=	$(2.24809)(10^5)$	lb
	=	$(2.24809)(10^2)$	k
	=	$(1.12405)(10^2)$	ton
1 dyne	=	$(1.00000)(10^{-5})$	N
	=	$(1.01971)(10^{-3})$	g_f
	=	$(1.01971)(10^{-6})$	kg_f
	=	$(1.01971)(10^{-9})$	t_f
	=	$(2.24809)(10^{-6})$	lb
	=	1.00000	erg/cm
1 g_f	=	$(9.80665)(10^{-3})$	N
	=	$(9.80665)(10^2)$	dynes
	=	$(1.00000)(10^{-3})$	kg_f
	=	$(1.00000)(10^{-6})$	t_f
	=	$(2.20462)(10^{-3})$	lb
	=	$(1.10231)(10^{-6})$	ton
1 kg_f	=	9.80665	N
	=	$(9.80665)(10^5)$	dynes
	=	$(1.00000)(10^3)$	g_f
	=	$(1.00000)(10^{-3})$	t_f
	=	2.20462	lb
	=	$(2.20462)(10^{-3})$	k
	=	$(1.10231)(10^{-3})$	ton
1 t_f	=	$(9.80665)(10^3)$	N
	=	$(1.00000)(10^8)$	dynes
	=	$(1.00000)(10^6)$	g_f
	=	$(1.00000)(10^3)$	kg_f
	=	$(2.20462)(10^3)$	lb
	=	2.20462	k
	=	1.10231	ton
1 lb	=	4.44822	N
	=	$(4.44822)(10^5)$	dynes
	=	453.592	g_f
	=	0.453592	kg_f
	=	$(4.53592)(10^{-4})$	t_f
	=	$(1.00000)(10^{-3})$	k
	=	$(5.00000)(10^{-4})$	ton

APPENDIX

1 k (= kip)	=	$(4.44822)(10^3)$	N
	=	$(4.44822)(10^8)$	dynes
	=	$(4.53592)(10^5)$	g_f
	=	$(4.53592)(10^2)$	kg_f
	=	$(4.53592)(10^{-1})$	t_f
	=	$(1.00000)(10^3)$	lb
	=	$(5.00000)(10^{-1})$	ton
1 ton	=	$(8.89644)(10^3)$	N
	=	$(8.89644)(10^8)$	dynes
	=	$(9.07184)(10^5)$	g_f
	=	$(9.07184)(10^2)$	kg_f
	=	$(9.07184)(10^{-1})$	t_f
	=	$(2.00000)(10^3)$	lb
	=	2.00000	k

STRESS

$1\ N/m^2 = 1\ Pa$	=	$(1.00000)(10^{-4})$	N/cm^2
	=	$(1.01972)(10^{-2})$	g_f/cm^2
	=	$(1.01972)(10^{-5})$	kg_f/cm^2
	=	$(1.01972)(10^{-1})$	kg_f/m^2
	=	$(1.01972)(10^{-4})$	t_f/m^2
	=	$(1.45038)(10^{-4})$	$lb/in.^2$
	=	$(2.08854)(10^{-2})$	lb/ft^2
	=	$(2.08854)(10^{-5})$	k/ft^2
	=	$(1.04427)(10^{-5})$	ton/ft^2
1 kg_f/cm^2	=	$(9.80665)(10^4)$	N/m^2
	=	9.80665	N/cm^2
	=	$(1.00000)(10^4)$	kg_f/m^2
	=	$(1.00000)(10)$	t_f/m^2
	=	$(1.42233)(10)$	$lb/in.^2$
	=	$(2.04817)(10^3)$	lb/ft^2
	=	2.04817	k/ft^2
	=	1.02409	ton/ft^2
1 t_f/m^2	=	$(9.80665)(10^3)$	N/m^2
	=	$(1.00000)(10^3)$	kg_f/m^2
	=	$(1.00000)(10^{-1})$	kg_f/cm^2
	=	$(1.02408)(10^{-1})$	ton/ft^2
	=	$(2.04817)(10^{-1})$	k/ft^2
	=	$(2.04817)(10^2)$	lb/ft^2
	=	1.42233	psi

1 lb/in.²	=	(6.89476) (10³)	N/m²
	=	(6.79476) (10⁻¹)	N/cm²
	=	(7.03069) (10)	g_f/cm²
	=	(7.03069) (10⁻²)	kg_f/cm²
	=	(7.03069) (10²)	kg_f/m²
	=	(7.03069) (10⁻¹)	t_f/m²
	=	(1.44000) (10²)	lb/ft²
	=	(1.44000) (10⁻¹)	k/ft²
	=	(7.20000) (10⁻²)	ton/ft²
1 lb/ft²	=	(4.78803) (10)	N/m²
	=	(4.78803) (10⁻³)	N/cm²
	=	(4.88242) (10⁻¹)	g_f/cm²
	=	(4.88242) (10⁻⁴)	kg_f/cm²
	=	4.88242	kg_f/m²
	=	(4.88242) (10⁻³)	t_f/m²
	=	(6.94444) (10⁻³)	lb/in.²
	=	(1.00000) (10⁻³)	k/ft²
	=	(5.00000) (10⁻⁴)	ton/ft²
1 k/ft²	=	1000	lb/ft²
	=	0.500000	ton/ft²
	=	(4.78803) (10⁴)	N/m²
	=	(4.88242) (10⁻¹)	kg_f/cm²
	=	(4.88242) (10³)	kg_f/m²
	=	4.88242	t_f/m²
	=	6.94444	psi
	=	(1.00000) (10⁻⁶)	lb/ft²
	=	(5.00000) (10⁻¹)	ton/ft²
1 ton/ft²	=	2000	lb/ft²
	=	(9.57605) (10⁴)	N/m²
	=	(9.76485) (10⁻¹)	kg_f/cm²
	=	(9.76485) (10³)	kg_f/m²
	=	9.76485	t_f/m²
	=	(1.38888) (10)	psi
	=	(2.00000) (10³)	lb/ft²
	=	2.00000	k/ft²

UNIT WEIGHT

1 N/m³	=	(1.01972) (10⁻⁴)	g_f/cm³
	=	(1.01972) (10⁻¹)	kg_f/m³
	=	(1.01972) (10⁻⁴)	t_f/m³

	=	(6.36586) (10^{-3})	lb/ft^3
	=	(3.18293) (10^{-6})	ton/ft^3
1 g$_f$/cm^3	=	(9.80665) (10^3)	N/m^3
	=	(1.00000) (10^{-3})	kg$_f$/cm^3
	=	(1.00000) (10^3)	kg$_f$/m^3
	=	1.00000	t$_f$/m^3
	=	(6.24277) (10)	lb/ft^3
	=	(3.12139) (10^{-2})	ton/ft^3
1 kg$_f$/m^3	=	9.80665	N/m^3
	=	(1.00000) (10^{-3})	g$_f$/cm^3
	=	(1.00000) (10^{-6})	kg$_f$/cm^3
	=	(1.00000) (10^{-3})	t$_f$/m^3
	=	(6.24277) (10^{-2})	lb/ft^3
	=	(3.12139) (10^{-5})	ton/ft^3
1 t$_f$/m^3	=	(9.80665) (10^3)	N/m^3
	=	(1.00000) (10^3)	kg$_f$/m^3
	=	(6.24277) (10)	lb/ft^3
	=	(3.12139) (10^{-2})	ton/ft^3
1 lb/ft^3	=	(1.57092) (10^2)	N/m^3
	=	(1.60189) (10)	kg$_f$/m^3
	=	(1.60189) (10^{-2})	t$_f$/m^3
	=	(5.00000) (10^{-4})	ton/ft^3
1 ton/ft^3	=	(3.14184) (10^5)	N/m^3
	=	(3.20379) (10^4)	kg$_f$/m^3
	=	(3.20379) (10)	t$_f$/m^3
	=	(2.00000) (10^3)	lb/ft^3

1 at = 1.0 kg$_f$/cm^2 = 736 torr =
 = 1 technical atmosphere

1 atm = 1.0335 kg$_f$/cm^2 = 760 torr =
 = 1 physical atmosphere

TEMPERATURE

$$1 \,[K] = 1\,[C] = (1.8)\,[F]$$
$$1\,[F] = (0.555)\,[C] = (0.555)\,[K]$$

Absolute zero: $0\,°K = -273.15\,°C = -459.67\,°F$

$$\begin{aligned}
T_C &= (5/9)(T_F - 32°) \\
&= T_K - 273.15° \\
T_K &= T_C + 273.15° \\
&= (T_F + 459.67°)/1.8 \\
T_F &= (9/5)(T_C) + 32° \\
&= (T_C + 17.8°)(1.8)
\end{aligned}$$

HEAT

1 BTU	= 1 B	= $(1.05435)(10^3)$ Nm (or Joule, J.)
		= 0.252 Cal = 252 cal
1 Cal	= 1000 cal = 3.968 B	
1 cal	= 0.003968 B	
1 Nm	= 1 J = 1 ws	
	= $(2.77778)(10^{-4})$ wh	
	= $(2.77778)(10^{-7})$ kwh	
	= $(2.38846)(10^{-4})$ Cal	
	= $(2.38846)(10^{-1})$ cal	
	= $(9.47741)(10^{-4})$ B	
	= $(3.77672)(10^{-7})$ PSh	
	= $(3.72505)(10^{-7})$ HPh	

1 Cal/(m)(h)(°C)	=	(0.67197) B/(ft)(h)(°F)
	=	(8.064) B-in./(ft²)(h)(°F)
	=	$(2.778)(10^{-3})$ cal/(cm)(s)(°C)
	=	$(1.163)(10^{-2})$ w/(cm)(°C)
1 cal/(cm)(s)(°C)	=	(360) Cal/(m)(h)(°C)
	=	(241.9) B/(ft)(h)(°F)
	=	$(2.9)(10^3)$ B-in./(ft²)(h)(°F)
	=	(4.18680) w/(cm)(°C)
1 B/(ft)(h)(°F)	=	(1.48817) Cal/(m)(h)(°C)
	=	$(4.134)(10^{-3})$ cal/(cm)(s)(°C)

	=	(12) B-in./(ft²) (h) (°F)
	=	(1.730) (10⁻²) w/(cm) (°C)
1 B/(ft²) (h) (°F/in.)	=	(0.124) Cal/(m) (h) (°C)
	=	(3.447) (10⁻⁴) cal/(cm) (s) (°C)
	=	(8.34) (10⁻²) B/(ft) (h) (°F)
	=	(1.44314) (10⁻³) w/(cm) (°C)
1 w/(cm) (°C)	=	(85.9845) Cal/(m) (h) (°C)
	=	(0.2388) cal/(cm) (s) (°C)
	=	(57.81477) B/(ft) (h) (°F)
	=	(6.92933) B/(ft²) (h) (°F/in.)

APPENDICES 6 AND 7

DYNAMIC VISCOSITY TABLES FOR WATER
DYNAMIC VISCOSITY CORRECTION FACTOR TABLES FOR WATER

Prepared by A. R. Jumikis

In analyzing soil freezing experiment and permeability data, there is a need for dynamic viscosity values of water in the freezing range of temperatures, as well as for values of viscosity for whole degrees and for every decimal of degree.

The viscosity values found in various publications are not usually given for such close temperatures. Therefore, it was expedient to prepare such viscosity tables, which would be useful for work in connection with studies on freezing soil systems and on soil moisture migration. The viscosity and the viscosity correction tables were calculated and prepared based on N. E. Dorsey's data found in his book entitled *Properties of Ordinary Water-Substance* (New York, Reinhold Publishing Corporation, 1940), the viscosity values for most part of which are based on *International Critical Tables for Numerical Data, Physics, Chemistry and Technology*, published for the U.S. National Research Council by the McGraw-Hill Book Co., Inc., New York, 1929, vol. 5, p. 10.

The published viscosity values were plotted to a large scale on millimeter paper, the viscosity ordinates were connected by a curve, and the viscosity values for each decimal of a degree of temperature were scaled on such a curve and tabulated. The work on freezing soil systems was sponsored by the National Science Foundation, Washington, D.C., and performed in the Soil Mechanics and Foundation Engineering Laboratory of the Department of Civil Engineering at Rutgers—The State University, New Brunswick, New Jersey.

APPENDIX 6

Dynamic Viscosity of Water, $\eta[\text{g}/(\text{cm sec})]$ Table

T°C	0	0.1	0.2	0.3	0.4	0.5	0.6	0.7	0.8	0.9
−10	0.0260	—	—	—	—	—	—	—	—	—
−9	250	0.0251	0.0252	0.0253	0.0254	0.0255	0.0256	0.0257	0.0258	0.0259
−8	240	241	242	243	244	245	246	247	248	249
−7	230	231	232	233	234	235	236	237	238	239
−6	220	221	222	223	224	225	226	227	228	229
−5	0.02140	0.02142	0.02146	0.02151	0.02156	0.02162	0.02168	0.02175	0.02183	0.02191
−4	2050	2060	2067	2075	2083	2092	2100	2110	2120	2130
−3	1979	1980	1987	2000	2007	2015	2022	2030	2036	2044
−2	1910	1915	1922	1930	1935	1943	1950	1956	1965	1972
−1	1840	1847	1853	1860	1866	1873	1880	1887	1895	1902
0	0.01790	0.01795	0.01799	0.01803	0.01808	0.01813	0.01818	0.01824	0.01830	0.01835
0	0.01790	0.01785	0.01778	0.01773	0.01766	0.01760	0.01754	0.01748	0.01742	0.01736
1	0.01730	0.01724	0.01718	0.01712	0.01706	0.01700	0.01694	0.01687	0.01682	0.01675
2	1670	1663	1657	1651	1647	1640	1635	1629	1623	1617
3	1610	1605	1600	1594	1588	1583	1578	1573	1568	1564
4	1560	1555	1551	1546	1543	1539	1535	1531	1528	1524
5	1520	1515	1510	1505	1501	1493	1487	1484	1478	1475

Dynamic Viscosity of Water, $\eta[g/(cm\ sec)]$ Table (continued)

T°C	0	0.1	0.2	0.3	0.4	0.5	0.6	0.7	0.8	0.9
6	0.01470	0.01466	0.01463	0.01458	0.01454	0.01450	0.01446	0.01442	0.01438	0.01434
7	1430	1426	1422	1418	1414	1410	1406	1402	1398	1394
8	1390	1387	1382	1377	1373	1370	1365	1361	1357	1353
9	1350	1345	1342	1338	1334	1330	1326	1322	1318	1313
10	1308	1305	1301	1298	1293	1290	1285	1282	1278	1275
11	0.01271	0.01268	0.01264	0.01261	0.01258	0.01254	0.01350	0.01246	0.01243	0.01240
12	1236	1233	1229	1226	1222	1218	1215	1212	1208	1205
13	1202	1199	1196	1193	1190	1186	1183	1180	1176	1173
14	1171	1166	1164	1161	1157	1154	1151	1148	1145	1143
15	1140	1136	1133	1131	1128	1125	1122	1119	1116	1113
16	0.01111	0.01107	0.01104	0.01101	0.01098	0.01095	0.01093	0.01090	0.01087	0.01084
17	1082	1078	1076	1073	1070	1068	1065	1062	1059	1056
18	1055	1051	1048	1046	1043	1040	1038	1035	1033	1030
19	1029	1025	1022	1020	1016	1014	1012	1009	1007	1006
20	1005	1003	1000	0998	0995	0993	0991	0988	0986	0983

Dynamic Viscosity of Water, η [g/(cm sec)] Table (continued)

T°C	0	0.1	0.2	0.3	0.4	0.5	0.6	0.7	0.8	0.9
21	0.00981	0.00979	0.00976	0.00974	0.00972	0.00969	0.00967	0.00965	0.00962	0.00960
22	958	956	953	951	949	947	945	942	940	938
23	936	933	931	929	927	925	922	920	918	916
24	914	912	910	908	906	904	902	900	898	896
25	894	892	890	888	886	884	882	880	788	786
26	0.00874	0.00872	0.00870	0.00868	0.00866	0.00864	0.00862	0.00860	0.00858	0.00856
27	854	852	850	848	846	845	843	841	839	837
28	836	834	832	830	828	826	824	823	821	819
29	818	816	814	812	811	809	807	806	804	802
30	801	799	798	796	794	792	791	788	787	785
31	0.00784	0.00782	0.00781	0.00779	0.00777	0.00776	0.00774	0.00772	0.00771	0.00769
32	768	766	764	763	861	760	758	757	755	754
33	752	751	749	748	746	745	743	742	740	739
34	737	736	734	733	731	730	728	726	725	724
35	722	720	719	718	716	715	714	712	711	709
36	0.00708	0.00707	0.00705	0.00704	0.00703	0.00701	0.00700	0.00699	0.00697	0.00696
37	695	693	692	691	689	688	687	685	684	683
38	681	680	679	677	676	675	673	672	670	669
39	668	667	666	665	663	662	661	660	658	657
40	656	—	—	—	—	—	—	—	—	—

Appendix 7

Dynamic Viscosity Correction Factors, η_T/η_{20} for Water

Calculated and prepared by A. R Jumikis

T°C	0.0	0.1	0.2	0.3	0.4	0.5	0.6	0.7	0.8	0.9
-10	2.58706	—	—	—	—	—	—	—	—	—
-9	2.48756	2.49751	2.50746	2.51741	2.52736	2.53731	2.54726	2.55721	2.56716	2.57711
-8	2.38805	2.40796	2.40796	2.41791	2.42786	2.43781	2.44776	2.45771	2.46766	2.47761
-7	2.28855	2.30845	2.30845	2.31840	2.32835	2.33830	2.34825	2.35820	2.36815	2.37810
-6	2.18905	2.19900	2.20895	2.21890	2.22885	2.23880	2.24875	2.25870	2.26865	2.27860
-5	2.12935	2.13134	2.13532	2.14029	2.14527	2.15124	2.15721	2.16417	2.17213	2.18009
-4	2.03980	2.04975	2.05671	2.06467	2.07263	2.08159	2.08955	2.09950	2.10945	2.11940
-3	1.96915	1.97014	1.97711	1.99004	1.99701	2.00497	2.01194	2.01990	2.02587	2.03383
-2	1.90049	1.90547	1.91243	1.92039	1.92537	1.93333	1.94029	1.94626	1.95522	1.96218
-1	1.83084	1.83781	1.84378	1.85074	1.85671	1.86368	1.87064	1.87761	1.88557	1.89253
0	1.78109	1.78606	1.79004	1.79402	1.79900	1.80398	1.80895	1.81492	1.82089	1.82587
0	1.78109	1.77611	1.76915	1.76417	1.75721	1.75124	1.74527	1.73930	1.73333	1.72736

Dynamic Viscosity Correction Factors, η_T/η_{20} for Water (continued)

T°C	0.0	0.1	0.2	0.3	0.4	0.5	0.6	0.7	0.8	0.9
1	1.72139	1.71542	1.70945	1.70348	1.69751	1.69154	1.68557	1.67860	1.67363	1.66666
2	1.66169	1.65472	1.64875	1.64278	1.63880	1.63184	1.62686	1.62089	1.61492	1.60895
3	1.60199	1.59701	1.59203	1.58606	1.58009	1.57512	1.57014	1.56517	1.56019	1.55621
4	1.55223	1.54328	1.54328	1.53830	1.53532	1.53134	1.52736	1.52338	1.52039	1.51641
5	1.51243	1.50746	1.50248	1.49751	1.49353	1.48557	1.47960	1.47661	1.47064	1.46766
6	1.46268	1.45870	1.45572	1.45074	1.44676	1.44278	1.43880	1.43482	1.42084	1.42686
7	1.42288	1.41890	1.41492	1.41094	1.40696	1.40298	1.39900	1.39502	1.39104	1.38706
8	1.38308	1.38009	1.37512	1.37014	1.36616	1.36318	1.35820	1.35422	1.35024	1.34626
9	1.34328	1.33830	1.33532	1.33134	1.32736	1.32338	1.31940	1.31542	1.31094	1.30646
10	1.30149	1.29850	1.29452	1.29104	1.28656	1.28358	1.27860	1.27562	1.27164	1.26865
11	1.26467	1.26169	1.25771	1.25472	1.25174	1.24776	1.24378	1.23980	1.23681	1.23383
12	1.22985	1.22686	1.22288	1.21990	1.21592	1.21194	1.20895	1.20597	1.20199	1.19900
13	1.19601	1.19303	1.19004	1.18706	1.18407	1.18009	1.17711	1.17412	1.17014	1.16716
14	1.16507	1.16019	1.15820	1.15522	1.15124	1.14825	1.14527	1.14228	1.13930	1.13731
15	1.13432	1.13034	1.12736	1.12537	1.12238	1.11940	1.11641	1.11343	1.11044	1.10746

Dynamic Viscosity Correction Factors, η_T/η_{20} for Water (continued)

T°C	0.0	0.1	0.2	0.3	0.4	0.5	0.6	0.7	0.8	0.9
16	1.10547	1.10149	1.09850	1.09552	1.09253	1.08955	1.08756	1.08457	1.08159	1.07860
17	1.07661	1.07263	1.07064	1.06766	1.06467	1.06268	1.05970	1.05671	1.05373	1.05074
18	1.04975	1.04577	1.04278	1.04079	1.03781	1.03482	1.03283	1.02985	1.02787	1.02487
19	1.02388	1.01990	1.01691	1.01492	1.01094	1.00895	1.00696	1.00399	1.00199	1.00099
20	1.00000	0.99761	0.99502	0.99303	0.99034	0.98805	0.98587	0.98328	0.98109	0.97810
21	0.97611	0.97363	0.97114	0.96915	0.96716	0.96437	0.96218	0.96019	0.95721	0.95522
22	0.95323	0.95124	0.95124	0.94626	0.94427	0.94228	0.94029	0.93731	0.93532	0.93333
23	0.93134	0.92835	0.92636	0.92437	0.92238	0.92039	0.91741	0.91542	0.91343	0.91144
24	0.90945	0.90746	0.90547	0.90348	0.90149	0.89950	0.89751	0.89552	0.89353	0.89154
25	0.88955	0.88756	0.88756	0.88358	0.88159	0.87960	0.87761	0.87562	0.87363	0.87164
26	0.86965	0.86766	0.86567	0.86368	0.86169	0.85970	0.85771	0.85572	0.85373	0.85174
27	0.84975	0.84776	0.84577	0.84378	0.84179	0.84079	0.83880	0.83681	0.83482	0.83283
28	0.83184	0.82985	0.82786	0.82587	0.82388	0.82189	0.81990	0.81890	0.81691	0.81492
29	0.81393	0.81194	0.80995	0.80796	0.80696	0.804497	0.80298	0.80199	0.80000	0.79800
30	0.79701	0.79502	0.79402	0.79203	0.79004	0.78805	0.78706	0.78407	0.78308	0.78208

APPENDICES 6 AND 7

Dynamic Viscosity Correction Factors, η_T/η_{20} for Water (continued)

$T°C$	0.0	0.1	0.2	0.3	0.4	0.5	0.6	0.7	0.8	0.9
31	0.78009	0.77810	0.77711	0.77512	0.77313	0.77213	0.77014	0.76815	0.76716	0.76517
32	0.76517	0.76218	0.76019	0.75920	0.75721	0.75621	0.75422	0.75323	0.75124	0.75024
33	0.74825	0.74526	0.74527	0.74427	0.74228	0.74129	0.73930	0.73830	0.73631	0.73532
34	0.73333	0.73233	0.73034	0.72935	0.72736	0.72636	0.72437	0.72238	0.72139	0.72039
35	0.71840	0.71641	0.71542	0.71442	0.71243	0.71144	0.71044	0.70845	0.70746	0.70547
36	0.70447	0.70348	0.70149	0.70049	0.69950	0.69751	0.69651	0.69552	0.69353	0.69253
37	0.69154	0.68955	0.68855	0.68756	0.68557	0.68457	0.68358	0.68159	0.68059	0.67960
38	0.67761	0.67661	0.67562	0.67363	0.67263	0.67164	0.66965	0.66865	0.66666	0.66567
39	0.66467	0.66368	0.66268	0.66169	0.65970	0.65870	0.65771	0.65671	0.65472	0.65373
40	0.65273	—	—	—	—	—	—	—	—	—

ABOUT THE AUTHOR

Alfreds Richards Jumikis was born in Riga, Latvia. Since 1952, Dr. Jumikis has been a member of the Department of Civil and Environmental Engineering of the College of Engineering of Rutgers, The State University of New Jersey. Professor Jumikis had previously taught at the University of Delaware, and at the University of Latvia in Riga. He holds earned academic degrees of Dr. Eng. Sc., Dr. techn., and Dr.-Ing. from the University of Latvia, the Technical University in Vienna, and the University of Stuttgart, respectively.

Dr. Jumikis is the author of more than one hundred publications, among them 17 books and booklets on shale rocks, soil mechanics, foundation engineering, thermal soil mechanics, and a book on Glossary of Terms in Thermal Soil Mechanics.

Professor Jumikis has lectured and presented papers at national and international conferences.

Dr. Jumikis is a member of several professional engineering and scientific societies, and is the recipient of three National Science Foundation grants for his research in thermal soil mechanics.

Alfreds R. Jumikis is a registered professional engineer (P. E.), and consults on problems in geotechnics.

INDEX

AUTHOR INDEX

A

Abbet, R. W., 428
Abel, J. R., 430
Airy, G. B., 339-341, 356
Alexander, L. G., 260
Alley, R. P., 256
Andreae, C., 254
Andrus, F. M., 427
Archambault, G., 514
Attewell, P. B., 72

B

Bacon, R., Sir, 78
Baidyuk, B. V., 255
Bain, G. W., 427
Baldwin, B., 71
Barkan, D. D., 260
Barnes, H. T., 85, 89
Barney, K. R., 255, 427
Barson, J. M., 263
Båth, M., 261, 386, 427
Berg, G. V., 427
Bergman, M., 428
Berkey, C. P., 428
Bernatzik, W., 296
Birch, F., 255
Birkenmeyer, M., 460, 466
Bjerrum, L., 428, 433
Blume, J. A., 434
Blyth, G. H., 72
Bolni, H. W., 428
Bolt, B. A., 428
Borst, R. L., 261
Boussinesq, J. V., 228, 255
Bowen, R., 296
Brace, W. F., 255
Bray, J. W., 433, 467, 500, 511
Brekke, T. L., 255, 296
Brock, D., 261
Brown, E. T., 513
Bullen, K. E., 261
Burwell, E. B., 255

C

Calyxite, 110
Cambefort, H., 255, 428
Cancani, A., 379-382
Carlson, R. W., 433
Casagrande, A., 138, 255, 287, 296, 397, 428
Cecil III, O. S., 296
Cedergren, H. R., 513
Chae, Y. S., 433
Challinor, J., 559
Chapman, D. R., 559
Clark, Jr., S. P., 255
Cleaves, A. B., 72
Cleveland, G. B., 428
Clouh, G. W., 466
Coates, D. F., 72, 261, 559
Collin, A., 428, 489, 511
Connolly, J. E., 90
Cook, G. W., 72, 257
Considine, D. M., 559
Cording, E. J., 261, 445-447, 466
Coulomb, C. A., 161-166, 193, 194, 197, 200, 201, 236, 255, 339, 340, 345, 350, 361, 443, 460, 463, 482, 483, 501
Coulter, H. W., 428
Crieger, W. P., 401, 428

D

Darcy, H. P. G., 137, 139, 140, 255
Deere, D. U., 62-64, 71
Dimas, J., 261, 296, 405, 428
Dodson, E. L., 469
Donath, F. A., 261
Donovan, N. C., 428
Dowrick, D. J., 433
Dreyer, W., 72
Duckworth, W. H., 205, 255
Dumont-Villares, A. D., 89
Duncan, N., 72
Duvall, W. I., 72, 200, 259, 305, 310, 330, 335, 430

E

Eckel, E. B., 428
Eichinger, A., 431
Eide, O., 428, 433
Emery, C. L., 38, 71, 428

Engle, H. M., 428
Erickson, H. B., 291, 296
Eshbach, O. W., 158, 255
Espinosa, A. F., 428
Euler, L., 315
Everell, M. D., 255
Evison, F. F., 429

F

Fadum, R. E., 138, 255
Fairhurst, C., 250, 255, 256, 261, 514
Farmer, I. W., 72, 80, 89, 168, 180, 202, 256, 441-443, 466
Feld, J., 370, 429
Fellenius, B., 433
Fellenius, W., 511
Fenner, R., 89, 363, 364
Folberth, P. J., 433
Franklin, J. A., 433
Frey, J. Q., 71
Freitas, de, M. H., 72
Fröhlich, O. K., 433, 511
Fumagally, E., 261

G

Gaziev, E. G., 433
Gildersleeve, W. K., 90
Glen, J., 85, 89
Glossop, R., 296
Gold, L. W., 85, 89
Goodier, J. N., 260, 335, 353
Goodman, R. E., 256, 513
Gordon, G., 89, 429
Graham, A. C., 559
Gralewska, A., 485, 513
Gregory, C. E., 261
Gretener, P, E., 261
Griffith, A. A., 161, 168, 256
Griffith, J. H., 131, 155, 256
Griffith, D. H., 261
Griggs, D. T., 182, 261, 369, 429
Grobbelaar, C., 261
Gumensky, D. B., 429
Gutenberg, E. G., 429, 433

H

Haimson, B. C., 256
Hall, J. R., Jr., 262
Handin, J., 182, 513
Hanna, T. H., 467
Hansen, W. R., 429

Hard, H. C., 256
Harlow, G., 559
Harris, C. M., 511
Hartman, H., 467
Harza, L. F., 433
Hayes, J., 247-249, 258
Heck, N. H., 386, 429, 511
Heflin, L. H., 467
Heim, A., 303, 310
Heiskanen, W. A., 261
Heitfeld, K. H., 297
Hellblom, K., 467
Hencky, H., 81, 89
Herget, G., 255
Henning, D., 263
Herzog, M., 468
Heuze, F. E., 256, 513
Hino, K., 433
Hix, Jr., C. F., 256
Hobbs, D. W., 256
Hobst, L., 467
Hocking, G., 513
Hoek, E., 433, 467, 485, 500, 511, 513
Holman, W. W., 256
Holmsen, P., 429
Hooke, R., 79, 80, 159, 173, 174, 314, 320, 331, 339
Hori, M., 430
Horn, W. L., 428
Horvath, J., 256, 307, 310
Hosking, A. D., 468
Hoskins, E. R., 257
Hunt, V. D., 559
Hvorslev, M. J., 94, 256

I

Iacopi, R., 429
Idriss, I. M., 431

J

Jaeger, C., 35, 72, 257, 429
Jaeger, Ch., 198, 257, 409
Jaeger, J. C., 72, 257, 511
Jäger, B., 468
Jähde, H., 288, 291, 292, 296
Jampole, S., 90
Janbu, N., 511
Jessberger, H. L., 90
John, K. W., 513
Johnson, A. M., 72, 257
Johnson, A. W., 296
Joosten, H. J., 290, 291, 296

AUTHOR INDEX

Jörstad, F. A., 255
Judd, W. R., 72, 73, 89, 124, 258, 559
Jumikis, A. A., 71, 258, 304
Jumikis, A. R., 71, 89, 256-258, 296, 304, 307, 335, 353, 364, 429, 430, 467, 511, 512, 559

K

Kapp, M. S., 467
Kapur, K. K., 434
Kármán, Th., 82, 89, 182, 258
Karol, R. H., 297
Kastner, H., 335, 341, 344, 353, 363, 364
Kehle, R. O., 261
Keller, W. D., 261
Kendrick, T. D., 386, 430
Kenney, N. T., 433
Kenwort, C. E., 71
Kiersch, G. A., 35, 258, 424, 426, 430
King, R. F., 261
Kjellman, W., 433
Koenig, H. W., 297
Komarnitskii, N. I., 261, 272, 273, 297, 513
Kovari, K., 258
Krebs, E., 258
Krsmanović, D., 261
Krynine, D. P., 72, 124, 258, 559
Kulp, J. L., 42, 71
Kurdi, I. I., 467

L

Ladanyi, B., 514
Ladd, G. E., 430, 512
Lama, R. D., 260, 261
Lamé, G., 260, 312, 317
Lang, T., 467
Lee, W. H. K., 262
Leeman, E. R., 247-249, 258
Legget, R. F., 72, 262, 514
Lemke, R. W., 430
Lepedes, D. N., 559
Lizzi, F., 505-508, 512
Lo, K. Y., 430, 434
Lomnitz, C., 262
Londe, P., 258, 297, 430, 433, 467, 468, 500, 512, 514
Look, A. D., 297
Lorenz, H., 262
Love, A. E. H., 262
Loverdo, De., 90
Lugeon, M., 140, 258, 292, 297, 399, 400
Lukajic, B., 430

M

Macdonald, G. A., 428
Manghnani, M. H., 258
Marcello, C., 258, 433
Martin, D. R., 468
Mary, M., 408, 433
Mattner, R. H., 468
Maurseth, R. D., 434, 514
McCormack, R. K., 256
McCoy, B. O., 401, 428
Meinerz, F. A., 261
Mencl, V., 258, 434, 512, 513
Mercalli, G., 258, 378-382, 384
Middlebrooks, T. A., 434
Migliaccio, R. R., 428
Miller, R. P., 62, 63, 71
Minard, J. P., 256
Mindlin, R. D., 335
Mises, R. von., 82, 90
Mohr, F., 262
Mohr, Otto, 161, 163-167, 193, 194, 197, 201, 258, 339, 340, 345, 350, 361, 482, 483, 484, 512
Mohs, F., 156, 157
Moore, J. T., 297
Moos, A., 258
Morfeldt, C. O., 369, 430
Muhs, H., 263
Müller, L., 72, 258, 262, 434, 514
Mumpton, F. A., 107, 108, 258, 259
Muscat, M., 262

N

Nádai, A., 79, 81, 90, 259
Nasiatka, Th. M., 35
Nelsen, J. C., 467
Nesbit, R. H., 255
Neumann, F., 378-381, 432
Newmark, N. M., 262, 434, 514
Nichols, T. C., 430
Nothwood, T. D., 434

O

Oakeshott, G. B., 434
Obert, L., 72, 200, 259, 262, 305, 310, 330, 335, 430
Ormsby, W. C., 107, 108, 259
Ortlepp, W. D., 468
Ostermeyer, H., 467
Ôura, H., 85, 90

P

Packham, G. R., 297
Palmer, J. H. L., 434
Patton, F. D., 259
Pavlović, M., 262
Pender, E. P., 468
Pequinot, C. A., 35
Persen, L. N., 262
Peters, J., 559
Petterson, K. E., 512
Petzny, H., 514
Pierzinski, D. C., 432
Piteau, D. R., 468
Poisson, S. D., 159, 174, 187-189, 193, 233, 244, 304, 307, 314, 318, 326, 329-331, 334, 356, 362
Popov, S. I., 272, 273
Pough, F. H., 559
Prandtl, L., 81, 90
Price, N. J., 157, 259
Prinz, M., 559
Proctor, R. V., 90, 364, 432
Puls, L. G., 259, 431
Pynnonen, R. O., 297

Q

Quervain, de, F., 199, 258

R

Rainer, J. H., 434
Ranke, A., 467
Rayleigh (J. W. Strutt)., 262
Real, C. R., 432
Rechitski, V. I., 433
Redlinger, J. F., 469
Rice, C. M., 559
Richart, Jr., F. E., 262, 363, 364
Richey, J. E., 398, 399, 431, 512
Richter, C. F., 383, 384, 429, 431
Rinne, J. E., 431
Ritschie, A. M., 512
Roberts, A. F., 259, 262
Robertson, E. C., 259
Rocha, M., 244, 246, 259, 262, 271, 297, 431
Rodrigues, F. P., 262
Rolfe, S. T., 263
Roš, M., 431
Rosenblueth, E., 262, 514
Rowe, P. W., 259
Runcorn, S. K., 559

S

Saint-Venant, de B., 82, 90
Saluja, S. S., 260
Sanger, F. J., 90
Savage, W. Z., 430
Savini, T., 261, 296, 428
Scharpe, C. F. S., 431
Scheidegger, A. E., 263
Schleicher, F., 468
Schmidt, A. E., 469
Schnitter, G., 425, 431
Schousboe, J., 468
Schreiber, E., 258
Schriever, R., 428, 511
Schroter, G. A., 434, 514
Schütz, J. R., 72
Schultze, E., 263
Scott, R. F., 428
Seed, H. B., 431
Seltz-Petrash, A., 71, 469
Serafim, J. L., 259, 431
Sharma, P. V., 259
Sharpe, C. F. S., 514
Sieberg, A., 379-382, 384, 431
Simmonds, A. W., 297, 431
Singer, F. L., 259
Singh, D. P., 263
Singh, S., 428
Sir Roger Bacon, 78
Slater, H., 72
Slusarchuk, W. A., 257
Smith, D. P., 391, 431
Smith, G. R., 90
Smith, N., 35
Smith, R., 469
Snow, D. T., 259
Snyder, J. L., 263
Soga, N., 258
Sozen, M. A., 431
Stacey, F. D., 263
Stagg, K. G., 72, 232, 259
Stearns, D. W., 513
Steinbrugge, K. V., 390, 431, 432
Stewart, G. C., 90
Stini, J., 72, 76, 90
Stokes, G. I., 71
Sturm, E., 71
Swinzow, G. K., 88, 90
Széchy, K., 72, 90, 202, 260, 335, 363, 364, 432

AUTHOR INDEX

T

Tait, R. G., 469
Talobre, J. A., 72, 303, 310
Taylor, H. T., 469
Ter-Stepanian, G., 263
Terzaghi, K., 71, 260, 263, 292, 335, 363-365, 400, 432, 512
Thomas, L. J., 260
Timoshenko, S., 260, 335, 353
Toppozoda, T. R., 432
Trefethen, J. M., 72
Tresca, H., 82, 90, 161, 162, 164, 165, 260
Tsui, Y., 466

U

Ugolini, F. C., 71
Uvarov, E. B., 559

V

Van, K., 256
Varnes, D. J., 559
Vening, M. F. A., 261
Vigier, G., 514
Viner-Brady, E. V., 434, 514
Voight, B., 434
Vollmer, E., 559
Vormeringer, R., 514
Vutukuri, V. S., 260, 261

W

Wahlstrom, E. E., 72
Walters, R. C. S., 72

Ware, K. R., 469
Weber, E., 469
Wenner, F., 263
Wesseloh, J., 415, 432
Weyermann, W., 261, 296, 428
Wiegel, R. L., 432
White, T. L., 90, 364, 432
White, W. S., 71, 432
Wilhoyt, E. E., Jr., 71
Wilkinson, T. A., 432
Williams, C. I., 448-450, 468
Windes, S. L., 259
Windolf, G., 71, 469
Winkler, H. G. F., 103, 260
Wittke, W., 513
Wöhlbier, H., 263
Wood, A. M., 434, 514
Wood, H. O., 378, 381, 432
Woods, R. D. 262

Y

Young, D., 159, 173, 187, 188, 203, 205, 233
Yuoen, C. M. K., 430

Z

Zajic, J., 467
Zaruba, Q., 513
Zemanek, J., 260
Zignoli, V., 398, 432
Zischinsky, U., 513

SUBJECT INDEX

A

About the author, 580
Abrasion, 123
Absolute strain, 248
Absolute stress, 247
Absorption, 123
Abutment, elastic displacement at, 333, 334
Acidic precipitation, 145
Acidic rocks, 46, 186
Acids, 144, 145, 290
Acknowledgments, 21-24, 25-27
Active faults, 275
Adits, 312
Adjusted compressive strength, 199
Advances in rock mechanics, 32, 33
Advantages of
    dilatometer test, 244, 245
    flat-jack testing, 246
    in-situ testing of rocks, 190, 228
    laboratory testing of rocks, 115
    pressure chamber (tunnel) test, 233
    rock bolting, 448
    tie-back soil anchors, 464, 466
    underground openings, 68, 311
Aeolian (eolian), 529
Aeolotropic material, 38
Aerial photographs, 95-98
    of Longarone, Italy, 423
    of Round Valley, New Jersey, 95, 97, 98
    of surface drainage pattern, 96
    of trace of San Andreas fault zone, 9, 394
    of Tuve landslide, Sweden, 416
    of Vita Sikudden slide, Sweden, 114
Age, 31, 451
    bronze, 31
    of the Earth, 42, 44
    Stone, 31
Airphoto mosaic of Round Valley, New Jersey, 95, 97, 98
Airy's stress function, 339-341, 356
Alaska earthquake, 422
Alaska pipeline, 385
Albula tunnel, 149
Allegheny Portage Tunnel, Pa, 32
Alpine tunnels, 149
    overburden thickness of, 149
    temperature in, 149
Alteration, hydrothermal, of rocks, 143
American Cyanamid, 291
    AM-9 chemical grout, 291
American rocks,
    coefficient of thermal expansion of, 155
    dry unit weight of, 131
    porosity of, 131, 133

American Society for Testing and Materials (ASTM), 32, 33
American tunnels, 32
Amphibole, 529
Amphibolite, 451
    metavolcanic, 451
Amplitude, 160
Analcime, 107
Analysis of stability of rock slopes, 493
Analysis of stability of soil slopes, 486-489, 491-493
Anaverde formation, 394
Anchor force, determination of, 457-466
Anchoring, 437
    requirements of, 440
    of rock slopes, 478-482, 498-501
    of a rupture wedge, 439
    of a U-shaped structure, 438
    of a weak zone to a stronger one of rock, 438
Anchors, pre-tensioning of, 458
Anchors, soil, 460-466
    pressure-grouted, 440-466
    anchor coefficients, 465
Anchor system, 440
Ancient slip planes, 266
Ancient tunnels, 31
Andesite, 45, 46, 530
Angle of friction, 123, 160, 223
Angle of inclination ω, of anchor P_{min}, 477, 499
Angle of residual shear, ϱ_r, 217
Angle of rupture (shear), α, 163-165
Angle of shear resistance, 217, 236
Angular unconformity, 59
Anhydrite, 530
Anisotropic deformation, 160
Anisotropic material, 38
Anisotropy, 38, 62, 172, 305, 530

of rocks, 38, 62, 172, 184, 233
  directional, 62
Anticline, 276-280
  cracks in, 277
  position of tunnel in, 278
  tunnel through, 278
Apatite, 47, 157
Appenine tunnel, 145, 149
Apparent (bulk) specific gravity, 123, 127, 128
Appearance of deformed rock specimens, 182, 183
Appendices, 515-580
  about the author, 580
  conversion factors for units of measurement, 563-571
  dynamic viscosity tables for water, 572-575
  dynamic viscosity correction tables for water, 572, 576-579
  glossary of terms, 529-559
  Greek alphabet, 517
  key to signs and notations, 519-527
  rock defects, 561, 562
  SI units, 563
    use of in geotechnical engineering, 563
Application (use) of rock materials, 64-70
Apron, 530
Aqueous tunnels, 312
Aquifers, 266
Archeozoic era, 42
Arching, 325
Arch roof (rock), pointed, 268
Arcosic sand, 48
Argillaceous sandstones, 99
Arlberg tunnel, 149
Artificial freezing of soil and rock, 86, 87
Artificially induced earthquakes, 376
ASCE Rock Mechanics Seminar, 62
Aseismic design building codes, 95, 394
Assumed stress fields, 307, 308
Assumptions made in the theory of elasticity, 79-81, 188
ASTM standards, 32, 33
Attapulgite, 105
Attitude, 172, 423, 531
Auburn tunnel, 32
Augite, 46
Avalanches, rock, 422

B

Barrage, 282, 283
Barre, Vermont, quarry, 48, 49, 65, 66

Basalt, 45-47, 131, 151, 155, 181, 202, 223
Basalt lava, 86
Basement foundation for the World Trade Center in New York, N.Y., 455, 457
Batholite, 48
b/d-ratio, 199
Bearing capacity of rocks, 202, 225-248
Bedrock, 31, 40, 95
Bedding planes, 54, 95, 265-267, 271, 282, 423
Bending of beams, 204
Bending strength, 191, 203
Bending stress, 205
Bentonite, 102, 104
Bent rails, 386
  Guatemala, 386
b/h-ratio, 199
Biaxial stress components, 339
Biaxial stress field, 307, 308, 322
Biotite (black mica), 47
Black granite, 47
Black mica, 47
Block flow, 474, Bluff, 532
Bologna-Florence tunnel, 142
  fire in, 145
Bolted rock elements, 441
Bolting of rocks, 437, 440
Borehole, 97
  calyx, 97, 109, 111, 112
  EX, NX, 99, 100, 247
Borehole deformation meter, 234
Borehole deformation test, 234
Borehole examination, 113
  large boreholes vs. small ones, 113
Borehole permeability test, 251
  bottom hole, 251
  Lugeon, 140, 292, 400
  packer, 251-253
  periscope, 113
Borehole strain measurement, 244, 247, 248
Borings, 97
  calyx, 97
  depth and spacing of, 100, 101
  their effect on disclosure on subsurface conditions, 101
  large-diameter, 109
  shot-core drilling method, 109, 110
  small-diameter, 109, 110
Boss, 48
Boundary of elastic-plastic zone, 342, 343
Boussinesq's elasticity solution, 228
"Box" shear strength tests, 213, 214
Brazilian tensile strength test, 205

SUBJECT INDEX 593

Breccia, 51
Brick caisson, 31
Bridge pier over a fault, 274
Brittle fracture of rocks, 183
Brittle-ductile transition, 177
Brittle materials, 177
Brittle splitting, 197
Bronze Age, 31
Brunswick Triassic shale, 53, 54, 97, 124, 126, 127, 128, 184, 185, 200, 211
Buckling of columns, 204
Building codes, aseismic, 95
Bulk (apparent) specific gravity, 129, 131
Buoyant (submerged) unit weight, 131
Bursts, rock, 374, 486
By-passing pressure, 318

C

Cable jacking test, 232
Cainozoic era (= Cenozoic era), 42
Caisson, brick, 31
Calcareous sandstones, 52
Calcite, 54, 157
Calcium carbonate, 52
Calcium chloride, 290
Calcium sulfate, 144
Calkspar, 151
California earthquakes, 384
   representative magnitudes of, 385
   table of, 385
California major faults, 392
Calorie (kilo, large), 150
Calyx, 99, 109-112
Calyx rock cores, 99, 100
Calyx rock samples, 111, 112
Calyxite, 110
Cambrian period, 42, 43
Canadian Shield, 369
Cancani-Sieberg earthquake intensity scale, 379-382
Cancer, rock, 146
Canelles dam, 404-408
Canyon, Madison, 240
Carbon dioxide, 143, 144
Carboniferous period, 42, 43
Carbon monoxide, 144
Carbonate, 52
Carbonate rocks, 52
Casings,
   AX, BX, EX, NX, 99
Caverns, 67, 456
Cavities, 74, 124, 266, 281
   solution in, 397, 398

Cedar Springs dam, California, 286
Cenozoic era, 42
Chabazite, 107-109
   scanning electron micrograph of, 108
Chamber (tunnel) test, 232, 233
Channel Tunnel Study Group, 72
   reports of the, 72
Chemical action, 143
   effects of, 124
Chemical grouts, 291
   AM-9, 291
Chemical injection, 290, 291
Chemical solidification of soil, 290, 291
   silication, 291
   strength of, 291
Chemical weathering, 143
Chezy's coefficient, 139
Choia Valley, 393
Circularly curved rupture surfaces, 486-489
Circular openings, underground, 299, 319, 324
   stresses in rock about, 299
   stress distribution diagrams around horizontal section of, 324, 328
   vertical, 357, 358, 361
Classification of rocks, 44, 62, 63, 77
   on the basis of strength, 44
   engineering, 44, 62, 63
   by genesis, 44
   geological or lithological, 44, 62
   in-situ, 62, 63
   intact, 44, 62, 63
   according to strength and modulus ratio, 63
Clastic rock, 52
Clastic texture, 52
Clay minerals, 105-107
Clays, 104, 266
Clay seams, 268
   movement (sliding) over, 268, 269
Clays, grouting, 290
Cleavage planes, 266
Closing remarks about stability of rock slopes, 485
Codes, 95, 394
Codes for aseismic design, 95, 394
Coefficient of
   contraction, 149
   displacement m, 229
   expansion, 149
   friction, 160, 222, 223
      dynamic, 222
   lateral pressure, λ_o, at rest, 305-308, 323, 324, 326, 327, 333

values for rocks, 305
passive earth pressure (earth resistance) K_p, 484
permeability 137-139, 291, 292
thermal diffusivity, 149, 152, 153
thermal expansion and contraction, 153-155
   linear, 153
   of American rocks, 155
   volumetric, 154
   weakness of rocks, 272-273
Cohesion, 123, 195, 207, 215
Columnar structure, 124, 125
Colloid, 102
Colosseum, 145
Committee on Rock Mechanics, 73
Comparison of earthquake intensity and magnitude, 384
   after Richter, 383
Competent ground, 41
Competent rock, 40
Complete state of stress, 248
Composition, mineral, of rock, 123, 124
Compressibility of in-situ rock, 226
Compression bending strains, 205
Compression modulus, 176
Compression and decompression of rocks, 172
Compression strength tests, 191-194, 234
   on cubes, 197
   on disks, 211, 212
   in the laboratory, 191-201
   large scale, 234
   in-situ, 232, 234
   small-scale, 191-201
   triaxial, 195-197
   unconfined, 191-195
   ultimate, 193, 198
Compression testing machine, 118
Compressive strength of rocks, 54, 62, 202
   adjusted, 199
   in-situ, 226
   laboratory, 114, 115, 118, 196, 207-221
   static, 191-201
   triaxial, 195
   ultimate 174, 193
   unconfined, 54, 62, 191-195
      dry, 202
Comstock mine (USA), 368
Concentration of
   force lines, 319
   stresses, 319
      in the lower part of a slope, 485

Concept of
   plastic zone in rock, 337
   stability of slopes, 475-501
   stress, 243
Concrete dams, 280, 283
Conductance, electrical, 146
Conductivity,
   temperature, 149
   thermal, 149, 152, 153
Cones, 185, 195
Confining pressure, 198
Confinement of rock, 304
Conformity, 57
Conglomerate, 51
Construction material, rock as, 31, 47, 48, 52, 61, 62, 64, 65, 74
Continuum, 73
Controlled-strain shear testing device, 216
Control of water, 397
Constants, elasticity, 159, 174, 187, 188
Conversion factors for units of measurement, 563-571
Cooling, 132
   quick, 132
   slow, 132
Core drilling, 110
Cores, rock, 99, 110, 113-115
Coring, 100, 110
Corundum, 157
Coulee, 86
Coulomb's passive earth pressure (resistance) coefficient, K_p, 484
Coulomb's shear strength diagram, 344
Coulomb's shear strength envelope, 197, 201
Coulomb's shear strength equation, 162, 200, 236
Coulomb's shear strength line, 162, 163, 340, 350
Coulomb's criterion of failure, 162
Coulomb-Mohr
   failure criterion, 161-164
   plasticity condition, 339
   strength diagram, 344, 345, 483
Coulomb-Mohr-Tresca failure criterion, 161, 165
Cracks, 74, 172, 265
   in anticline, 277
   haircracks, 172, 265
   in syncline, 277
Crataceous period, 42, 43
Creep, 121, 123, 173, 180, 181, 182, 191, 414

SUBJECT INDEX

accelerated, 182
definition of, 181
plastic displacement, 180
primary (transient), 182
secondary (steady state), 182
of slopes, 414
tertiary, 182
tests, 182
time-dependent deformations, 172, 173, 180, 182
Creep strains, 181
Criteria of rock failure, 161-168
  Coulomb's, 161
  Coulomb's-Mohr, 161-164
  Griffith's, 161
  maximum shear stress, 161
  maximum tensile stress for rock, 161
  Mohr's, 161
  shear strength, 193, 194
  Tresca's, 161, 162, 165
  various, 162
Criterion for failure impending of sliding of a rock block, 500
Critical stress, 319, 330
Critical stress concentration, 330
Critical wedge, support of, 447
Cross-sections of underground openings, 324, 325
Crushed zones, 266, 273
"Crushing drama," 426, 427
Cube compression testing, 197, 216-221
  shale cubes, 218, 219
Curved rupture (shear, sliding) surfaces in rock, 439
  in soil, 160, 240, 439, 490
Curves, stress-strain, 173, 176, 177
Cutting edge of limestone, 31
Cyclic deformation graph, 229, 230
Cyclic loading and unloading, 176, 229, 230
Cylinder, thick-walled, 312, 313

D

Dams, 282, 403
  across a fault, 274
  Boulder (Hoover), 403, 404
  Canelles, 404-408
  Cedar Springs, 286
  Grand Coulee, 86
  Grande Dixence, 283
  Hale's Bar, 397
  Hoover (Boulder), 403, 404
  Malpasset, 240

failure of, 240
massive, 281
Oroville, 451, 452
performance of, 280
Power Authority of the State of New York, 110
  (Robert Moses Power Dam), 110
Darcy's law of permeability, 137, 139, 140
Debris flow, 414
Decompression of rocks, 172
Deere's lectures on rock mechanics, 62
Deep-seated rupture surface, 461, 462
Defects of rocks, 172, 190, 226, 265, 561, 562
Definition of
  creep, 181
  deformation of rocks, 169
  elasticity, 158, 159
  geotechnics (géotechnique), 40
  intact rock, 62
  permeability, 136, 250
  plasticity, 160
  rock (by Emery), 38
  rock mechanics, 73
  rock slopes, 471, 472
  San Andreas fault zone, 391
  shear strength, 207
  soil (by Terzaghi), 38
  tensile strength of rock, 203
  thixotropy, 102
Deformability factors, 227
Deformation meter, borehole, 234
Deformation modulus, 176
Deformations, 170
  elastic, 170
  factors affecting, 72
  inelastic, 227, 229
  permanent, 159, 175
  plastic, 170, 175, 187
  radial, 331
  time-dependent, 172, 173, 180, 182
  viscous, 170
Deformations of rock-core specimens,
  gneiss, pegmatite, 184, 185
  Triassic (Brunswick) shale, 184, 185
  Vermont marble, 184, 185
Deformations of rocks, 169-187, 233
Deformations of rock slopes, 373, 374
  of excavation floor in a cut, 374, 375
Degree of elasticity of rock, 175
Degree of plasticity of rock, 176
Degree of saturation, S, 136
Degree of weathering, 172

Delaware-Chesapeake Canal bank slide, 411, 413
Density, 123
Depth of borings, 100, 101
Derivation of plasticity condition in rock, 337
Design in rock, requirements for, 371, 372
Destructive earthquakes, 386-389
Determination of anchor force, 458, 459
Determination of modulus of elasticity of rocks, 186, 191
Determination of rock-bolt support, 445-448
Development of rock mechanics, 32-35
Devices for testing of
  rock cubes, 218, 219
  rock disks, 210-213
Devonian Age, 43
Diabase, 45, 47, 124, 125, 131, 155, 181, 202, 223, 305
Diagrams,
  phase, 136
  shear-stress-displacement, 216
  strain hardening, 176
  stress distribution around a tunnel, 324, 328
  stress-strain, 173, 176-180
  for ductile materials, 176
  stress-strain for rocks, 177-180
  time-dependent deformation (creep), 123, 173, 180-182, 191, 414
Diameter-to-height ratio, d/h, 199
Diametrical deformation calculation, 244
Diamond, 157
Diastrophism, 536
Dielectric constant, 146
Difference between strength of materials and rock mechanics, 82-84
Differential equation, Euler's, 315
Diffusivity, thermal, 149, 152, 153
Dihedral block, 408
Dikes, 47, 48
Dilatancy, 123
Dilatometer tests, 244, 245
Diminution of rock, 122
Dioctahedral illite, 54
Diorite, 45, 47
Dip, 172, 474
Dipping beds (strata), 126
Direct "box" shear strength of rocks, 213, 214
Direct shear strength test of rocks, 62, 191, 209, 210, 213-220

along a forced inclined plane, 215, 216
Direct strain measurement, 244
  absolute, 248
Disadvatages of laboratory testing of rocks, 195, 196
  by dilatometer, 245
  by flat-jacking, 246
  in pressure chamber (tunnel), 233
Disadvantages of pressure-grouted soil anchors, 466
Disadvantages of rock bolting, 448
Disadvantages of tie-back soil anchors, 466
Disadvantages of triaxial testing, 195, 196
Disconformity, 57
Discontinuity, 475, 477, 493-498
  geological, in rock, 54, 74, 396, 403, 475, 493-498
  planar, 493-498
  pre-existing, 476, 493-498
  in rock, 203, 396
  sliding on, 475, 493-498
  slope stability problem, 476, 478, 479, 493-498
Discontinuum, 73, 74
Discussion of equations, 328-330
  on stresses in elastic and plastic zones, 344-347
  on stresses around vertical shafts, 358-361
Disks, shale rock, for testing in punch shear, 210, 212
Displacements, 331
  coefficient m, of, 229
  radial, 316
Dissolution of soluble mineral components, 119, 144
Distribution of shear stresses in a solid-shaft specimen, 220
Distribution of stresses around circular openings, 324
Disturbance zone around a tunnel, 347, 348
Division of time, 41-44
Dolerite, 47
Dolomite, 52, 131, 151, 187, 202, 223, 305
Domain,
  elastic, 175
  inelastic, 173, 175
  partly elastic-plastic, 175
  plastic, 175, 176
Double shear test, 209, 210
Drainage, 119, 140
  subsurface, 397
  surface, 397

SUBJECT INDEX

Drainage pattern, 95, 96
Drilled-and-shot pilot bore, 453
Drilling fluids, 102-106
Drills, rock core, 116, 117
 laboratory, 116, 117
Dry/wet factor of shale, 213
Duckworth's tensile strength of ceramic beams, 205
Ductile behavior, 174, 183
Ductile materials, 173
Ductile rocks, 177
Ductile rupture, 177
Durability of rocks, 156, 158
Dynamic or absolute viscosity η, 140, 572-579
Dynamic deformability tests, 227
Dynamic friction, 222
Dynamic properties of rocks, 222, 223
Dynamic strength of rocks, 123
Dynamic tests, 186
Dynamic viscosity,
 correction tables for water, 140, 572, 576-579

E

Earth bank, uplift of, 402
Earth pressure coefficient, passive, K_p, 484
Earthquake effects, 501-504
 on earth retaining walls, 502, 503
 on stability of rock and soil slopes, 504
Earthquake factors, 395, 396
Earthquake intensity scales, 378-382
 Cancani scale, 379-382
 comparison of intensity and magnitude, 384
 description of, 378-386
 Mercalli scale, 379-382
 table of 379-382
Earthquake magnitude, 382-384
 Richter scale, 383
Earthquakes, 375, 501
 artificially induced, 376
 destructive, 377, 386, 387
 intensity of, 379-382
 magnitude of, 382-383
 seismic forces, 501, 503, 504
 tectonic, 376
 volcanic, 376
Earthquakes,
 Alaska, 385, 422
 Anchorage, 385
 California, 385
 Daly City, 385

El Asnam, Algeria, 389
Guatemala, 386
Homestead Valley, Cal., 385
Imperial Valley, Cal., 385
Kern County, Cal., 385
Lisbon, Portugal, 386
Long Beach, Cal., 384
Romania, 1977, 389
San Fernando, Cal., 385
San Francisco, Cal., 1906, 384
Santa Barbara, Cal., 385
Scopje, Yugoslavia, 389
Seward, Alaska, 385
Valdez, Alaska, 385
Earth resistance coefficient, K_p, 484
Earth retaining walls, 502, 503
 earthquake effects on, 502, 503
Effect of depth and spacing of borings on disclosure of subsurface conditions, 100, 101
Effect of pore water on principal stresses, 200
Effect of precipitation on stability of rock slopes, 482
Effect of temperature on strength of rocks, 369
Effective normal stresses, 200
Effective shear stress, 216
Effective stress circle, 201
Effervescence, 52, 53, 109, 145
Elastic constants, 191
Elastic deformation, 319
 in sound rock, 319, 331
  around a circular opening upon excavation, 319, 331
Elastic displacements, 332-334
 at abutment, 333, 334
 at crest, 332
Elastic material, 170, 179
Elastic-plastic boundary, 342
Elastic-plastic domain (zone), 338
Elastic strains, 175, 177, 181, 315
Elastic stress analysis in rock, about openings, 311
Elastic stress condition s around a vertical shaft, 359
Elastic stresses, by Lamé, 317
Elastic substance, 170
Elastic tangential tensile stresses, 333, 334
 zone of 333
Elastic waves, 227
Elastic zones in rock, 338, 361
 radial stresses, 323
 tangential stresses in, 325, 333

Elasticity, 123, 158, 159, 186, 233
　theory of, 79-81
Elasticity constants, 159, 174, 187, 188
Elasticity equations, 314
Elasticity modulus 159, 177, 186-188, 203, 205, 233, 304
　of rocks, 123, 159, 186-188, 203, 304
Elasticity of rocks, 156, 158, 159, 175, 186-188, 233, 244, 304
Electrical logging, 146
Electrical parameters, 134
Electrical properties
　of rocks, 123, 124, 146, 186
　of water, 146
Electrolytes, 143
Electron micrographs, scanning, 107, 108
Elements of a thick-walled cylinder, 313
Emery's definition of rock, 38
Emulsion, 289
End surfaces of rock specimens, quality of, 184
Energy, 179
　heat, 147
　strain, 171, 172
　wave, 186
　work, 171
Engineering classification of intact rock, 62
　of rock in-situ, 62
Engineering geology, congresses on, 34
Engineering materials, 31
　rock as, 31
Engineering problems associated with work in rock, 75-78, 365, 366
Engineering properties of rocks, 122, 123, 146, 147, 155, 172, 190, 222, 223, 255, 265
　en masse properties of, 225, 226
Envelope of failure, 166, 167
　Coulomb's shear strength, 162-165
　　curvilinear, 166, 167
　　straight line, 163-166
Eocene epoch, 42, 43
Eolian (aeolian), 529
Epoch, 42, 43
Equipment
　for preparation of rock test specimens, 116-118
　for testing of rocks, 116-118, 211, 218, 219
Eras, 42, 43
Erionite, 107-109
　scanning electron micrograph of, 108
Euler's differential equation, 315

Examination of boreholes, 113
EX boreholes, 247
EX casing, 99
Excavation walls in rock, 474
　stable, 474, 475
　unstable, 474, 475
Exfoliation of rock, 147
Existing planes of weakness, 266
Expansion, thermal, 149, 153-155
　linear, 153, 154
　volumetric, 154
Exploration methods in rock, 93
　geological, 94
　geophysical, 94, 120, 121
　hydrological, 94, 118-120
　subsurface, 94, 97-102
　thermal, 94, 121, 122
Exposure of San Andreas fault zone, 393
Exsolution of gas, 144, 145
Extent of plastic zone in rock, 341
Extrusive (volcanic) rocks, 45, 46, 370

F

Factors
　affecting deformation, 172
　contributing to rock slope failure, 472, 473
　conversion, for units of measurement, 563-571
　of deformability, 227
　dry/wet, of shale, 213
　seismic, 172
Factors of safety,
　analytically, 476, 479, 484, 487, 489, 495-497, 500, 504
　graphically, 498
　of slopes, 476, 479, 484, 487-489, 495-497, 500, 504
Factors, of softening, 200
Failure criteria for rocks, 161-168
Failure of Malpasset dam, 240, 408-410
Failure of rocks,
　of marble, 184, 185
　of pegmatite gneiss, 184, 185
　plastic, 183
　of shale, 185, 212, 220
　in shear, 183, 274
　in tension, 183
Failure of rock slopes, 472-475
Fall line, 452
False hanging loose rock, 438
Faulted anticline, 280
Faulted syncline, 280

SUBJECT INDEX

Faulting in Guatemala, 386
  bent rails, 386
Faults, 54, 74, 266, 269, 272
  hazards from, 375, 390
  influx of water from, 271
  principal kinds of, 275
  significant of the California region, 392
  trace of, 391
  zone of, 266, 391
Features of rock mechanics, 74
Feldspars, 46, 47, 48, 157
Field Act of California, 396
Fire in tunnel, 145
Fissured rock, 39
  grouting of 186, 285
  injection of, 281, 282, 289
Fissures, 54, 74, 172, 265
  lithogenetic, 142
  tectonic, 142
  weathering, 142
Flat-jack testing, 245, 246
Flexural strength test of rocks, 191, 203
Fluorite, 157
Flysch, 57, 59
Folded rock structure in limestone, 124, 126
Folding of a rock stratum, 276
Folds, 266, 276, 277
  anticline, 276-280
  syncline, 276-280
Fondedile, S. p. A., 504, 505
Force fields, 301, 302, 306
  gravitational, 306
Force lines, concentration of, 319
Force majeure, 539
Force polygons, 461, 463, 498
Forced direct shear test, 217-220
Forced inclined plane, 215, 216
Forced shear plane, 215, 218, 219
Forces,
  anchor, 457-466
  on dams, 283
  geological, 283
  hydrodynamic, 489
  hydrostatic, 283
  seismic, 283
  seepage, 489
  self-weight of structure, 283
  tectonic, 319
  weight of rock, 283
Forewords, 5-8
Forms of underground openings, 268, 319, 322, 324, 325

Forms of slip lines, 351-353
Foundation models on sand, obliqued loaded, 240, 439, 490
Foundation work
  for the World Trade Center, New York, 455, 457
Fractures, 124, 161, 265, 267, 269
Free-hanging rock, 326
Freezing of ground against rock slides, 86
  in tunnel engineering, 86
Freezing of rock, 86, 158
Freeze-thaw test, 158
Friction, 221
  angle of, 223
  coefficient of, 222, 223
  dynamic, 222
  static, 222
Frost action, 123, 472
Fusion, latent heat of, 149, 151, 152

G

Gabbro, 45-47, 131, 155, 187, 202, 223, 305
Gabions, 510
Galleries, 312
Gases, exsolution of, 144, 145
Gases in tunnels, 122, 144, 145
Gauges, strain, 115, 116, 186, 193, 247
Gel, 291
General notes about slip lines, 348-350
Generatrix, 196
Genesis, rock classification by, 44
Geology, 40
Geological classification of rocks, 44, 62
Geological defects of rocks, 172, 190, 226, 265, 561, 562
Geological discontinuity, 54, 74, 396, 403
  in stability problems of rock slopes, 475, 476, 478, 479, 493-498
Geological exploration of rocks, 94-97
Geological hazards, 393, 398
Geological structures, 94, 124
Geologic time scale, 41-44
Geologist, 95
Geologist's report, 95
Geophysical exploration of rock, 120, 121, 146
Geotechnical engineering, 40, 74, 84
Geotechnics (Géotechnique), 40, 74, 84
Geothermal gradient, 147, 148, 368
Geothermal step, 148, 368
Geotherms, 148
"Giant granite," 50

Glossary of terms, 529-559
Gneiss, 60, 127, 128, 131, 151, 155, 158, 184, 188, 200, 202, 223, 305, 371, 451
Gold ore mines, Johannesburg, 369
Göta Valley, 417
Göteborg, 416, 417, 454
Gotthard, St., tunnel, 149
Gouge, 124, 272, 367, 390, 391, 403, 411, 495
Gradient,
  geothermal, 147, 148, 368
  thermal, 121
Grand Coulee dam, 86
Grande Dixence dam, 283
Granite, 45, 47-50, 66, 115, 131, 155, 158, 187, 202, 203, 305
  quarry, 48, 66
Granite City of Aberdeen, Scotland, 48
Granitoid rock, 47
Graph showing cyclic deformation, 229, 230
Gravitational force field, 306
Graywacke, 53
Great Apennine tunnel, 145
Greek alphabet, 517
Griffith's criterion of tensile failure, 161, 168
Groundbreak, 490
Ground, competent, 41
Grout, chemical, 290, 291
Grout curtain, 287, 397, 404
Grouting, 282, 284-288, 398, 404
  effectiveness of, 288
  need for, 288
  of rock, 186, 285
  shortcomings of, 288
Guatemala earthquake, 386
Guatemala, 386
Gunite, 293
Gypsum, 144, 157

H

Haircracks, 172, 265
Hale's Bar dam, 397
Hanging rock, 275, 438
Hardness,
  of minerals, 106, 156, 157
  Mohs' scale of, 156, 157
  of rocks, 122, 156-158
  of water, 143
Hazards,
  from faults, 375, 390
  from water, 396

  geological, 398
  in granite rock, 400
  from rock, 398
  in rock engineering, 369, 372, 375-378
h/d-ratio, 191, 198
Heat, 147-153
Heat capacity, 149, 150
  of frozen rock or soil, 150
  of minerals and rocks, 150
  of unfrozen rock or soil, 150
Heat, specific, 150
Heim's hypothesis, 303
Hematite, 54
Heterogeneity of rocks, 38, 41, 184
Historical notes, 31
Holders of rock specimens, 218, 219
Homogeneous rocks, 38, 127, 128, 155
Hookean material, 173
Hooke's law, 79-81, 159, 173, 174, 314, 320, 331, 339
Hoop stresses, 542
Hoover dam, 403, 404
Hornblende, 47, 48, 151
Hot (magmatic) water, 143
Humus, 143
Hydraulic fracturing of rock, 248, 250, 397
Hydrochloric acid (HC1), 52, 106, 109, 195
  identification test for detecting of the presence of calcite in a rock, 53, 145
Hydrological exploration of rocks, 118-120
Hydrostatic stress field, 307, 308
Hydrostatic uplift, 400, 404, 438
  pressure diagram of,
    under a dam, 401, 402
    in earth bank, 402
Hydrothermal alteration of rock, 143
Hysteresis, 179, 180

I

Ice, 84, 118, 119, 148
  average ultimate unconfined compressive strength of, 85
  utilization of, 86
Ice age, 42
Ice mechanics, 84
Icing on pavements, 118, 119
ICOLD Committee, 409, 410
Ideal elasticity, 159
Ideal plastic material, 170, 338
Igneous rocks, 40, 44-48, 50, 127, 131, 155, 187, 202, 223, 369
Imbrication, joint patterns of, 271, 272

SUBJECT INDEX

Impact resistance strength, 123
Impact work, specific, 222
Imparting precompression stress on bolted rock elements, 441
Inclined forced shear plane, 215
Inclined planar discontinuity, 493-498
Incompetent rock, 61, 311
Index of toughness, 222, 223
Indirect (Brazilian) tensile strength test, 205
Induced (secondary) stresses, 303
Inelastic deformation, 227, 229
Inelastic domain, 173, 175
Inelasticity, 305
Inelastic material, 179
Inelastic rocks, 305
Inelastic strain, 176
Inferiorities of rocks, 367, 368
Influence of rock stability on dams and foundations, 283, 284
Influx of water from joints and faults, 142, 271
   into excavations, 142
   into underground openings, 94, 142, 271
Initial state of stress, 172
Initial stresses, 165
Initial stress field, 83, 307, 309
Initial tangent modulus, 62, 180
Injection of rock, 281, 282, 289
In-situ
   classification of rocks, 62
   properties of rocks, 225, 226
   shear strength tests, 231, 235-241
   state of stress, 306
   strength of rocks, 118, 189, 190
   tension test of rock, 235
   testing of rocks, 190, 228
      objectives of, 228
      against leakage, 397
Intact rocks, 41, 62, 63
   classification of, 62, 63
   shear parameters, 224
Intensity of earthquakes, 378-382
Interbeds, 266, 267
Interferometry, ultrasonic, 223
Internal friction of rocks, 123, 160, 223
Internal (residual) stresses within a rock mass, 132, 226, 242
International Conferences on Soil Mechanics and Foundation Engineering, 33
International Congresses on Engineering Geology, 34
International Congresses on Rock Mechanics, 33, 271, 320

International Society for Rock Mechanics, 196
   Laboratory tests on rocks, 196
Interstitial water, 557
Introduction, 31-35
Intrusive (plutonic) rocks, 45, 47, 48, 50, 124
Irreversible deformation (strain), 175, 176
Isotropic elasticity, 159
Isotropic material, 39, 61

J

Jack-and-plate loading, 228, 229
Jacking test, 231
Jigs, 216, 217, 219
Johnstown flood, Pa., 424
Joint pattern, 119
Joints, 54, 119, 124, 190, 265, 269
Jointed rock, 119
   closed, 270
   continuous, 270
   discontinuous, 270
   influx of water from, 271
   open, 124, 270
Joosten's principle of chemical injection, 290, 291

K

Kaolin, 143
Kaolinite, 54
Karawanken tunnel, 149
Karst regions and topography, 119, 120
Key to signs and notations, 519-527
Keypunch console, 116, 118
Kinematic viscosity, ν, 140

L

Laboratory static compressive strength, 191-201
Laboratory testing of rocks, 114-118, 196, 207-221
   by International Society for Rock Mechanics, 196
   summary on, 196, 224, 225
Lamé elastic stresses, 317
Lamé equations for a thick-walled cylinder, 315
Laminations, 265, 267
Landslides, 410, 411, 423
   in California, 413
   at Longarone in the Piave Valley, 425
   at Tuve, Sweden, 415, 416
   Vaiont Reservoir, 423, 424

Large-scale tests, 234, 238, 239
  compressive strength, 234
  shear strength, 238, 239
Latent heat of fusion, 149, 151, 152
Lateral pressure coefficient, λ_0, 305
  for rocks, 305
Lava, 127, 132
  basalt, 86
Layered shale structure, 124, 126
Leakage, 140, 396
  testing of, 397
Leaky tunnels, 134
length of tunnels, 149
Limestone, 31, 40, 52, 57, 59, 120, 121,
    124, 131, 143, 151, 155
      158, 187, 202, 223, 305
  folded rock structure of, 124, 126
Limonite, 544
Linear-elastic material, 174, 178
Liquefaction, 377
  Lisbon earthquake 1755, 386
  Lisbon, 386
Lithogenetic fissures, 142
Lithological classification of rocks, 44, 62
Lithology, 62
Load-bearing capacity, 397
Loading plates, forms of, 229
Loading of rock pillars, 192
Loading-unloading-loading tests, 175-177,
    229, 230
Location of California significant faults,
    392
Location of tunnels,
  in anticline, 278-280
  in syncline, 278-280
Log, 133
Logarithmic spiral tables, 352
Logarithmically spiraled rupture surface
  curves in sand, 160, 240, 490
Logarithmically spiraled slip lines, 351, 352
Logging, electrical, 146
Longitudinal resonance test, 186
Longitudinal shear displacement, 208
Lötschberg tunnel, 149
Lugeon unit, 292
Lugeon water pressure test, 140, 292
Lugeon's criterion for groutability, 292
Lugeon's dictum, 400

M

Macroscale properties of rocks, 123, 124
Madame Barrage, 282, 283
Madison Canyon, 240

Magma, 127, 132
Magmatic water (hot water), 143
Magnesium, 52, 144
Magnetite, 47
Magnetometric surveys, 120
Magnitude of earthquakes, 384, 385
  Richter Magnitude Scale, 384, 385
Major principal stress, 163, 334, 358
Malpasset dam failure, 240, 408
Manhattan schist, 371
Manifestation for the need of rock mechanics, 34
Mantle rock, 41
Map, soil and rock, 95, 97, 98
  of part of Hunterdon County, New
    Jersey, 95, 97, 98
Mapping of rock exposures, 95
Marble, 61, 131, 151, 155, 158, 184, 185,
    188, 202, 223, 305
  Vermont, failure of, 184, 185
Markings on mantle surfaces of tested rock
  and soil specimens, 349
Marsh gas, 144
Mass heat capacity, 150, 151
Mass wasting, 545
Materials, various, 38, 61, 159, 170, 173,
    174, 177-180, 338
Matrix, rock, 172
Maximum shear stress, 161, 162
  criterion of, 161, 162
Maximum tensile stress criterion, 161, 162
Measurements, units of, 563
  conversion factors, 565-571
  SI units, 563
  use of in geotechnical engineering, 563
Measurements of strain,
  in dilatometer test, 244
Mechanical defects of rocks, 265, 281
Mechanical properties of rocks, 143,
    156-160
Mechanical slippage, 227
Meerschaum (sea foam), 105, 106
Mercally intensity scale, earthquake,
    378-382
Mesozoic era, 42, 43
Metamorphic rocks, 40, 44, 45, 60, 61,
    127, 128, 131, 151, 155, 158, 184, 185,
    187, 188, 202, 223, 451
Metamorphism, 60
Methane gas, 144
Metavolcanic, 451
Methods of
measuring strain, 243-248

SUBJECT INDEX

rock exploration, 93
shear testing of rocks in the laboratory, 207, 209-221
in-situ testing of rocks, 228-253
slices, 491-493
Methods of remedy against rock slides, 477-482, 498-501, 504-510
Metro,
subway system of Paris, 86
subway of Washington, D. C., 68, 70, 451, 453
Mica, 47, 48, 50, 129, 151
black (biotite), 47
white (muscovite), 129
Mica schist, 149, 371
Micrographs, electron, scanning, 107, 108
Microscale properties of rocks, 123
Mineralogical composition of rocks, 123, 124
Minerals, Mohs' hardness scale of, 156, 157
Mines, gold, at Johannesburg, 369
Minor principal stress, 163, 358
Modes of deformation
of idealized substances, 170
of rocks, 204
in tension, 204
Modes of rock failure in unconfined compression, 184, 192
Modes of shear failure in rock, 207, 208
Modes of stress distribution in rock, 363
Modulus of compression (or deformation), 176, 545
Modulus of elasticity, 123, 159, 173, 176, 186, 203, 205, 233, 244, 304, 545
of confined rock, 304
initial, 68, 180
of in-situ rock, 120
of rocks, 159, 186
determination of, 186
secant, 180
tangent, 180
true, 177
Young's, 123, 159, 187, 203, 233
50%, 179, 180
Modulus ratio, E_{t50}/σ_{ult}, 62, 63
Modulus of rigidity, 545
Modulus of rupture, 205
Moho, 546
or Mohorovičić disconformity, 546
or discontinuity, 546
Mohr-Coulomb failure condition, 161, 163-165, 340, 483

Mohr's plasticity condition, 339, 340
Mohr's theory, 340, 482-484
Mohs' scale of hardness of minerals, 157,
Moisture content, 124, 133, 135
Mono-system rock, 39
Mont Blanc tunnel, 32, 370, 398, 399
Mont Cenis tunnel, 32, 149
Mont Toc Mountain, 423
Montmorillonite, 103, 104, 391
scanning electron micrograph of, 104
Moroccan Meseta, 57
Mosaic, uncontrolled, airphoto, 98
of Round Valley, New Jersey, 98
Mount Rushmore, 65, 66
Movement of rock strata over clay seams, 268
Movement of water through rock, 396
Mud flows, 414
Multiple-body rock system, 39, 74
Munsell color coordinates, 54
Muscovite (white mica), 129
Mutual dependence of stress and strain, 169
m-values of displacement coefficients, 229

N

Natrium = sodium, 102
Natura non facit saltus, 346
Need for rock exploration, 93
Need for rock mechanics, 33
Net, wire, 452
Neutral pressure (stress), 65, 200
Niagara Waterfalls, 417, 419, 420
American Falls, 419
Horseshoe Falls, 418
Nitrogen, 145
Nonconformity, 57
Normal stress (pressure), 165, 200
effective, 200
neutral, 200
No-load shear strength (= cohesion), 193
Nonelastic material, 179, 180
Normal fault, 275
Notations, key to signs and symbols, 519-527
NX borehole, 247

O

Objectives of in-situ testing of rocks, 228
Objectives of rock mechanics, 76, 77
Obliquely loaded foundation models on sand, 240, 439, 490

Obsidian, 45, 50, 223
Oil storage caverns in rock in Sweden, 455, 456
Olivine, 46
Openings, underground, 67, 68, 268, 299, 311, 312, 337, 454
"Opponent water," 396
Ordovician period, 43
Organizations studying rocks, 32
Oroville dam, 451, 452
Orthoclase, 46, 48, 157
Orientation (attitude) of rocks, 172
Other related references, 72, 260-263, 433, 444, 468-469, 513, 514
Outcrop, 42, 95
Overbreak, 122, 270
Overburden, 41, 147, 149, 319
  thickness of, some alpine tunnels, 147, 149, 331
Overcoring technique, 247, 248

P

Packers, 251, 252, 253
Paleozoic era, 42, 43
Pali radici (root piles), 504-507
Palisades, along Hudson River, N.J., 124, 125
PAN-AM building, 371
Panama Canal slides, 411, 412
Parameters,
  electrical, 134
  shear strength, 207, 214
    residual, 216, 217
  stress, 340
  test, 214
Paris Metro system, 86
Parthenon, 146
Partings, 265, 266
Passive earth pressure coefficient, K_p, 484
Patterns
  of joint imbrication, 271, 272
  of rupture surfaces in dry sand, 240, 439, 490
Peak shear strength, 127
Peeling of rock surface, 122
Pegmatite, 45, 50
Pegmatite gneiss, failure of, 184, 185
Perfectly elastic material, 170, 179
Period, 42, 43
Permacrete, 87
Permafrost, 86
Permanent set, 159
Permeability, 123, 124, 136-142, 156, 190, 250, 291, 292

Lugeon's test, 140, 292
  of rocks, 140, 141, 292
Permeable gouge, 124
Petrification, 291
Petrographic structure of rocks, 172
Phase diagram, 136
Phenocrysts, 547
Physical constants of soils, 135
Physical properties of rocks, 123-155
Pillar, 192
Plagioclase, 46-48
Planes of
  discontinuity, 493-498
  rupture (shear), 163
    forced shear, 215
    inclined, 215
  weakness, 208, 237, 240, 265, 266
Plastic creep, 180-182
Plastic deformations (strains), 160, 177, 187, 337
Plastic domain, 176
Plastic equilibrium, 351
Plastic failure, 183
Plastic flow conditions, 170, 180
Plastic material, ideal, 170, 338
Plastic properties of rocks, 191
Plastic substance, 170
Plastic yield criterion by Horvath, 307, 310
Plastic zones, 337, 338, 341
Plasticity, 80, 81, 156, 159, 160, 176
Plasticity condition in rock, 337
  derivation of, 337
  in Coulomb-Mohr presentation, 339, 340
Platelets, 58, 62
  shale, 58
Plate loading test, 228, 229
Plutonic (intrusive) rocks, 45
Poisson's number, $m = 1/\mu$, 123, 159, 174, 188, 189, 304, 305, 318, 329, 330, 334, 356
Poisson's ratio, $\mu = 1/m$, 75, 123, 159, 174, 187, 188, 189, 193, 233, 244, 304, 305, 307, 314, 326, 331
  values for rocks, 75, 123, 187-189, 307
Poly-body rock system, 39
Popping, 370
Pore volume, 132
Pore-water pressure, 200, 201, 240
  effect on normal and principal stresses, 201
Porosity, 123, 131, 132, 281
  of rocks, 123, 131-133, 198, 281
  of soils, 132

SUBJECT INDEX

Porphyry, 50, 125, 127, 128
Position of horizontal seismic force, F_s, 503, 504
Position of tunnels, 279
Potential planes of weakness, 266
Potential rupture wedge, 439
Precambrian age, 451
Precambrian rocks, 42, 44
Precipitation, effect of, on stability of slopes, 482
Precompression stress on rock elements, 441, 442
Preface, 19, 20
Pressure at rest, 306-308
Pressure chamber (tunnel) test, 232, 233
Pressure,
  by-passing, 318
  effective, 200
  geostatic, 472
  hydrostatic, 380
  neutral, 65, 200
  pore-water, 200, 201, 240
Pressure coefficient at rest, λ_o, lateral, for rocks, 305-308, 323, 324, 326, 327, 333
Pressure-grouted soil anchors, 460-464
Prestressing, 442, 457-460
Primary stress condition in the elastic zone, 331, 355, 356
Primary stress field,
Primary stresses, 123, 318, 322, 331, 334
  in elastic zone, 332
  principal, 332
  in sound rock, 123, 302
Principal stresses, 334
  primary, 334
  secondary, 334
Principal kinds of faults, 275
Principle of shot-core drilling, 110
Principles of direct shear, 209, 238, 239
Principles in rock classification, 44, 62-64
Problems in rock mechanics, 75, 76, 365, 366
Profile, 102
Proportional limit, 174
Properties of rocks, 122, 123, 222, 265
  dynamic, 190, 222, 223
  elastic, 172
  electrical, 146
  in-situ, 225, 226
  macroscale, 122, 123, 265
  mechanical, 155
  microscale, 122, 123, 265
  physical, 122, 123-155

  plastic, 172
  radioactive, 147
  rheological, 172
  static, 190, 222
  technological, 123
  thermal, 147
Province of activity of the geologist, 76, 88, 93, 95
Pseudo-viscous flow, 182
"Public enemy No. 1" (water), 427
Punch shear tests, 210-212
Punching strength, 191
  of tested shale disks, 210, 212
Purely elastic deformations, 170
Purely plastic deformations, 170, 175, 187
Purely viscous deformations, 170
Pyroxene, 46, 47

Q

Quarry, 48, 49, 65, 66
Quartz, 46-48, 157, 451
Quartzite, 40, 60, 131, 155, 188, 202, 223
Quasi-elastic rock, 178, 180
Quaternary period, 43
"Quick clay," 415
"Quick clay," slides, 415
Quick condition, 86, 291
Quick cooling of lava, 132
Quick cooling of magma, 132
"Quicksand," 378

R

Radial deformation, 331
Radial displacement, 316
Radial stresses, 344
Radioactivity, 147
Radius of extent of plastic zone, 341
Railway embankment slide
  at Vita Sikudden, Sweden, 411, 413, 414
Reasons for studying
  small-size rock samples in the laboratory, 115
References, 35, 71, 89, 90, 254-260, 296, 297, 310, 335, 353, 427-432, 466-468, 511-513
  other related, 72, 260-263, 433, 444, 468, 469, 513, 514
Regolith, 41
Reinforcement of rock by anchoring, 437
Reinforcement of rock slopes, 477
Relaxation of stresses, 320, 373
Remedy against rock slides, 507-509
Repeated loading-unloading-loading test, 176, 229, 230

Requirements of anchorage in the design in rock, 371, 372, 440
Reservoir disaster, Vaiont, 423
Residual shear resistance (strength), 216, 217, 241
Residual stresses, 80, 486
  tectonic, 486
Resistance
  to chemical influence, 123
  to frost action, 123
  to impact, 123
  to weathering, 123
Resistance of rocks to abrasion, 123
Resistance of rocks to drilling, 99, 100
Resistivity, electrical, of rocks, 146
Resonance, longitudinal, test of, 186
Retaining wall, 502, 503
Reticulated root piles, 504-509
Reverse fault, 575
Rheological phenomenon, 80, 173, 181
Rheological properties of rocks, 172
Rhyolite, 45, 50
Richmond water supply tunnel shaft in Brooklyn, N.Y., 86, 87
Richter's earthquake magnitude scale, 382
  comparison with intensity scale, 384
Rift Valley, 550
Rip-rap, 64, 134
Robert Moses Power Dam on Barnhart Island, N.Y., 110
"Rock of Ages," 48, 49, 66
Rock anchors, 437, 438, 477-481
Rock block, sliding of, 475, 493-498
Rock bolt support, 443-445
  determination of, 445-448
Rock bolt, Williams, 448-451, 468
Rock bolting, 451, 452
Rock burst, 374
Rock cancer, 146
Rock caverns, 67, 451
Rock classification hammer, 116
Rock cores, 97, 98, 114, 115, 183
  calyx, 97, 112
  drills, 116, 117
  large diameter, 109, 111, 112
  small diameter, 97, 98, 113
  from St. Lawrence Power Project, 112
Rock coring, 100
Rock cubes, 216-220
  devices for testing of, 218, 219
  testing in shear, of, 216-219
Rock defects, 172, 190, 226, 265, 561, 562
Rock definition, 38, 40

Rock deformation, 169
Rock disks, 211
Rock engineering, 84
  hazards in, 276, 365-372
  problems in, 365, 366
  work in, 88
Rock as a construction (engineering) material, 31, 47, 48, 52, 61, 62, 64, 65, 75
Rock exfoliation, 147
Rock exploration methods, 93-97
Rock failure criteria, 161-168
Rock failure, some modes of, 207, 208
Rockfall, 417, 420, 474
  control by means of wire mesh, 452, 454, 509
Rock grouting, 186
Rock hazards, 369, 372, 390
Rock improvements, 282, 284-288, 398, 404
Rock-jointing characteristics, 273
Rock inferiorities, 367, 368
Rock-making minerals, 46-48
Rock maps, 98
Rock matrix, 172
Rock mechanics, 32, 33, 35, 73, 74, 78, 79, 82-84
  definition of, 73
  development of, 32-35
  difference between strength of materials and, 82-84
  features of, 74
  international congresses on, 33
  need for, 33
  theoretical basis of, 78, 79
Rock mechanics organizations, 32
Rock mechanics problems, 75, 76
Rock mechanics research, 32
Rock Mechanics Seminars, ASCE, 62
Rock overburden thickness, 147, 149, 331
Rock physical properties, 48, 49
Rock pressures, 366, 367
Rock lateral pressure coefficient λ_o, 306-308
Rock quality designation, 64
Rock reinforcement, 437
Rock roof, 438
Rock resistance to drilling, 99, 100
Rock saws, 116, 117
Rock samples,
  calyx, 97
  small diameter, 113, 115
Rock slides, 421, 422, 493

SUBJECT INDEX

methods of remedy against, 507-509
Rock slope failures, 475-478
Rock slopes, 471-486, 493
  anchoring of, 477-481, 498-501
  definitions of, 471, 472
  deformations of, 373, 374
  stabilization by means of reticulated root piles, 504-507
  treatment by means of gabions and wire mesh, 510
Rock stabilization, 435-469
Rock strength, 156
Rock stresses,
  internal, 132, 226, 242
  natural, 123
  residual, 80, 486
  tectonic, 486
Rock structure, 124
Rock substance, 41
Rock systems, 39
Rock temperature, 148
Rock testing equipment, 116-118, 211, 218 219
Rock testing in the laboratory, 191
  in-situ, 191
Rock texture, 124, 127, 128
Rock tunneling for the Rosslyn subway station, Virginia, 453
Rock weaknesses, summary of, 281-284
Rock wedges, 447
Rocks,
  classification of, 44
  clastic, 52
  competent, 40
  deformation of, 169-186
  description of, 44
  durability of, 156, 158
  dynamic properties of, 222, 223
  extrusive, 46
  igneous, 40, 46
  inelastic, 179, 305
  intact, 41
  internal friction of, 223
  intrusive, 45, 46
  mechanical defects of, 265, 281
  mechanical properties of, 156-160, 265
  metamorphic, 44, 45, 60
  modulus of elasticity, 123, 186
  nonhomogeneity of, 38, 127, 128
  petrographic structure, 172
  plastic flow of, 170, 180
  plastic properties of, 172
  plutonic, 45, 46

pressure coefficient, lateral, λ_o, 305-308, 323, 324, 326, 327, 333
rheological properties of, 172
rupture of, 230
sedimentary, 51
soft, 305
strength properties of, 118, 189, 190
stress-strain diagrams, 177-179
thermal properties of, 147-155
volcanic, 46
water in, 118
workability of, 122
Romania earthquake, 1977, 389
Roof in rock, 438
Root piles, 504-507
Rosettes, strain gauge, 115, 247, 249
Rosslyn subway station, Virginia, 68, 70, 451, 453
Round Valley, N.J., reservoir, 95, 97, 98
  airphoto mosaic of, 97
Rupture, modulus of, 205
Rupture angle, 163-165
Rupture of earth slope of the Delaware-Chesapeake Canal bank, 411, 413
Rupture plane, 163
Rupture of rocks, 230
Rupture of slope, 413
Rupture surfaces, 240
  circularly curved, in rock and in soil, 439
  in dry sand, 240
  logarithmically spiraled, 160, 240, 490
Ruptured rock wedges, 439

S

Safety factor, 476, 479, 484, 487-489, 495-498, 500, 504
San Andreas fault, Cal., 95, 96, 390-392
San Andreas fault zone, 96, 392-394
  definition of, 391
  exposure of, 393
Sand,
  arcosic, 48
  failure in shear, 439
  quartz, 151
Sandstones, 40, 52, 53, 127, 131, 151, 155, 157, 158, 187, 202, 203, 223, 305
  argillaceous, 99
  calcareous, 52
  siliceous, 52
Saturated flow, 139
Saturation moisture content, 124

Saturation unit weight, 130
Scanning electron micrographs of, 107, 108
  montmorillonite, 103, 104, 391
  sepiolite, 105, 106
  zeolite, 10, 107, 109
    analcime, 107
    chabazite, 107, 108
    erionite, 107-109
Scarp, 551
Schist, 60, 61, 131, 155, 188, 223, 305, 371
  chlorite, 61, 451
  Manhattan, 371
  mica, 149, 371
  talc, 57
Schistose gneisses, 200, 451
Schistose rock, 266
Scopje earthquake, 389
Sea defense, 65, 67
Sea foam (Meerschaum), 105, 106
Seams, 266, 267
Secant modulus, 180
Secondary stresses, 331, 334
Secondary stress condition in rock, 311, 321, 332
  elastic, 332, 356-363
  plastic, 361
  around vertical shafts, 361
Sedimentary rocks, 44, 45, 51, 52, 131, 155, 187, 202, 223, 281, 370
Seepage, 119, 396
  Darcy's law, 137, 139, 140
Seepage force, 489
Seismic effects on rock slopes, 423
Seismic exploration methods, 120
Seismic factors, 172
Seismic forces, 502-504
Semi-elastic rock, 178, 180
Seminars, ASCE, on Rock Mechanics, 62
Señor Rock, 282
Sepiolite, 105, 106
Shaft, vertical, 97, 220, 221
Shale, 53-58, 97, 124, 127, 128, 131, 176, 187, 202, 220, 223, 266, 305, 371
Shale core specimens
  after testing for their unconfined compressive strength, 185
Shale disks, 211, 212
Shale structure, layered, 58, 124, 126
Shale, Triassic, 53-57, 97, 124, 126, 127, 128, 184, 185, 200, 211
  calcite veins in, 55
  cube specimens, 220
  disks of, 211

dry/wet factor, 213
for punch shear strength testing, 211, 212
Shear angle, 163-165, 217, 236
Shear box, 214
Shear, direct, principles of, 209
Shear displacement, 216
Shear in a fault, 167
Shear failure in rock, some modes of, 187, 207, 208
Shear planes, forced, 215
Shear resistance, 236
  residual, 216, 217
Shear strength, 236
  Coulomb's line of, 162, 163
  definition of, 204
  direct, box, 213-215
  direct, with normal stress absent, 210
    with normal stress present, 213
  direct, of rock cubes, 216
  in double shear, 210
  in-situ, 235-241
  peak, 217
  punch shear, 210, 212, 213
  of rocks, 202, 205
  in single shear, 210
  unconfined, 54, 62, 191-195
Shear strength parameters, 207, 214
Shear strength tests, 210, 213, 215-217
  in the laboratory, 210, 213, 215-217
  large-scale, principles of, 235-241
    in-situ, 235-241
    with one jack, 231, 236, 237
  in an underground opening, 231, 235-237
Shear stress distribution in a solid shaft, 220
Shear stress parameters, residual, 216, 217
Shot-core drilling, 109, 110
Shotcrete, 293-295, 451
Stabilized rock slopes, 294, 295, 451
SI units of measurement, 563
Significant faults of the California region, 392
Signs and notations, key to, 519
Sodium silicate, 290
Siliceous sandstones, 52
Silt, 53
Silurian period, 43
Simplon tunnel, 149
Sinai Peninsula, 31
Single-body rock system, 39
Sinkholes, 120, 121, 281
Skopje earthquake 1963, 389

SUBJECT INDEX

Slate, 60, 61, 131, 155, 188, 202, 223
Slides,
    Delaware-Chesapeake Canal, 411, 413
    Panama Canal, 411, 412
    quick-clay, 415, 417
    rock, 421
    Tuve, Sweden, 415, 416
    Vita Sikudden, Sweden, 411, 413, 414
Sliding over a geological discontinuity, 216, 268, 269, 475, 493-498
Sliding surfaces, rough and wavy, 237, 241
Sliding tests in situ, 237-241
Slip lines, 348-350, 351-353
Slippage, mechanical, 227
Slip surfaces, 160
Slope failure, 472-475
Slope stability problem over a discontinuity, 475
Slopes, 486
    definition of, 471, 472
    failure of, 472
    stability analyses of, 475-477, 482, 486-489, 491-493
Small rock samples versus large ones, 113, 115
Sodium (natrium) bentonite suspension, 102
Sodium chloride, 144
Sodium silicate, 290
Soft cushion of our conscience, 76
Softening factor, 200
Soft rocks (inelastic), 305
Soil,
    definition of, 38
    physical constants of, 135
Soil anchors, pressure-grouted, 460-466
Soil freezing, 86
Soil maps, 98
Soil mechanics, 34, 40
Soil and rock map of a part of Hunterdon County, New Jersey, 98
Solidification of soil chemically, 290, 291
Solifluction, 415
Spacing of boring, 100, 101
Spalling, 122
Specific gravity, 106, 123, 127-129
Specific heat, 150
Specific impact work, 222
Specimens for unconfined compression, 114
Splitting of core specimens, 184
Spirals, logarithmic, 160, 351
Springing of a vault, 324

Stability analyses of rock slopes, 144, 475-477, 482, 485
Stability analyses of soil slopes, 486
Stabilization of landslides and rock slopes by means of reticulated root piling, 504
Stabilization of rock, 435-469
    by means of anchoring, 477-481
    by means of grouting, 282, 284-288, 398, 404
Stabilization of soil chemically, 290, 291
Standard sizes for casings, rods, core barrels, and holes, 99, 100
State of stress in-situ, 306
State of stress within a rock mass, 172
    complete, 248
    initial, 172
Static compression test on cubes, 197-220
Static laboratory compressive strength, 191-201
Static tests, 186
St. Gotthard tunnel, 142, 149
St. Lawrence Power Project, New York, 112
    large diameter rock core, 112
Stone Age, 31
Strain, absolute, 248
Strain measurements in rock, 244, 247, 248
Strain recovery, 175
Strain scanner, 116, 118
Strain gauges, 116, 186, 247
    rosettes, 115, 247
Strain hardening, 175, 176
Strains, 175-177, 181
Stratification, 51, 265, 267, 270
Stratigraphical column, 41-44
Strength, 156
    bending, 191, 203
    compressive, 54
    dynamic, 123, 190
    flexural, 191, 203
    in-situ, 118, 189, 190
    laboratory, 191-201
    no-load (= cohesion), 193
    punch, 191, 210, 212
    shear, 210, 213, 215-217
    static, 123, 190, 191
    tangential, 193
    tensile, 202-206
Strength parameters, 207, 214, 216, 217
Strength properties of rocks, 122, 189, 191
Stress analysis, elastic, in rock, 311, 318-326
Stress, biaxial, 339

Stress circle, Mohr's, 163-167, 339, 340, 351, 483
Stress concentrations, critical, 330
Stress conditions around a vertical shaft, 355-363
Stress diagram around a tunnel, 328
Stress ellipsoid, 306
Stress equations, 326
Stress fields, 83, 301, 307-309, 322
Stress function, Airy's, 339-341
Stress relaxation, 320, 373
Stress and strain, mutual dependence of, 169, 173
Stress-strain diagrams, 169, 173, 174, 176, 177, 183, 337, 338
   for ductile materials, 176
   for an ideal, plastic material, 170, 338
   for rocks, 174, 177-180, 183
Stresses, 243
   absolute, 247
   biaxial, 339
   effective, 200, 216
   initial, 165, 309
   internal, 242
   neutral, 65, 200
   pore-water, 200, 201, 240
   precompression on bolted rock elements, 440-443
   primary, 123, 322, 331, 334
   radial, 315, 344
   residual, 80, 486
   secondary, 311, 321, 332
   tangential, 344
   tectonic, 305
   thermal, 191, 315
   ultimate, 174
   yield, 309
Stresses,
   concentration in a slope, 485
   in elastic rock, 299, 355-363
   in elastic zone, 344-347
   in plastic zone, 344-347
Stresses around horizontal, circular openings in rock, 299
   internal, 80
Stresses in a thick-walled cylinder, 312
Stresses around vertical shafts, 355-363
Strike, 172
Strike-slip fault, 275
Structure, layered, of shale, 124, 126
Structure of rocks, 123, 124
Structures across faults, 274
Subsurface drainage, 397

Subsurface water, 368
Sulfates, 145
Sulfur dioxide, 145
Sulfur trioxide, 145
Sulfuric acid, 143
Summary on laboratory testing of rocks, 224, 225
Summary on rock mass weaknesses, 281-284
Summary about slip lines, 353
Support of critical wedge, 447
Support of gravity load, 447
Surcharge pressure on tunnels, 278
Surface drainage, 397
Surface, rupture, 160, 240, 439, 490
Surface tension, 137
Swelling pressure, 133, 144
Syenite, 45, 50
Syncline, 276-280
System, 44
System's permeability, 139

T

Table of contents, 9-18
Table of geologic time scale, 43
Taj Mahal, 146
Talc, 129, 157
Tangent (initial) modulus of elasticity, 62, 180
Tangential strength, 193
Tangential stresses, 315, 330
Tauern tunnel, 149
Tectonic earthquakes, 376
Tectonic fissures, 142
Tectonic forces, 319
Tectonic stresses, 306, 486
Tectonism, 282
Temperature, 368, 369
   in Alpine tunnels, 149
   in mines, in tunnels, 148, 368, 369
Temperature conductivity, 149
Temperature conversion factors, 570
Temperature variation, annual, daily, monthly, 148
Tensile failure, Griffith's, 161, 168
Tensile strength of ceramic beams, by Duckworth, 205
Tensile strength of rock, 191, 201-203
Tensile strength of rock, Brazilian test, 206
Tensile test in-situ, 235
Tension zones, 319, 325
Terms, glossary of, 529-559
Terzaghi's definition of soil, 38

SUBJECT INDEX

Test parameters, 214
Test specimens, 115, 192
Testing of rocks, 32
  laboratory, 115, 191
  objectives of, 115
  in-situ, 120, 121, 191
Tests,
  borehole, 234
  compression, 191-194, 234
  on cubes, 197
  dilatometer, 244
  dynamic, 186, 191
  freeze-thaw, 158
  hydrofracturing, 248, 250, 397
  for leakage, 397
  in-situ, 191
  jacking, 231
  laboratory, 191
  plate loading, 228, 229
  pressure chamber (tunnel), 232, 233
  resonance, 186
  shear, 210, 213, 215-217
  time-dependent, 196
  torsion, 191, 219, 241
  triaxial, 195
  ultrasonic velocity, 186
  unconfined compression, 194
Texture, 5, 123-125
Texture of rocks, 52, 123-125, 127, 128
Theoretical basis for rock mechanics, 78
Theoretical basis for stress analysis in rock, 318-326
Theoretical course of stress distribution around a vertical shaft, 361
Theory of elasticity, 79-81, 188
Theory, Heim's, 303
Theory of plasticity, 81, 82
Thermal conductivity, 149, 152, 153
Thermal diffusivity, 149, 152, 153
Thermal expansion and contraction coefficients, 153-156
Thermal exploration of rocks, 121, 122
Thermal properties of rocks, 123, 124, 147-155
Thermal stabilization of ground, 86
  of the Metro subway system of Paris, 86
  for the shaft for the Richmond water supply tunnel, N.Y., 86, 87
  in tunnels, 86
Thermal strains and stresses, 191
Thermo-osmotic phenomenon 134
Thickness of overburden, 147, 149, 331
Thickness of zone of disturbance, 347, 348

Thick-walled cylinder, 312, 313
Thixitropy, 102,
  simple test of, 102, 103
Tie-backs, 461-466
Time-dependent deformations (creep), 172, 173, 180, 182, 196
Time scale, geologic, 41, 43
Topaz, 157
Torsion tests, 191, 219, 241
Toughness index, 222, 223
Toula glacier, 399
Trace of San Andreas fault zone, 391, 393
Trachite, 45, 50
Trap rock, 45, 47, 65, 66
Tresca failure criterion, 161, 162
Triassic age, 43
Triassic (Brunswick, New Jersey) shale, 53, 54, 97, 124, 126-128, 184, 185, 200, 211
Triaxial compression test, 183, 195-197
True modulus of elasticity, 177
Tuff, 305
Tunnels, 148, 149
  Albula, 149
  Allegheny Portage, 32
  American, 32
  ancient, 31
    near Naples, 31
  in anticline, 278-280
  Apennine, 145, 149
  aqueous, 86, 312
  Arlberg, 149
  Auburn, 32
  Bologna-Florence (Great Apennine), 142, 145
  Karawanken, 149
  leaky, 134
  length of, 149
  Lötschberg, 149
  Mont Blanc, 32, 370
  Mont Cenis, 32, 149
  Naples (ancient), 32
  beneath Oroville dam, 451
  railway tunnel in Pennsylvania, 32
  of the Rosslyn Metro subway station in rock, Virginia, 68, 70, 451
  St. Gotthard, 142, 149
  Symplon, 149
  in syncline, 287-280
  Tauern, 149
  vehicular, 312
  Washington, D.C., Metro, 68, 70, 451, 453
Tunnel-rock sytem in a biaxial stress field, 322

Tuve landslide, Sweden, 415, 416

U

Ultimate strength, 193, 198
Ultrasonic interferometry, 223
Unconfined compression test, 194
Unconfined compressive strength, 54, 62, 191-195
Unconformity, 57, 59
    angular, 57, 59
Underground caverns in rock, their various uses of, 32, 67-70, 311, 451-456
Uniaxial compression, 183
Uniaxial stress field, 307, 308
Unit weight of some American rocks, 131
Unit weight, 123, 129, 131
    buoyant (submerged), 131
    dry soil, 129
    moist soil, 130
    of rocks, 130, 131
    of water, 131
Units of measurement, 563
    conversion factors for, 565-571
    SI units, 563
Unlined tunnel, 328
Uplift in earth bank, 402
Uplift, hydrostatic, 400, 401, 438
Uses of rocks, 64-67
Use of SI units in geotechnical engineering, 563
U-shaped structure, 438
Utility line across a fault, 274
Ut tensio sic vis, 79, 173, 320, 339

V

Vaiont Reservoir disaster, 240, 398, 423, 424
Valle de la Caodos, 399
Value of rock mechanics, 78
Veins, calcite, 55
Vermont marble, 184, 185
Vertical shafts, 357, 359
Very fissured poly-body rock system, 39
Viscosity, absolute (dynamic), 140
    tables for water, 572
    correction factor tables for water, 576
    kinematik, 140
Viscous deformations, 170
Viscous material, 175
Viscous substance, 170
Vita Sikudden slide, Sweden, 411, 413, 414
Void ratio, 123, 132, 133

Voids in rock, 132, 266, 281
Volcanic earthquakes, 376
Volcanic (extrusive) rocks, 45, 46
Volumetric heat capacity, 150
Volumetry, 136

W

Washington, D.C., Metro tunnel, 68, 70 451, 453
Water, artesian, 530
    electrical properties of, 146
Water, gravitational, 137
Water control, 119, 397
Water film, 134
Water glass, 290
Water, hardness of, 143
Water, hot (magmatic), 143
Water,
    influx from joints underground openings, 142, 271
    interstitial, 557
    leakage, 396
    phreatic, 548, 558
    "public enemy No. 1," 427
    in rock, 133, 134
    seepage, 396
    from shale, frozen to ice along a road, 119
    subsurface, 368
    surface tension of, 137
    in tunnel under Oroville dam, 452
    unit weight of, 129-131
Waterfalls, Niagara, 417
Water movement through rock, 119, 396
Water problems, 119, 398
Waves,
    elastic, 227
    energy, 186
    primary, 186
    secondary, 186
Weaknesses of rocks, 226, 265, 266, 272, 281
    coefficient of, 272
    as a function of rock jointing characteristics, 273
    planes of, 208, 216, 265, 266
    summary on, 281-284
Weathering fissures, 142
Weathering of rocks, 123, 143, 158, 172
Wedge, critical, 447
Weight of rock, 123, 129-131
White mica (muscovite), 129

Williams rock bolt, 448-451, 468
Wire net (mesh), 452, 454, 509
Work energy, 171
Work in rock engineering, 88, 89
Workability of rock, 122
World Trade Center, New York,
  basement foundation of the, 455, 457

Y

Yield point, 173, 174
Yield strength, 173, 174
Yield stress, 174, 309
Young's modulus of elasticity, 123, 159, 173, 187, 188, 203, 205, 233

Z

Zeolite, 106, 107, 109
  analcime, 107
  chabazite, 107, 108
  erionite, 107-109
Zones,
  crushed, 266, 273
  of disturbance in rock, 347, 348
  earthquake, 393
  elastic, 338, 361
  of elastic tangential tensile stresses, 333
  fault, 266
  plastic, 337, 338, 361
  of San Andreas fault, 391
  tension, 319
  of weakness, 266

Whenever Structural Safety is the Point... Kern Precision Instruments are the Right Choice

For measuring distances up to 2500 m:

Mekometer ME 3000

Electro-optical Precision Distance Meter with the extremely high accuracy of $\pm(0.2\ mm+1\ ppm)$ and a range of 2.5 km. Digital distance display to 0.1 mm.
Universal application: structural deformation measurements, large area slip and displacement measurements, precision layout work and fundamental surveying.

The Mekometer used for dam control measurements

For measuring length variations within a distance range of 50 m:

Distometer ISETH

Precision instrument for accurate determination of length variations by means of Invar wires. Measuring accuracy ± 1 ppm; length of the Invar wire 1—50 m; measuring range for length variations 100 mm.
Special advantages: lengths of any inclination including vertical may be measured; simple layout of the measuring arrangement.
Application: structural deformation measurements.

The Distometer ISETH used for tunnel wall deformation measurement

Kern & Co. Ltd.
Mechanical, Optical and Electronic Precision Instrument
CH-5001 Aarau, Switzerland
Telex 981 106

Please send me your detailed documentation on:

☐ Mekometer ME 3000
☐ Distometer ISETH

Name:

Occupation:

Address:

BUEHL & FAUBEL

Test Equipment and Machines for testing Asphalt, Bitumen, Conrete, Cement, Aggregates, Soils and Rock Materials

Isotopic Equipment for measurements of density and moisture and for the determination of bituminous content

Mobile Trailer Laboratories, Container-Laboratories

Non-destructive measuring devices for the thickness of asphalt layers

BUEHL & FAUBEL

Appareils d'essais pour faire des épreuves de Mécanique des sols, de Mécanique des roches, de l'Asphalte, du Bitume, du Béton, du Ciment et des Agrégats

Appareils á source radio-active et indicateur pour déterminer in situ la densité et la teneur eneau du sol et la densité de la couche bitumineux et des autres pour déterminer la teneur en bitume

Laboratoires mobiles en remorque à quatres roues et Labor-Container tous les deux aménagés

Attirails pour mesurer la profondeur des revêtements de route à la mode non-destructif

BUEHL & FAUBEL

BUEHL & FAUBEL

Prüfgeräte und Maschinen
zur Untersuchung von Asphalt,
Bitumen, Beton, Zement,
Zuschlagstoffen, Boden- und
Felsmaterialien

Isotopengeräte für Dichte-,
Feuchtigkeits- und
Bitumengehaltsbestimmung

Fahrbare Laborwagen,
Container-Laborwagen

Zerstörungsfreie
Schichtdickenmeßgeräte für
Asphaltstraßenbeläge

BUEHL & FAUBEL
Ges.m.b.H.

TEST-EQUIPMENT FOR CIVIL ENGINEERING

A-1237 Vienna/Austria · P.O. Box 20 · Zangerlestraße 49
Tel. Nr.: 88 69 22 - 88 69 23 · Telex Nr.: 13 52 85 - 13 61 89

Foundations in Tension
— Ground Anchors —

by Prof. Dr. **T. H. Hanna**,
University of Sheffield, England

1982, 600 pp. 345 figs, 55 tables, 920 refs.
price: US$ 58.00

Series on Rock and Soil Mechanics, Vol. 6
ISBN 0-87849-044-2   ISSN 0080-9004

Available in the U.S. and Canada through McGraw-Hill Book Company, New York.

CONTENTS: 1. Tensile Foundations and Ground Engineering — 2. Soil and Rock Anchors — 3. Anchor Corrosion and Its Protection — 4. Transfer of Load in Anchor and Associated Problems — 5. The Load Testing of Anchors — 6. The Analysis and Design of Anchored Structures — 7. Ground Anchor Use and the Performance of Anchored Structures — 8. Miscellaneous Aspects of Ground Anchors and Their Use.

An educational approach has been followed throughout the book to show how the engineer may benefit from the use of ground anchors, how ground anchors behave and how an engineer tackles a design where anchors are used. It will be clear from the text that many changes have taken place in anchoring techniques and design methods over the past 15 years.

The book should be of direct interest to the practising Civil, Structural or Mining Engineer who is involved with the design, construction or maintenance of ground sturctures. Research engineers, teaching staff and graduate students should also benefit from it.

Trans Tech Publications

P.O. Box 266 · D-3392 Clausthal-Zellerfeld · West Germany
16 Bearskin Neck · Rockport, MA 01966 · USA

DYWIDAG
Thread Bars

- indestructible continuous threads
- deformations allow anchorages and couplers to thread onto the thread bar at any point
- simple rugged anchorage and coupling
- most efficient shear bond
- load range of single bars from 190 to 1250 kN bundles up to 2835 kN ultimate and more

for Rock Anchors Soil Anchors Rock Bolts

- for permanent use
  ☐ with double corrosion protection
  ☐ fully bonded or free elasticity
- bond to rock
  ☐ with cement grout
  ☐ with resin mortar
  ☐ with expansion shell

DYWIDAG-SYSTEMS INTERNATIONAL GmbH
Postfach 81 02 68 · D-8000 München 81
Erdinger Landstraße 1
Phone 089/92 67-1 · Telex 05-216 195 dsi d

DYWIDAG-SYSTEMS INTERNATIONAL USA, Inc.
Beaverbrook Road 107, Lincoln Park, NJ 07035
Phone 001-201-628-8700
Telex 134 232 dsi USA lnpk

OVERSEAS OFFICES AND REPRESENTATIVES: WORLDWIDE

T 120

DYWIDAG SYSTEMS INTERNATIONAL

Handbook on Mechanical Properties of Rocks

Testing Techniques and Results
Volumes I, II, III and IV

by Dr. **R. D. Lama,** CSIRO Division of Applied Geomechanics, Australia and Prof. **V. S. Vutukuri,** University of New South Wales, Australia

Volume I (1974); Volume II, III and IV (1978) 1746 pp, 860 figs, 292 tables, 2090 refs

Price: Each volume: US$ 65.00
Price for complete set, Vols. I—IV: US $ 248.00
Series on Rock & Soil Mechanics
Vol. 2, No. 1; Vol. 3, Nos. 1—3
ISBN 0-87849-031-0 (set); ISSN 0080-9004

CONTENTS:

Vol. 1: 1. Specimen Preparation for Laboratory Tests — 2. Compressive Strength of Rock — 3. Tensile Strength of Rock — 4. Shear Strength of Rock — 4. Strength of Rock Under Triaxial and Biaxial Stresses — Appendix I: Stiff Testing Machines.

Vol. II: 6. Static Constants of Rock — 7. Dynamic Elastic Constants of Rock — Appendix II: Laboratory Mechanical Properties of Rocks.

Vol. III: 8. In Situ Testing of Rock — 9. Time-Dependent Properties of Rocks — Appendix III: In Situ Mechanical Propterties of Rock — Appendix IV: Crack Propagation Velocity in Rock.

Vol. IV: 10. Mechanical Behavior of Jointed Rock — 11. Classification of Rock — 12. Miscellaneous Properties of Rock — Appendix V: Stereographic Projections — Appendix VI: Definition of Some Rock Mechanics Terms (English, French, German) — Appendix VII: Imperial, Metric and SI Units.

"These four volumes can be regarded as the most up-to-date reference work on mechanical properties of rocks. They are of interest to scientists and engineers in geophysics, geological engineering, civil, mining and petroleum engineering."

Trans Tech Publications

P.O. Box 266 · D-3392 Clausthal-Zellerfeld · West Germany
16 Bearskin Neck · Rockport, MA 01966 · USA

SPECIALISTS IN ROCK MECHANICS
Measuring Techniques in Mining,
Civil Engineering and in Boreholes

MANUFACTURING OF IN-SITU TEST EQUIPMENT FOR:

- Plate Load Test
- Shear Test
- Flat Jack Test
- Point Load Tester
- Dilatometer Test
- Stress Determination
- Sonic Measurement
- TV Borehole Camera

PERFORMANCE OF IN-SITU TESTS WITH ABOVE EQUIPMENT

Interfels GmbH
Deilmannstr. 1
D-4444 Bad Bentheim
F. R. G.
Tel. 05922 / 72666
Telex 98919

Eastman Instruments GmbH
Carl-Zeiss-Str. 16
D-3005 Hemmingen 1
F. R. G.
Tel. 0511 / 413737
Telex 924011

Interfels Ges. m. b. H.
Schwarzstr. 27
A-5020 Salzburg
Austria
Tel. 06222 / 75104
Telex 633872

The Continuum Theory of Rock Mechanics

by Prof. Dr. **Cs. Assonyi** and Prof. Dr. **R. Richter**, Hungary

1979, 365 pp, 115 figs. US$ 58.00

Series on Rock and Soil Mechanics, Vol. 4
ISBN 0-87849-027-2
ISSN 0080-9004

A McGraw-Hill Civil Engineers' Book Club Selection

CONTENTS: Theoretical Bases of the Rheological Phenomena. 1. Stress and State of Stress — 2. State of Deformation — 3. Mechanical Equations of State — 4. Approximate Laws for Rock Material — 5. Basic Rheological Relations in Rock Mechanics — 6. Solutions of the Fundamental Rock Mechanics Equations. **Effect of Openings in Rheological Rock Surroundings** — 7. Primary State — 8. Formation of Mechanical Field Due to Driving Openings — 9. Mines with Circular Section for Hydrostatic State of Stress — 10. Mechanical Field Around a Drift for a Biaxial State of Deformation — 11. Drifts in a Primary Field of Arbitrary Orientation — 12. Co-action of the Binary System of Rock and Support.

Excerpts from the Preface

A single book, even if of several volumes, cannot possibly encompass the entire scope of theoretical and practical rock mechanics. In our selection of both the theoretical models to serve as a basis for analysis, and of the problems to be discussed, we were faced with a choice between a detailed discussion of the questions raised as opposed to an outline discussion of their most important aspects. In the second option, it is of course necessary to reduce both the number of theoretical models presented and the number of basic problems and phenomena covered by the discussion. As indicated by the title of this volume, it was this latter choice that we have opted for.

Trans Tech Publications

P.O. Box 266 · D-3392 Clausthal-Zellerfeld · West Germany
16 Bearskin Neck · Rockport, MA 01966 · USA

The Pressuremeter and Foundation Engineering

by Dr. **F. Baguelin**, Dr. **J. F. Jézéquel**,
Laboratoire Central des Ponts et Chaussées, France
and Prof. Dr. **D. H. Shields**,
University of Manitoba, Canada

1978, 624 pp, 314 figs, 280 refs, US $ 58.00

Series on Rock and Soil Mechanics Vol. 2, No. 4
ISBN 0-87849-019-1   ISSN 0080-9004

A McGraw-Hill Civil Engineers' Book Club Selection

Excerpts from the Preface
"The design and construction of foundations require a thorough knowledge of the behaviour of soils and rocks in the field. Since even elaborate laboratory tests on large subsurface samples can at best only approximate the field conditions, in-situ tests are often preferable. The pressuremeter is probably the most versatile in-situ testing device available at present for investigating static and cyclic strength and deformation of solids and rocks."

"Unfortunately, this important technique for subsurface exploration and foundation design has not been sufficiently appreciated outside France where pressuremeters have been widely used with great success for about twenty years. It is, therefore, timely that this first book on the subject of the pressuremeter and foundation engineering be published, and the authors have done a valuable service in sharing their special experience with the geotechnical profession at large. The readers will find in this definitive book a comprehensive treatment of the various practical aspects and fundamental principles of pressuremeter tests under a great variety of field conditions."

"Especially noteworthy are the detailed chapters on the interpretation and evaluation of test results in subsurface explorations and their application to estimates of the bearing capacity and settlement of shallow and deep foundations."

"In this way pressuremeter tests can lead to safe and economical solutions to many geotechnical problems, as shown in this warmly recommended book."

G. G. Meyerhoff

Trans Tech Publications
P.O. Box 266 · D-3392 Clausthal-Zellerfeld · West Germany
16 Bearskin Neck · Rockport, MA 01966 · USA

Design and Construction of Dry Docks

by Prof. Dr. **B. K. Mazurkiewicz**, Gdynia, Poland

1980, 381 pp, 549 figs, 28 × 21 cm

US $ 68.00 (hard cover), (ISBN 0-87849-028-0)
US $ 38.00 (soft cover), (ISBN 0-87849-036-1)

CONTENTS: 1. Introduction — 2. Function and Main Dimensions — 3. Development of Dry Dock Construction — 4. Floor and Side Walls — 5. Structural Design of Dry Docks — 6. Dry Dock Structural Calculations — 7. Cranes for Dry Docks — 8. Flooding and Dewatering of a Dry Dock — 9. Dock Entrances and Dock Gates — 10. Construction of Dry Docks.

Book Review:

"This is a valuable text book that combines an authoritative review of design parameters and construction procedures with a wealth of information on existing dry docks around the world."

"The text is set out clearly and the subject matter well organised. It is presented in a clear and direct style of English, and the reader ist not conscious that it is a translation, expect perhaps in an occasional paragraph in the early chapters. The 381 pages are furnished with 549 excellent photographic and line illustrations."

"The theoretical approach is admirably counterbalanced by descriptions of the application of the design principles in existing docks."

"These minor criticisms on peripheral subjects do not detract from the considerable value of this book, which includes a copious bibliography, to anyone involved in Dry Dock design."

Excerpts from a book review prepared by R. O. Campbell, Sir William Halcrow & Partners (Project Engineer for Dubai Dry Dock) and published in **"The Dock and Harbour Authority"**, *December 1980, p. 262.*

Trans Tech Publications

P.O. Box 266 · D-3392 Clausthal-Zellerfeld · West Germany
16 Bearskin Neck · Rockport, MA 01966 · USA